Lecture Notes in Artificial Intelligence 6332

Edited by R. Goebel, J. Siekma

Subseries of Lecture Notes

Bernhard Pfahringer   Geoff Holmes
Achim Hoffmann (Eds.)

Discovery Science

13th International Conference, DS 2010
Canberra, Australia, October 6-8, 2010
Proceedings

 Springer

Series Editors

Randy Goebel, University of Alberta, Edmonton, Canada
Jörg Siekmann, University of Saarland, Saarbrücken, Germany
Wolfgang Wahlster, DFKI and University of Saarland, Saarbrücken, Germany

Volume Editors

Bernhard Pfahringer
Geoff Holmes
The University of Waikato, Department of Computer Science
Private Bag 3105, Hamilton 3240, New Zealand
E-mail:{bernhard, geoff}@cs.waikato.ac.nz

Achim Hoffmann
The University of New South Wales, School of Computer Science and Engineering
Sydney 2052, Australia
E-mail: achim@cse.unsw.edu.au

Library of Congress Control Number: 2010935451

CR Subject Classification (1998): I.2, H.3, H.4, H.2.8, J.1, F.1

LNCS Sublibrary: SL 7 – Artificial Intelligence

ISSN   0302-9743
ISBN-10   3-642-16183-9 Springer Berlin Heidelberg New York
ISBN-13   978-3-642-16183-4 Springer Berlin Heidelberg New York

springer.com

© Springer-Verlag Berlin Heidelberg 2010
Printed in Germany

Typesetting: Camera-ready by author, data conversion by Scientific Publishing Services, Chennai, India
Printed on acid-free paper   06/3180

Preface

This volume contains the papers presented at the 13th International Conference on Discovery Science (DS 2010) held in Canberra, Australia, October 6–8, 2010.

The main objective of the Discovery Science (DS) conference series is to provide an open forum for intensive discussions and the exchange of new ideas and information among researchers working in the area of automating scientific discovery or working on tools for supporting the human process of discovery in science. It has been a successful arrangement in the past to co-locate the DS conference with the International Conference on Algorithmic Learning Theory (ALT). This combination of ALT and DS allows for a comprehensive treatment of the whole range, from theoretical investigations to practical applications. Continuing in this tradition, DS 2010 was co-located with the 21st ALT conference (ALT 2010). The proceedings of ALT 2010 were published as a twin volume (6331) of the LNCS series.

The international steering committee of the Discovery Science conference series provided important advice on a number of issues during the planning of Discovery Science 2010. The members of the steering committee were Alberto Apostolico, Setsuo Arikawa, Hiroki Arimura, Jean-Francois Boulicaut, Vitor Santos Costa, Vincent Corruble, Joao Gama, Achim Hoffmann, Tamas Horvath, Alipio Jorge, Hiroshi Motoda, Ayumi Shinohara, Einoshin Suzuki (Chair), Masayuki Takeda, Akihiro Yamamoto, and Thomas Zeugmann.

We received 43 full-paper submissions out of which 25 long papers were accepted for presentation and are published in this volume. Each submission was allocated three reviewers from the program committee of international experts in the field. In total 125 reviews were written discussing in detail the merits of each submission. The selection of papers was made after careful evaluation of each paper based on originality, technical quality, relevance to the field of Discovery Science, and clarity.

This volume consists of two parts. The first part contains the papers accepted for presentation at the conference.

The second part contains the invited talks of ALT 2010 and DS 2010. Since the talks were shared between the two conferences, for the speakers invited specifically for ALT 2010, only abstracts are contained in this volume, while the full papers are found in the twin volume, LNCS 6331 (the proceedings of ALT 2010). The following invited speakers presented their work: Peter Bartlett *Optimal Online Prediction in Adversarial Environments*, Ivan Bratko *Discovery of Abstract Concepts by a Robot*, Alexander Clark, *Towards General Algorithms for Grammatical Inference*, Rao Kotagiri *Contrast Pattern Mining and Its Application for Building Robust Classifiers*, and Manfred Warmuth *The Blessing and the Curse of the Multiplicative Updates*.

We are deeply indebted to the program committee members as well as their subreferees who had the critically important role of reviewing the submitted papers and contributing to the intense discussions which resulted in the selection of the papers published in this volume. Without this enormous effort, ensuring the high quality of the work presented at Discovery Science 2010 would not have been possible.

We also thank all the authors who submitted their work to Discovery Science 2010 for their efforts. We wish to express our gratitude to the invited speakers for their acceptance of the invitation and their stimulating contributions to the conference. Furthermore, we wish to thank the *Air Force Office of Scientific Research, Asian Office of Aerospace Research and Development* as well as the *Artificial Intelligence Journal* for their financial support contributing to the success of this conference.

Finally, we wish to thank everyone who helped to make Discovery Science 2010 a success: the DS steering committee, the ALT conference chairs, invited speakers, the Publicity Chair for Discovery Science 2010, Albert Bifet, for the well-designed web presence, and last but not least the Local Arrangements Chair for Discovery Science 2010, Eric McCreath, and the Local Arrangements Chair for Algorithmic Learning Theory, Mark Reid, and their team of supporters who worked very hard to make both conferences a success.

July 2010   Bernhard Pfahringer
                                                         Geoff Holmes
                                                        Achim Hoffmann

Organization

Organization Committee

Steering Committee Chair	Einoshin Suzuki
Conference Chair	Achim Hoffmann
Program Chairs	Bernhard Pfahringer
	Geoffrey Holmes
Publicity Chair	Albert Bifet
Local Arrangements Chair	Eric McCreath

Program Committee

Akihiro Yamamoto	Kyoto University, Japan
Albert Bifet	University of Waikato, New Zealand
Albert Yeap	AUT, New Zealand
Alipio Jorge	University of Porto, Portugal
Alneu de Andrade Lopes	University of Sao Paulo, Brazil
Andre Carvalho	University of Sao Paulo, Brazil
Antoine Cornuejols	AgroParisTech, France
Antonio Bahamonde	University of Oviedo, Spain
Bettina Berendt	Katholieke Universiteit Leuven, Belgium
Carlos Soares	University of Porto, Portugal
Colin de la Higuera	University of Nantes, France
Concha Bielza	Technical University of Madrid, Spain
Daisuke Ikeda	Kyushu University, Japan
Daniel Berrar	University of Ulster, Ireland
David Dowe	Monash University, Melbourne, Australia
Dino Pedreschi	Pisa University, Italy
Donato Malerba	University of Bari, Italy
Einoshin Suzuki	Kyushu University, Japan
Filip Zelezny	Technical University Prague, Czech Republic
Gerhard Widmer	Johannes Kepler University, Austria
Gerson Zaverucha	Universidade Federal do Rio de Janeiro, Brazil
Guillaume Beslon	INSA Lyon, France
Hendrik Blockeel	Katholieke Universiteit Leuven, Belgium
Hideo Bannai	Kyushu University, Fukuoka, Japan
Hiroshi Motoda	AOARD and Osaka University, Japan
Hisashi Kashima	University of Tokyo, Japan
Ian Watson	University of Auckland, New Zealand
Inaki Inza	University of Basque Country, Spain
Ingrid Fischer	University of Konstanz, Germany
Irene Ong	UW-Madison, USA

Jaakko Hollmen   Helsinki University of Technology, Finland
Janos Csirik   University of Szeged and RGAI Szeged,
       Hungary
Jean-Francois Boulicaut   INSA Lyon, France
Jesse Davis   University of Washington, USA
Joao Gama   University of Porto, Portugal
Jose Luis Borges   University of Porto, Portugal
Kevin Korb   Monash University, Australia
Kouichi Hirata   Kyushu Institute of Technology, Japan
Kristian Kersting   Fraunhofer IAIS and University of Bonn,
       Germany
Kuniaki Uehara   Kobe University, Japan
Ljupco Todorovski   Josef Stefan Institute, Slovenia
Luis Torgo   University of Porto, Portugal
Maarten van Someren   University of Amsterdam, Netherlands
Maguelonne Teisseire   Cemagraf Montpellier, France
Makoto Haraguchi   Hokkaido University, Japan
Marta Arias   Universitat Politècnica de Catalunya, Spain
Mario J. Silva   Universidade de Lisboa, Portugal
Masayuki Numao   University of Osaka, Japan
Michael Berthold   University of Konstanz, Germany
Michael May   Fraunhofer IAIS Bonn, Germany
Mohand-Said Hacid   University Claude Bernard Lyon 1, France
Nada Lavrač   Jožef Stefan Institute, Slovenia
Nuno Fonseca   University of Porto, Portugal
Patricia Riddle   University of Auckland, New Zealand
Paulo Azevedo   Universidade do Minho, Portugal
Pedro Larranaga   Polytechnic University of Madrid, Spain
Peter Andreae   Victoria University of Wellington,
       New Zealand
Peter Christen   Australian National University, Australia
Ross King   University of Wales, UK
Simon Colton   Imperial College London, UK
Sriraam Natarajan   UW-Madison, USA
Stan Matwin   University Ottawa, Canada
Stefan Kramer   Technische Universitat München, Germany
Stephen Marsland   Massey University, New Zealand
Szymon Jaroszewicz   National Institute of Telecommunications,
       Poland
Takashi Washio   Osaka University, Japan
Tamas Horvath   University of Bonn and Fraunhofer IAIS,
       Germany
Tapio Elomaa   Tampere University of Technology, Finland
Vincent Corruble   Universite Pierre et Marie Curie, France
Vitor Santos Costa   University of Porto, Portugal
Will Bridewell   CSLI Stanford, USA

Additional Reviewers

Carlos Ferreira   Elena Ikonomovska
Cristiano Pitangui   Ken-ichi Fukui

Table of Contents

Sentiment Knowledge Discovery in Twitter Streaming Data 1
  Albert Bifet and Eibe Frank

A Similarity-Based Adaptation of Naive Bayes for Label
Ranking: Application to the Metalearning Problem of Algorithm
Recommendation 16
  Artur Aiguzhinov, Carlos Soares, and Ana Paula Serra

Topology Preserving SOM with Transductive Confidence Machine 27
  Bin Tong, ZhiGuang Qin, and Einoshin Suzuki

An Artificial Experimenter for Enzymatic Response Characterisation ... 42
  Chris Lovell, Gareth Jones, Steve R. Gunn, and Klaus-Peter Zauner

Subgroup Discovery for Election Analysis: A Case Study in Descriptive
Data Mining 57
  Henrik Grosskreutz, Mario Boley, and Maike Krause-Traudes

On Enumerating Frequent Closed Patterns with Key in Multi-relational
Data 72
  Hirohisa Seki, Yuya Honda, and Shinya Nagano

Why Text Segment Classification Based on Part of Speech Feature
Selection 87
  Iulia Nagy, Katsuyuki Tanaka, and Yasuo Ariki

Speeding Up and Boosting Diverse Density Learning 102
  James R. Foulds and Eibe Frank

Incremental Learning of Cellular Automata for Parallel Recognition of
Formal Languages.. 117
  Katsuhiko Nakamura and Keita Imada

Sparse Substring Pattern Set Discovery Using Linear Programming
Boosting 132
  *Kazuaki Kashihara, Kohei Hatano, Hideo Bannai, and
  Masayuki Takeda*

Discovery of Super-Mediators of Information Diffusion in Social
Networks 144
  *Kazumi Saito, Masahiro Kimura, Kouzou Ohara, and
  Hiroshi Motoda*

Integer Linear Programming Models for Constrained Clustering........ 159
  Marianne Mueller and Stefan Kramer

Efficient Visualization of Document Streams........................ 174
  Miha Grčar, Vid Podpečan, Matjaž Juršič, and Nada Lavrač

Bridging Conjunctive and Disjunctive Search Spaces for Mining a New
Concise and Exact Representation of Correlated Patterns 189
  Nassima Ben Younes, Tarek Hamrouni, and Sadok Ben Yahia

Graph Classification Based on Optimizing Graph Spectra 205
  Nguyen Duy Vinh, Akihiro Inokuchi, and Takashi Washio

Algorithm for Detecting Significant Locations from Raw GPS Data..... 221
  *Nobuharu Kami, Nobuyuki Enomoto, Teruyuki Baba, and
  Takashi Yoshikawa*

Discovery of Conservation Laws via Matrix Search 236
  Oliver Schulte and Mark S. Drew

Gaussian Clusters and Noise: An Approach Based on the Minimum
Description Length Principle 251
  Panu Luosto, Jyrki Kivinen, and Heikki Mannila

Exploiting Code Redundancies in ECOC............................. 266
  Sang-Hyeun Park, Lorenz Weizsäcker, and Johannes Fürnkranz

Concept Convergence in Empirical Domains 281
  Santiago Ontañón and Enric Plaza

Equation Discovery for Model Identification in Respiratory Mechanics
of the Mechanically Ventilated Human Lung........................ 296
  *Steven Ganzert, Josef Guttmann, Daniel Steinmann, and
  Stefan Kramer*

Mining Class-Correlated Patterns for Sequence Labeling 311
  Thomas Hopf and Stefan Kramer

ESTATE: Strategy for Exploring Labeled Spatial Datasets Using
Association Analysis 326
  Tomasz F. Stepinski, Josue Salazar, Wei Ding, and Denis White

Adapted Transfer of Distance Measures for Quantitative
Structure-Activity Relationships 341
  *Ulrich Rückert, Tobias Girschick, Fabian Buchwald, and
  Stefan Kramer*

Incremental Mining of Closed Frequent Subtrees 356
  Viet Anh Nguyen and Akihiro Yamamoto

Optimal Online Prediction in Adversarial Environments 371
  Peter L. Bartlett

Discovery of Abstract Concepts by a Robot 372
  Ivan Bratko

Contrast Pattern Mining and Its Application for Building Robust
Classifiers 380
  Kotagiri Ramamohanarao

Towards General Algorithms for Grammatical Inference............... 381
  Alexander Clark

The Blessing and the Curse of the Multiplicative Updates 382
  Manfred K. Warmuth

Author Index 383

Sentiment Knowledge Discovery in Twitter Streaming Data

Albert Bifet and Eibe Frank

University of Waikato, Hamilton, New Zealand
{abifet,eibe}@cs.waikato.ac.nz

Abstract. Micro-blogs are a challenging new source of information for data mining techniques. Twitter is a micro-blogging service built to discover what is happening at any moment in time, anywhere in the world. Twitter messages are short, and generated constantly, and well suited for knowledge discovery using data stream mining. We briefly discuss the challenges that Twitter data streams pose, focusing on classification problems, and then consider these streams for opinion mining and sentiment analysis. To deal with streaming unbalanced classes, we propose a sliding window Kappa statistic for evaluation in time-changing data streams. Using this statistic we perform a study on Twitter data using learning algorithms for data streams.

1 Introduction

Twitter is a "what's-happening-right-now" tool that enables interested parties to follow individual users' thoughts and commentary on events in their lives—in almost real-time [26]. It is a potentially valuable source of data that can be used to delve into the thoughts of millions of people as they are uttering them. Twitter makes these utterances immediately available in a data stream, which can be mined using appropriate stream mining techniques. In principle, this could make it possible to infer people's opinions, both at an individual level as well as in aggregate, regarding potentially any subject or event [26].

At the official Twitter Chirp developer conference in April 2010 [28], the company presented some statistics about its site and its users. In April 2010, Twitter had 106 million registered users, and 180 million unique visitors every month. The company revealed that 300,000 new users were signing up per day and that it received 600 million queries daily via its search engine, and a total of 3 billion requests per day based on its API. Interestingly, 37 percent of Twitter's active users used their phone to send messages.

Twitter data follows the data stream model. In this model, data arrive at high speed, and data mining algorithms must be able to predict in real time and under strict constraints of space and time. Data streams present serious challenges for algorithm design [3]. Algorithms must be able to operate with limited resources, regarding both time and memory. Moreover, they must be able to deal with data whose nature or distribution changes over time.

B. Pfahringer, G. Holmes, and A. Hoffmann (Eds.): DS 2010, LNAI 6332, pp. 1–15, 2010.

The main Twitter data stream that provides all messages from every user in real-time is called Firehose [16] and was made available to developers in 2010. To deal with this large amount of data, and to use it for sentiment analysis and opinion mining—the task considered in this paper—streaming techniques are needed. However, to the best of our knowledge, data stream algorithms, in conjunction with appropriate evaluation techniques, have so far not been considered for this task.

Evaluating data streams in real time is a challenging task. Most work in the literature considers only how to build a picture of accuracy over time. Two main approaches arise [2]:

- **Holdout:** Performance is measured using on single hold-out set.
- **Interleaved Test-Then-Train or Prequential:** Each individual example is used to test the model before it is used for training, and accuracy is incrementally updated.

A common problem is that for unbalanced data streams with, for example, 90% of the instances in one class, the simplest classifiers will have high accuracies of at least 90%. To deal with this type of data stream, we propose to use the Kappa statistic, based on a sliding window, as a measure for classifier performance in unbalanced class streams.

In Section 2 we discuss the challenges that Twitter streaming data poses and discuss related work. Twitter sentiment analysis is discussed is Section 3, and the new evaluation method for time-changing data streams based on the sliding window Kappa statistic is proposed in Section 4. We review text data stream learners in Section 5. Finally, in Section 6, we perform an experimental study on Twitter streams using data stream mining methods.

2 Mining Twitter Data: Challenges and Related Work

Twitter has its own conventions that renders it distinct from other textual data. Consider the following Twitter example message ("tweet"): `RT @toni has a cool #job`. It shows that users may reply to other users by indicating user names using the character @, as in, for example, `@toni`. Hashtags (#) are used to denote subjects or categories, as in, for example `#job`. `RT` is used at the beginning of the tweet to indicate that the message is a so-called "retweet", a repetition or reposting of a previous tweet.

In the knowledge discovery context, there are two fundamental data mining tasks that can be considered in conjunction with Twitter data: (a) graph mining based on analysis of the links amongst messages, and (b) text mining based on analysis of the messages' actual text.

Twitter graph mining has been used to tackle several interesting problems:

- **Measuring user influence and dynamics of popularity.** Direct links indicate the flow of information, and thus a user's influence on others. There are three measures of influence: indegree, retweets and mentions. Cha et al. [5] show that popular users who have high indegree are not necessarily influential in terms of retweets or mentions, and that influence is gained through concerted effort such as limiting tweets to a single topic.

– **Community discovery and formation.** Java et al. [15] found communities using HyperText Induced Topic Search (HITS) [17], and the Clique Percolation Method [8]. Romero and Kleinberg [25] analyze the formation of links in Twitter via the directed closure process.
– **Social information diffusion.** De Choudhury et al. [7] study how data sampling strategies impact the discovery of information diffusion.

There are also a number of interesting tasks that have been tackled using Twitter text mining: sentiment analysis, which is the application we consider in this paper, classification of tweets into categories, clustering of tweets and trending topic detection.

Considering sentiment analysis [18,21], O'Connor et al. [19] found that surveys of consumer confidence and political opinion correlate with sentiment word frequencies in tweets, and propose text stream mining as a substitute for traditional polling. Jansen et al. [14] discuss the implications for organizations of using micro-blogging as part of their marketing strategy. Pak et al. [20] used classification based on the multinomial naïve Bayes classifier for sentiment analysis. Go et al. [12] compared multinomial naïve Bayes, a maximum entropy classifier, and a linear support vector machine; they all exhibited broadly comparable accuracy on their test data, but small differences could be observed depending on the features used.

2.1   The Twitter Streaming API

The Twitter Application Programming Interface (API) [1] currently provides a Streaming API and two discrete REST APIs. Through the Streaming API [16] users can obtain real-time access to tweets in sampled and filtered form. The API is HTTP based, and GET, POST, and DELETE requests can be used to access the data.

In Twitter terminology, individual messages describe the "status" of a user. Based on the Streaming API users can access subsets of public status descriptions in almost real time, including replies and mentions created by public accounts. Status descriptions created by protected accounts and all direct messages cannot be accessed. An interesting property of the streaming API is that it can filter status descriptions using quality metrics, which are influenced by frequent and repetitious status updates, etc.

The API uses basic HTTP authentication and requires a valid Twitter account. Data can be retrieved as XML or the more succinct JSON format. The format of the JSON data is very simple and it can be parsed very easily because every line, terminated by a carriage return, contains one object.

3   Twitter Sentiment Analysis

Sentiment analysis can be cast as a classification problem where the task is to classify messages into two categories depending on whether they convey positive or negative feelings.

Twitter sentiment analysis is not an easy task because a tweet can contain a significant amount of information in very compressed form, and simultaneously carry positive and negative feelings. Consider the following example:

> I currently use the Nikon D90 and love it, but not as much as the Canon 40D/50D. I chose the D90 for the video feature. My mistake.

Also, some tweets may contain sarcasm or irony [4] as in the following example:

> After a whole 5 hours away from work, I get to go back again, I'm so lucky!

To build classifiers for sentiment analysis, we need to collect training data so that we can apply appropriate learning algorithms. Labeling tweets manually as positive or negative is a laborious and expensive, if not impossible, task. However, a significant advantage of Twitter data is that many tweets have author-provided sentiment indicators: changing sentiment is implicit in the use of various types of emoticons. Hence we may use these to label our training data.

Smileys or *emoticons* are visual cues that are associated with emotional states [24,4]. They are constructed by approximating a facial expression of emotion based on the characters available on a standard keyboard. When the author of a tweet uses an emoticon, they are annotating their own text with an emotional state. Annotated tweets can be used to train a sentiment classifier.

4   Streaming Data Evaluation with Unbalanced Classes

In data stream mining, the most frequently used measure for evaluating predictive accuracy of a classifier is prequential accuracy [10]. We argue that this measure is only appropriate when all classes are balanced, and have (approximately) the same number of examples. In this section, we propose the Kappa statistic as a more sensitive measure for quantifying the predictive performance of streaming classifiers. For example, considering the particular target domain in this paper, the rate in which the Twitter Streaming API delivers positive or negative tweets may vary over time; we cannot expect it to be 50% all the time. Hence, a measure that automatically compensates for changes in the class distribution should be preferable.

Just like accuracy, Kappa needs to be estimated using some sampling procedure. Standard estimation procedures for small datasets, such as cross-validation, do not apply. In the case of very large datasets or data streams, there are two basic evaluation procedures: holdout evaluation and prequential evaluation. Only the latter provides a picture of performance over time. In prequential evaluation

Table 1. Simple confusion matrix example

	Predicted Class+	Predicted Class-	Total
Correct Class+	75	8	83
Correct Class-	7	10	17
Total	82	18	100

Table 2. Confusion matrix for chance predictor based on example in Table 1

	Predicted Class+	Predicted Class-	Total
Correct Class+	68.06	14.94	83
Correct Class-	13.94	3.06	17
Total	82	18	100

(also known as interleaved test-then-train evaluation), each example in a data stream is used for testing before it is used for training.

We argue that prequential accuracy is not well-suited for data streams with unbalanced data, and that a prequential estimate of Kappa should be used instead. Let p_0 be the classifier's prequential accuracy, and p_c the probability that a chance classifier—one that assigns the same number of examples to each class as the classifier under consideration—makes a correct prediction. Consider the simple confusion matrix shown in Table 1. From this table, we see that Class+ is predicted correctly 75 out of 100 times, and Class- is predicted correctly 10 times. So accuracy p_0 is 85%. However a classifier predicting solely by chance—in the given proportions—will predict Class+ and Class- correctly in 68.06% and 3.06% of cases respectively. Hence, it will have an accuracy p_c of 71.12% as shown in Table 2.

Comparing the classifier's observed accuracy to that of a chance predictor renders its performance far less impressive than it first seems. The problem is that one class is much more frequent than the other in this example and plain accuracy does not compensate for this. The Kappa statistic, which normalizes a classifier's accuracy by that of a chance predictor, is more appropriate in scenarios such as this one.

The Kappa statistic κ was introduced by Cohen [6]. We argue that it is particularly appropriate in data stream mining due to potential changes in the class distribution. Consider a classifier h, a data set containing m examples and L classes, and a contingency table where cell C_{ij} contains the number of examples for which $h(x) = i$ and the class is j. If $h(x)$ correctly predicts all the data, then all non-zero counts will appear along the diagonal. If h misclassifies some examples, then some off-diagonal elements will be non-zero.

We define

$$p_0 = \frac{\sum_{i=1}^{L} C_{ii}}{m}$$

$$p_c = \sum_{i=1}^{L} \left(\sum_{j=1}^{L} \frac{C_{ij}}{m} \cdot \sum_{j=1}^{L} \frac{C_{ji}}{m} \right)$$

In problems where one class is much more common than the others, any classifier can easily yield a correct prediction by chance, and it will hence obtain a high value for p_0. To correct for this, the κ statistic is defined as follows:

$$\kappa = \frac{p_0 - p_c}{1 - p_c}$$

If the classifier is always correct then $\kappa = 1$. If its predictions coincide with the correct ones as often as those of the chance classifier, then $\kappa = 0$.

The question remains as to how exactly to compute the relevant counts for the contingency table: using all examples seen so far is not useful in time-changing data streams. Gama et al. [10] propose to use a forgetting mechanism for estimating prequential accuracy: a sliding window of size w with the most recent observations, or fading factors that weigh observations using a decay factor α. As the output of the two mechanisms is very similar (every window of size w_0 may be approximated by some decay factor α_0), we propose to use the Kappa statistic measured using a sliding window. Note that, to calculate the statistic for an n_c class problem, we need to maintain only $2n_c + 1$ estimators. We store the sum of all rows and columns in the confusion matrix ($2n_c$ values) to compute p_c, and we store the prequential accuracy p_0. The ability to calculate it efficiently is an important reason why the Kappa statistic is more appropriate for data streams than a measure such as the area under the ROC curve.

5   Data Stream Mining Methods

We experimented with three fast incremental methods that are well-suited to deal with data streams: multinomial naïve Bayes, stochastic gradient descent, and the Hoeffding tree.

Multinomial Naïve Bayes. The multinomial naïve Bayes classifier is a popular classifier for document classification that often yields good performance. It can be trivially applied to data streams because it is straightforward to update the counts required to estimate conditional probabilities..

Multinomial naive Bayes considers a document as a bag-of-words. For each class c, $P(w|c)$, the probability of observing word w given this class, is estimated from the training data, simply by computing the relative frequency of each word in the collection of training documents of that class. The classifier also requires the prior probability $P(c)$, which is straightforward to estimate.

Assuming n_{wd} is the number of times word w occurs in document d, the probability of class c given a test document is calculated as follows:

$$P(c|d) = \frac{P(c) \prod_{w \in d} P(w|c)^{n_{wd}}}{P(d)},$$

where $P(d)$ is a normalization factor. To avoid the zero-frequency problem, it is common to use the Laplace correction for all conditional probabilities involved, which means all counts are initialized to value one instead of zero.

Stochastic Gradient Descent. Stochastic gradient descent (SGD) has experienced a revival since it has been discovered that it provides an efficient means to learn some classifiers even if they are based on non-differentiable loss functions, such as the hinge loss used in support vector machines. In our experiments we use an implementation of vanilla stochastic gradient descent with a fixed learning rate, optimizing the hinge loss with an L_2 penalty that is commonly applied to learn support vector machines. With a linear machine, which is frequently applied for document classification, the loss function we optimize is:

$$\frac{\lambda}{2}||\mathbf{w}||^2 + \sum [1 - (y\mathbf{xw} + b)]_+,$$

where w is the weight vector, b the bias, λ the regularization parameter, and the class labels y are assumed to be in $\{+1, -1\}$.

We compared the performance of our vanilla implementation to that of the Pegasos method [27], which does not require specification of an explicit learning rate, but did not observe a gain in performance using the latter. On the contrary, the ability to specify an explicit learning rate turned out to be crucial to deal with time-changing Twitter data streams : setting the learning rate to a value that was too small meant the classifier adapted too slowly to local changes in the distribution. In our experiments, we used $\lambda = 0.0001$ and set the learning rate for the per-example updates to the classifier's parameters to 0.1.

Hoeffding Tree. The most well-known tree decision tree learner for data streams is the *Hoeffding tree* algorithm [9]. It employs a pre-pruning strategy based on the Hoeffding bound to incrementally grow a decision tree. A node is expanded by splitting as soon as there is sufficient statistical evidence, based on the data seen so far, to support the split and this decision is based on the distribution-independent Hoeffding bound.

Decision tree learners are not commonly applied to document classification due to the high-dimensional feature vectors involved. Simple linear classifiers generally yield higher accuracy. Nevertheless, we include Hoeffding trees in our experiments on Twitter data streams to verify that this observation also holds in this particular context. Moreover, decision trees can potentially yield valuable insight into interactions between variables.

6 Experimental Evaluation

Massive **O**nline **A**nalysis (MOA) [2] is a system for online learning from examples, such as data streams. All algorithms evaluated in this paper were implemented in the Java programming language by using WEKA [13] and the MOA software.

In our experiments, we used the Twitter training datasets to extract features using text filters in WEKA. Each tweet was represented as a set of words. We extracted 10,000 unigrams using the default stop word list in WEKA. We used term presence instead of frequency, as Pang et al. [22] reported that term presence achieves better results than frequency on sentiment analysis classification. The resulting vectors are stored in sparse format.

6.1 The `twittersentiment.appspot.com` and Edinburgh Corpora

Twitter Sentiment (`twittersentiment.appspot.com`) is a website that enables visitors to research and track the sentiment for a brand, product, or topic. It was created by Alec Go, Richa Bhayani, Karthik Raghunathan, and Lei Huang from Stanford University. The website enables a visitor to track queries over time. Sentiment classification is based on a linear model generated using the maximum entropy method.[1] The Twitter Sentiment website provides an API to use the maximum entropy classifier: one can use it to determine the polarity of arbitrary text, retrieve sentiment counts over time, and retrieve tweets along with their classification.

The developers of the website collected two datasets: a training set and a test one, which were also used for sentiment classification in [12]. The training dataset was obtained by querying the (non-streaming) Twitter API for messages between April 2009 and June 25, 2009 and contains the first 800,000 tweets with positive emoticons, and the first 800,000 tweets with negative emoticons. The list of positive emoticons used was: :), :-), :), :D, and =). The negative emoticons used were: :(, :-(, and : (. The test dataset was manually annotated with class labels and consists of 177 negative tweets and 182 positive ones. Test tweets were collected by looking for messages that contained a sentiment, regardless of the presence of emoticons. Each tweet contains the following information: its polarity (indicating the sentiment), the date, the query used, the user, and the actual text.

The Edinburgh corpus [23] was collected over a period of two months using the Twitter streaming API. It contains 97 million tweets and requires 14 GB of disk space when stored in uncompressed form.[2] Each tweet has the following information: the timestamp of the tweet, an anonymized user name, the tweet's text, and the posting method that was used.

The corpus was collected between November 11th 2009 and February 1st 2010, using Twitter's streaming API. It is thus a representative sample of the entire stream. The data contains over 2 billion words and there is no distinction between English and non-English tweets. We only considered tweets in English and only those that contained emoticons.

---

[1] The software is available at `http://nlp.stanford/software/classifier.shtml`

[2] The corpus can be obtained at `http://demeter.inf.ed.ac.uk/`

Table 3. Total prequential accuracy and Kappa measured on the `twittersentiment.appspot.com` data stream

	Accuracy	Kappa	Time
Multinomial Naïve Bayes	75.05%	50.10%	116.62 sec.
SGD	82.80%	62.60%	219.54 sec.
Hoeffding Tree	73.11%	46.23%	5525.51 sec.

Table 4. Accuracy and Kappa for the test dataset obtained from `twittersentiment.appspot.com`

	Accuracy	Kappa
Multinomial Naïve Bayes	82.45%	64.89%
SGD	78.55%	57.23%
Hoeffding Tree	69.36%	38.73%

6.2 Results and Discussion

We performed two data stream experiments: one using the training dataset from `twittersentiment.appspot.com`, and another one with the Edinburgh Corpus. We also performed a classic train/test experiment based on each training set and the test set from `twittersentiment.appspot.com`.

First, we consider the data from `twittersentiment.appspot.com`. We performed a prequential evaluation, testing and then training, using the training stream of $1,600,000$ instances, half positives and half negatives. Figure 1 shows the learning curve for this stream measuring prequential accuracy and Kappa using a sliding window of size $1,000$. Table 3 reports the total prequential accuracy and Kappa. In this data stream the last $200,000$ instances are positive, as the data was collected to have the same number of positive and negative tweets: the rate of tweets using positive emoticons is usually higher than that of negative ones. We see at the end of the learning curve in Figure 1 that prequential accuracy still presents (apparently) good results, but that the value of Kappa is zero or below. This is an extreme example of a change in class distribution (one class disappears completely from the stream), which shows very clearly why Kappa is useful when the distribution of classes evolves over time. We see that the worst method in accuracy, Kappa, and time for this dataset is the Hoeffding Tree, supporting our hypothesis that tree learners are not appropriate in this context.

The second experiment uses the data from `twittersentiment.appspot.com` in a classic train/test set-up. Table 4 reports accuracy and Kappa for the test set. The results for accuracy for naïve Bayes are comparable to those in [12]. As SGD is very sensitive to change in the data distribution, we trained it on a randomized version of the training dataset for this particular test. Doing this improves its accuracy on the test set, but naïve Bayes is somewhat better. Note that the Hoeffding tree is the slowest of the three methods, as the current implementation does not use sparse instances as multinomial naïve Bayes and SGD do.

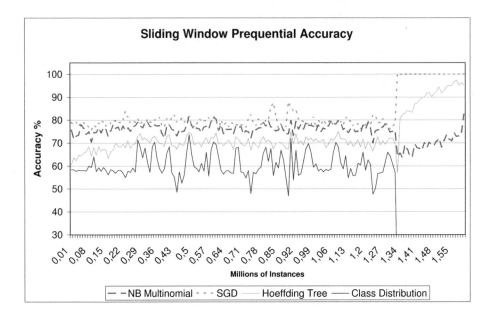

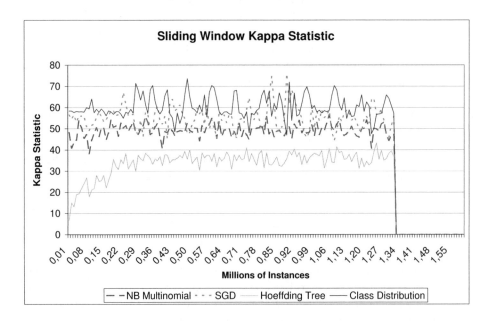

Fig. 1. Sliding window prequential accuracy and Kappa measured on the `twittersentiment.appspot.com` data stream. (Note: solid line shows accuracy in both graphs.).

Table 5. Total prequential accuracy and Kappa obtained on the Edinburgh corpus data stream

	Accuracy	Kappa	Time
Multinomial Naïve Bayes	86.11%	36.15%	173.28, sec.
SGD	86.26%	31.88%	293.98 sec.
Hoeffding Tree	84.76%	20.40%	6151.51 sec.

Table 6. Accuracy and Kappa for the test dataset obtained from `twittersentiment.appspot.com` using the Edinburgh corpus as training data stream

	Accuracy	Kappa
Multinomial Naïve Bayes	73.81%	47.28%
SGD	67.41%	34.23%
Hoeffding Tree	60.72%	20.59%

The `twittersentiment.appspot.com` data does not constitute a representative sample of the real Twitter stream due to the fact that the data was augmented to be balanced. Hence we now turn to the Edinburgh corpus. We converted the raw data following the same methodology as Go et al. [11,12]:

- Feature Reduction. Twitter users may use the @ symbol before a name to direct the message to a certain recipient. We replaced words starting with the @ symbol with the generic token USER, and any URLs by the token URL. We also replaced repeated letters: e.g., *huuuuuungry* was converted to *huungry* to distinguish it from *hungry*.
- Emoticons. Once they had been used to generate class labels, all emoticons were deleted from the input text so that they could not be used as predictive features.

Once this steps had been performed WEKA's text filter was used to convert the data into vector format.

The resulting data stream contains 324, 917 negative tweets and 1, 813, 705 positive ones. We observe that negative tweets constitute 15% of the labeled data and positive ones 85%. It appears that people tend to use more positive emoticons than negative ones.

Figure 2 shows the learning curve measuring prequential accuracy and Kappa using a sliding window of 1, 000, and Table 5 reports the total prequential accuracy and value of Kappa. We see in the learning curve of Figure 2 that accuracy is similar for the three methods, but this is not the case when one considers the Kappa statistic. Again, Kappa provides us with a better picture of relative predictive performance. In this stream, we see that multinomial naïve Bayes and SGD perform comparably.

Finally, we test the classifiers learned with the Edinburgh corpus using the test set from `twittersentiment.appspot.com`. Again, the training data was

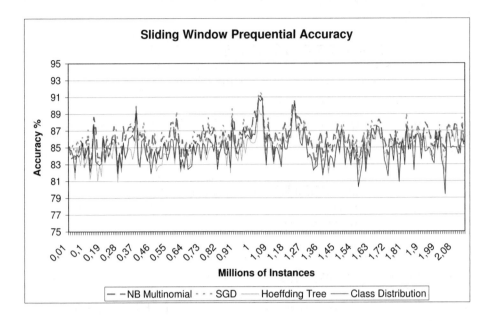

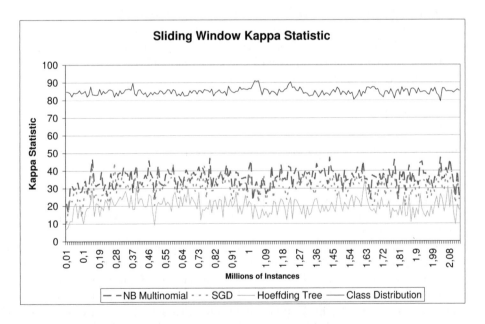

Fig. 2. Sliding window prequential accuracy and Kappa measured on data stream obtained from the Edinburgh corpus. (Note: solid line shows accuracy in both graphs.).

Table 7. SGD coefficient variations on the Edinburgh corpus

Tags	Middle of Stream Coefficient	End of Stream Coefficient	Variation
apple	0.3	0.7	0.4
microsoft	-0.4	-0.1	0.3
facebook	-0.3	0.4	0.7
mcdonalds	0.5	0.1	-0.4
google	0.3	0.6	0.3
disney	0.0	0.0	0.0
bmw	0.0	-0.2	-0.2
pepsi	0.1	-0.6	-0.7
dell	0.2	0.0	-0.2
gucci	-0.4	0.6	1.0
amazon	-0.1	-0.4	-0.3

randomized for SGD as in the case of the `twittersentiment.appspot.com` data. Table 6 shows the results. The value of Kappa shows that multinomial naïve Bayes is the most accurate method on this particular test set.

An advantage of the SGD-based model is that changes in its weights can be inspected to gain insight into changing properties of the data stream. Table 7 shows the change in coefficients for some words along the stream obtained from the Edinburgh corpus. The coefficients correspond to December 26th 2009, and February 1st 2010, respectively. Monitoring these coefficients, which determine how strongly absence/presence of the corresponding word influences the model's prediction of negative or positive sentiment, may be an efficient way to detect changes in the population's opinion regarding a particular topic or brand.

7   Conclusions

Twitter streaming data can potentially enable any user to discover what is happening in the world at any given moment in time. Because the Twitter Streaming API delivers a large quantity of tweets in real time, data stream mining and evaluation techniques are the best fit for the task at hand, but have not been considered previously. We discussed the challenges that Twitter streaming data poses, focusing on sentiment analysis, and proposed the sliding window Kappa statistic as an evaluation metric for data streams. Considering all tests performed and ease of interpretability, the SGD-based model, used with an appropriate learning rate, can be recommended for this data.

In future work, we would like to extend the results presented here by evaluating our methods in real time and using other features available in Twitter data streams, such as geographical place, the number of followers or the number of friends.

Acknowledgments

We would like to thank Alec Go, Lei Huang, and Richa Bhayani for very generously sharing their Twitter dataset with us. We would also like to thank Sasa Petrovic, Miles Osborne, and Victor Lavrenko for making their Twitter dataset publicly available.

References

1. Twitter API: (2010), `http://apiwiki.twitter.com/`
2. Bifet, A., Holmes, G., Kirkby, R., Pfahringer, B.: MOA: Massive Online Analysis Journal of Machine Learning Research, JMLR (2010),
   `http://moa.cs.waikato.ac.nz/`
3. Bifet, A., Holmes, G., Pfahringer, B., Frank, E.: Fast perceptron decision tree learning from evolving data streams. In: Proceedings of the 14th Pacific-Asia Conference on Knowledge Discovery and Data Mining, pp. 299–310 (2010)
4. Carvalho, P., Sarmento, L., Silva, M.J., de Oliveira, E.: Clues for detecting irony in user-generated contents: oh..!! it's "so easy";-). In: Proceeding of the 1st International CIKM Workshop on Topic-sentiment Analysis for Mass Opinion, pp. 53–56 (2009)
5. Cha, M., Haddadi, H., Benevenuto, F., Gummadi, K.P.: Measuring User Influence in Twitter: The Million Follower Fallacy. In: Proceedings of the 4th International AAAI Conference on Weblogs and Social Media, pp. 10–17 (2010)
6. Cohen, J.: A coefficient of agreement for nominal scales. Educational and Psychological Measurement 20(1), 37–46 (1960)
7. De Choudhury, M., Lin, Y.-R., Sundaram, H., Candan, K.S., Xie, L., Kelliher, A.: How does the data sampling strategy impact the discovery of information diffusion in social media. In: Proceedings of the 4th International AAAI Conference on Weblogs and Social Media, pp. 34–41 (2010)
8. Derenyi, I., Palla, G., Vicsek, T.: Clique percolation in random networks. Physical Review Letters 94(16) (2005)
9. Domingos, P., Hulten, G.: Mining high-speed data streams. In: Proceedings of the 6th ACM SIGKDD International Conference on Knowledge Discovery and Data Mining, pp. 71–80 (2000)
10. Gama, J., Sebastião, R., Rodrigues, P.P.: Issues in evaluation of stream learning algorithms. In: Proceedings of the 15th ACM SIGKDD International Conference on Knowledge Discovery and Data Mining, pp. 329–338 (2009)
11. Go, A., Bhayani, R., Raghunathan, K., Huangi, L.: (2009),
    `http://twittersentiment.appspot.com/`
12. Go, A., Huang, L., Bhayani, R.: Twitter sentiment classification using distant supervision. In: CS224N Project Report, Stanford (2009)
13. Hall, M., Frank, E., Holmes, G., Pfahringer, B., Reutemann, P., Witten, I.H.: The WEKA data mining software: an update. SIGKDD Explor. Newsl. 11(1), 10–18 (2009)
14. Jansen, B.J., Zhang, M., Sobel, K., Chowdury, A.: Micro-blogging as online word of mouth branding. In: Proceedings of the 27th International Conference Extended Abstracts on Human Factors in Computing Systems, pp. 3859–3864 (2009)

15. Java, A., Song, X., Finin, T., Tseng, B.: Why we twitter: understanding microblogging usage and communities. In: Proceedings of the 9th WebKDD and 1st SNA-KDD 2007 Workshop on Web Mining and Social Network Analysis, pp. 56–65 (2007)
16. Kalucki, J.: Twitter streaming API (2010),
    http://apiwiki.twitter.com/Streaming-API-Documentation
17. Kleinberg, J.M.: Authoritative sources in a hyperlinked environment. J. ACM 46(5), 604–632 (1999)
18. Liu, B.: Web data mining; Exploring hyperlinks, contents, and usage data. Springer, Heidelberg (2006)
19. O'Connor, B., Balasubramanyan, R., Routledge, B.R., Smith, N.A.: From tweets to polls: Linking text sentiment to public opinion time series. In: Proceedings of the International AAAI Conference on Weblogs and Social Media, pp. 122–129 (2010)
20. Pak, A., Paroubek, P.: Twitter as a corpus for sentiment analysis and opinion mining. In: Proceedings of the Seventh Conference on International Language Resources and Evaluation, pp. 1320–1326 (2010)
21. Pang, B., Lee, L.: Opinion mining and sentiment analysis. Foundations and Trends in Information Retrieval 2(1-2), 1–135 (2008)
22. Pang, B., Lee, L., Vaithyanathan, S.: Thumbs up? Sentiment classification using machine learning techniques. In: Proceedings of the Conference on Empirical Methods in Natural Language Processing, pp. 79–86 (2002)
23. Petrovic, S., Osborne, M., Lavrenko, V.: The Edinburgh Twitter corpus. In: #SocialMedia Workshop: Computational Linguistics in a World of Social Media, pp. 25–26 (2010)
24. Read, J.: Using emoticons to reduce dependency in machine learning techniques for sentiment classification. In: Proceedings of the ACL Student Research Workshop, pp. 43–48 (2005)
25. Romero, D.M., Kleinberg, J.: The directed closure process in hybrid social-information networks, with an analysis of link formation on Twitter. In: Proceedings of the 4th International AAAI Conference on Weblogs and Social Media, pp. 138–145 (2010)
26. Schonfeld, E.: Mining the thought stream. TechCrunch Weblog Article (2009),
    http://techcrunch.com/2009/02/15/mining-the-thought-stream/
27. Shalev-Shwartz, S., Singer, Y., Srebro, N.: Pegasos: Primal Estimated sub-GrAdient SOlver for SVM. In: Proceedings of the 24th International Conference on Machine learning, pp. 807–814 (2007)
28. Yarow, J.: Twitter finally reveals all its secret stats. BusinessInsider Weblog Article (2010), http://www.businessinsider.com/twitter-stats-2010-4/

A Similarity-Based Adaptation of Naive Bayes for Label Ranking: Application to the Metalearning Problem of Algorithm Recommendation

Artur Aiguzhinov[1,2],
Carlos Soares[1,2], and Ana Paula Serra[1,3]

[1] FEP - Faculdade de Economia da Universidade do Porto
[2] LIAAD-INESC Porto LA
[3] CEFUP - Centro de Economia e Finanças da Universidade do Porto
artur@liaad.up.pt, csoares@fep.up.pt, aserra@fep.up.pt

Abstract. The problem of learning label rankings is receiving increasing attention from several research communities. A number of common learning algorithms have been adapted for this task, including k-Nearest Neighbours (k-NN) and decision trees. Following this line, we propose an adaptation of the naive Bayes classification algorithm for the label ranking problem. Our main idea lies in the use of similarity between the rankings to replace the concept of probability. We empirically test the proposed method on some metalearning problems that consist of relating characteristics of learning problems to the relative performance of learning algorithms. Our method generally performs better than the baseline indicating that it is able to identify some of the underlying patterns in the data.

1 Introduction

Label ranking is an increasingly popular topic in the machine learning literature [9]. Label ranking studies the problem of learning a mapping from instances to rankings over a finite number of predefined labels. It is a variation of the conventional classification problem. In contrast to the classification setting, where the objective is to assign examples to a specific class, in label ranking we are interested in assigning a complete preference order of the labels to every example [5].

Several methods have been developed for label ranking, some of which consist of adapting existing classification algorithms (e.g., k-Nearest Neighbor [2], decision trees [15]). Some approaches (e.g., [5]) are based on probabilistic models for ranking, such as the Mallows model [10,5]. Other approaches take advantage of the possibility to compute the similarity/distance between rankings, unlike the traditional classification setting, where two classes are either the same or different [15].

B. Pfahringer, G. Holmes, and A. Hoffmann (Eds.): DS 2010, LNAI 6332, pp. 16–26, 2010.

In this paper, we follow the latter approach. We propose an adaptation of the naive Bayes (NB) algorithm for label ranking. Despite its limitations, NB is an algorithm with successful results in many applications [7]. Additionally, the Bayesian framework is well understood in many domains. For instance, we plan to apply this method on the problem of predicting the rankings of financial analysts. In the Financial Economics area, Bayesian models are widely used (e.g., the Black–Litterman model for active portfolio management [1]).

The main idea lies in replacing the probabilities in the Bayes theorem with the distance between rankings. This can be done because it has been shown that there is a parallel between the concepts of distance and likelihood [17].

The paper is organized as follows: section 2 provides the formalization of the label ranking problem; section 3 briefly describes the naive Bayes algorithm for classification; section 4 shows the adaptation of the NB algorithm for label ranking; section 5 explains the problem of metalearning, which will be the application domain for the empirical evaluation; section 6 presents empirical results; finally, section 7 concludes with the goals for future work.

2   Learning Label Rankings

Based on [16], a label ranking problem is defined as follows. Let $\mathcal{X} \subseteq \{\mathcal{V}_1, \dots, \mathcal{V}_m\}$ be an instance space of nominal variables, such that $\mathcal{V}_a = \{v_{a,1}, \dots, v_{a,n_a}\}$ is the domain of nominal variable a. Also, let $\mathcal{L} = \{\lambda_1, \dots, \lambda_k\}$ be a set of labels, and $\mathcal{Y} = \Pi_{\mathcal{L}}$ be the output space of all possible total orders[1] over $\mathcal{L}$ defined on the permutation space Π. The goal of a label ranking algorithm is to learn a mapping $h : \mathcal{X} \to \mathcal{Y}$, where h is chosen from a given hypothesis space $\mathcal{H}$, such that a predefined loss function $\ell : \mathcal{H} \times \mathcal{Y} \times \mathcal{Y} \to \mathbb{R}$ is minimized. The algorithm learns h from a training set $\mathcal{T} = \{x_i, y_i\}_{i \in \{1,\dots,n\}} \subseteq \mathcal{X} \times \mathcal{Y}$ of n examples, where $x_i = \{x_{i,1}, x_{i,2}, \dots, x_{i,m}\} \in \mathcal{X}$ and $y_i = \{y_{i,1}, y_{i,2}, \dots, y_{i,k}\} \in \Pi_{\mathcal{L}}$. Furthermore, we define $y_i^{-1} = \{y_{i,1}^{-1}, y_{i,2}^{-1}, \dots, y_{i,k}^{-1}\}$ as the order of the labels in example i. Given that we are focusing on total orders, y_i^{-1} is a permutation of the set $\{1, 2, \dots, k\}$ where $y_{i,j}^{-1}$ is the rank of label λ_j in example i.

Unlike classification, where for each instance $x \in \mathcal{X}$ there is an associated class $y_i \in \mathcal{L}^2$, in label ranking problems there is a ranking of the labels associated with every instance x and the goal is to predict it. This is also different from other ranking problems, such as in information retrieval or recommender systems. In these problems the target variable is a set of ratings or binary relevance labels for each item, and not a ranking.

---

[1] A total order is a complete, transitive, and asymmetric relation $\succ$ on $\mathcal{L}$, where $\lambda_i \succ \lambda_j$ indicates that λ_i precedes λ_j. In this paper, given $\mathcal{L} = \{A, B, C\}$, we will use the notation $\{A, C, B\}$ and $\{1, 3, 2\}$ interchangeably to represent the order $A \succ C \succ B$.

[2] Here, we use both y_i to represent the target class (label) in classification and the target ranking in label ranking to clarify that they are both the target of the learning problem. We will explicitly state the task we are dealing with when it is not clear from the context.

The algorithms for label ranking can be divided into two main approaches: methods that transform the ranking problem into multiple binary problems and methods that were developed or adapted to predict the rankings. An example of the former is the ranking by pairwise comparisons [9]. Some examples of algorithms that are specific for rankings are: the predictive clustering trees method [15], the similarity-based k-Nearest Neighbour for label ranking [2], the probabilistic k-Nearest Neighbour for label ranking [5] and the linear utility transformation method [8,6].

To assess the accuracy of the predicted rankings relative to the corresponding target rankings, a suitable loss function is needed. In this paper we compare two rankings using the Spearman correlation coefficient [2,16]:

$$\rho(\pi, \hat{\pi}) = 1 - \frac{6 \sum_{j=1}^{k} (\pi_j - \hat{\pi}_j)^2}{k^3 - k} \tag{1}$$

where π and $\hat{\pi}^3$ are, respectively, the target and predicted rankings for a given instance. Two orders with all the labels placed in the same position will have a Spearman correlation of $+1$. Labels placed in reverse order will produce correlation of -1. Thus, the higher the value of ρ the more accurate the prediction is compared to target. The loss function is given by the mean Spearman correlation values (eq. 1) between the predicted and target rankings, across all examples in the dataset:

$$\ell = \frac{\sum_{i=1}^{n} \rho(\pi_i, \hat{\pi}_i)}{n} \tag{2}$$

An extensive survey of label ranking algorithms is given by [16].

3   The Naive Bayes Classifier

We follow [11] to formalize the naive Bayes classifier. In classification, each instance $x_i \in \mathcal{X}$ is binded to class $y_i \in \mathcal{L}$. The task of a learner is to create a classifier from the training set $\mathcal{T}$. The classifier takes a new, unlabelled instance and assigns it to a class (label).

The naive Bayes method classifies a new instance x_i by determining the most probable target value, $c_{MAP(x_i)}^4$, given the attribute values that describe the instance:

$$c_{MAP(x_i)} = \arg\max_{\lambda \in \mathcal{L}} P(\lambda | x_{i,1}, x_{i,2}, \ldots, x_{i,m}) \tag{3}$$

where $x_{i,j}$ is the value of attribute j for instance i.

The algorithm is based on the Bayes theorem that establishes the probability of A given B as:

$$P(A|B) = \frac{P(B|A)P(A)}{P(B)} \tag{4}$$

---

[3] In the following, we will use y_i and π_i interchangeably to represent the target ranking.
[4] MAP – Maximum A Posteriori.

Thus, the Bayes theorem provides a way to calculate the posterior probability of a hypothesis.

Using (4), we can rewrite (3) as

$$c_{MAP(x_i)} = \underset{\lambda \in \mathcal{L}}{\arg\max} \frac{P(x_{i,1}, x_{i,2}, \ldots, x_{i,m} | \lambda) P(\lambda)}{P(x_{i,1}, x_{i,2}, \ldots, x_{i,m})}$$

$$= \underset{\lambda \in \mathcal{L}}{\arg\max} \, P(x_{i,1}, x_{i,2} \ldots x_{i,m} | \lambda) P(\lambda) \qquad (5)$$

Computing the likelihood $P(x_{i,1}, x_{i,2}, \ldots, x_{i,m} | \lambda)$ is very complex and requires large amounts of data, in order to produce reliable estimates. Therefore, the naive Bayes classifier makes one simple, hence, naive, assumption that the attribute values are conditionally independent from each other. This implies that the probability of observing the conjunction $x_{i,1}, x_{i,2}, \ldots, x_{i,m}$ is the product of the probabilities for the individual attributes: $P(x_{i,1}, x_{i,2}, \ldots, x_{i,m} | \lambda) = \prod_{j=1}^{m} P(x_{i,j} | \lambda)$. Substituting this expression into equation (5), we obtain the naive Bayes classifier:

$$c_{nb}(x_i) = \underset{\lambda \in \mathcal{L}}{\arg\max} \, P(\lambda) \prod_{j=1}^{m} P(x_{i,j} | \lambda) \qquad (6)$$

4   Adapting NB to Ranking

Consider the classic problem of the play/no play tennis based on the weather conditions. The naive Bayes classification algorithm can be successfully applied to this problem [11, chap. 6]. For illustration purposes, we extend this example application to the label ranking setting by replacing the target with a ranking on the preferences of a golf player regarding three tennis courts on different days (Table 1).

The last three columns in Table 1 represent the ranks of the tennis courts A, B and C.

Table 1. Example of tennis courts $\{A, B, C\}$ rankings based on the observed weather conditions

Day	Outlook	Temperature	Humidity	Wind	Ranks A B C
1	Sunny	Hot	High	Weak	1 2 3
2	Sunny	Hot	High	Strong	2 3 1
3	Overcast	Hot	High	Weak	1 2 3
4	Rain	Mild	High	Weak	1 3 2
5	Rain	Mild	High	Strong	1 2 3
6	Sunny	Mild	High	Strong	2 3 1

As described earlier, the difference between classification and label ranking lies in the target variable, y. Therefore, to adapt NB for ranking we have to adapt the parts of the algorithm that depend on the target variable, namely:

- prior probability, $P(y)$
- conditional probability, $P(x|y)$

The adaptation should take into account the differences in nature between label rankings and classes. For example, if we consider label ranking as a classification problem, then the prior probability of ranking $\{A, B, C\}$ on the data given in Table 1 is $P(\{A, B, C\}) = 3/6 = 0.5$, which is quite high. On the other hand, the probability of $\{A, C, B\}$ is quite low, $P(\{A, C, B\}) = 1/6 = 0.167$. However, taking into account the stochastic nature of these rankings [5], it is intuitively clear that the observation of $\{A, B, C\}$ increases the probability of observing $\{A, C, B\}$ and vice-versa. This affects even rankings that are not observed in the available data. For example, the case of unobserved ranking $\{B, A, C\}$ in Table 1 would not be entirely unexpected in the future considering a similar observed ranking $\{B, C, A\}$.

One approach to deal with stochastic nature characteristic of label rankings is to use ranking distributions, such as the Mallows model (e.g., [10,5]). Alternatively, we may consider that the intuition described above is represented by varying similarity between rankings.

Similarity-based label ranking algorithms have two important properties:

- they assign non-zero probabilities even for rankings which have not been observed. This property is common to distribution-based methods;
- they are based on the notion of similarity between rankings, which also underlies the evaluation measures that are commonly used. Better performance is naturally expected by aligning the algorithm with the evaluation measure.

Similarity and probability are different concepts and, in order to adapt NB for label ranking based on the concept of similarity, it is necessary to relate them. A parallel has been established between probabilities and the general Euclidean distance measure [17]. This work shows that maximizing the likelihood is equivalent to minimizing the distance (i.e., maximizing the similarity) in a Euclidean space. Although not all assumptions required for that parallel hold when considering distance (or similarity) between rankings, given that the naive Bayes algorithm is known to be robust to violations of its assumptions, we propose a similarity-based adaptation of NB for label ranking.

In the following description, we will retain the probabilistic terminology (e.g., prior probability) from the original algorithm, even though it does not apply for similarity functions. However, in the mathematical notation, we will use the subscript $_{LR}$ to distinguish the concepts. Despite the abuse, we believe this makes the algorithm easier to understand.

We start by defining $\mathcal{S}$ as a similarity matrix between the target rankings in a training set, i.e. $\mathcal{S}_{n \times n} = \rho(\pi_i, \pi_j)$. The prior probability of a label ranking is given by:

$$P_{LR}(\pi) = \frac{\sum_{i=1}^{n} \rho(\pi, \pi_i)}{n} \tag{7}$$

We say that the prior probability is the mean of similarity of a given rankings to all the others. We measure similarity using the Spearman correlation coefficient (1). Equation 7 shows the average similarity of one ranking relative to others. The greater the similarity between two particular rankings, the higher is the probability that the next unobserved ranking will be similar to the known ranking. Take a look a the Table 2 with the calculated prior probability for the unique rankings. We also added a column with prior probabilities considering the rankings as one class.

Table 2. Comparison of values of prior probability by addressing the label ranking problem as a classification problem or using similarity

π	$P(\pi)$	$P_{LR}(\pi)$
A B C	0.500	0.708
B C A	0.333	0.583
A C B	0.167	0.792

As stated above, the ranking $\{A, C, B\}$, due to its similarity to the other two rankings, achieves a higher probability. Note that since we measure prior probability of label ranking as a similarity between rankings, it would not add to one as the in case of probability for classification.

The similarity of rankings based on the value i of attribute a, $(v_{a,i})$, or conditional probability of label rankings, is:

$$P_{LR}(v_{a,i}|\pi) = \frac{\sum_{i:x_{i,a}=v_{a,i}} \rho(\pi, \pi_i)}{|\{i : x_{i,a} = v_{a,i}\}|} \tag{8}$$

Table 3 demonstrates the logic behind the conditional probabilities based on similarity. Notice that there are no examples with $Outlook = Sunny$ and a target ranking of $\{A, C, B\}$; thus, $P(Outlook = Sunny|\{A, C, B\}) = 0$. However, in the similarity approach, the probability of $\{A, C, B\}$ depends on the probability of similar rankings, yielding $P_{LR}(Outlook = Sunny|\{A, C, B\}) = 0.750$.

Table 3. Comparison of values of conditional probability by addressing the label ranking problem as a classification problem or using similarity

π	$P(Outlook = Sunny|\pi)$	$P_{LR}(Outlook = Sunny|\pi)$
A B C	0.33	0.500
B C A	1.00	0.750
A C B	0.00	0.750

Applying equation (6), we get the estimated posterior probability of ranking π:

$$P_{LR}(\pi|x_i) = P_{LR}(\pi) \prod_{a=1}^{m} P_{LR}(x_{i,a}|\pi) = \tag{9}$$

$$= \frac{\sum_{j=1}^{n} \rho(\pi, \pi_j)}{n} \left[\prod_{a=1}^{m} \frac{\sum_{j:x_{j,a}=x_{i,a}} \rho(\pi, \pi_j)}{|\{j : x_{j,a} = x_{i,a}\}|} \right]$$

The similarity-based adaptation of naive Bayes for label ranking will output the ranking with the higher $P_{LR}(\pi|x_i)$ value:

$$\hat{\pi} = \arg\max_{\pi \in \Pi_{\mathcal{L}}} P_{LR}(\pi|x_i) = \tag{10}$$

$$= \arg\max_{\pi \in \Pi_{\mathcal{L}}} P_{LR}(\pi) \prod_{a=1}^{m} P_{LR}(x_{i,a}|\pi)$$

5   Metalearning

The algorithm proposed in the previous section was tested on some metalearning problems. Algorithm recommendation using a metalearning approach has often been address as a label ranking problem [2,15]. Here, we provide a summary of a problem.

Many different learning algorithms are available to data analysts nowadays. For instance, decision trees, neural networks, linear discriminants, support vector machines among others can be used in classification problems. The goal of data analysts is to use the one that will obtain the best performance on the problem at hand. Given that the performance of learning algorithms varies for different datasets, data analysts must select carefully which algorithm to use for each problem, in order to obtain satisfactory results.

Therefore, we can say that a performance measure establishes a ranking of learning algorithms for each problem. For instance, Table 4 illustrates the ranking of four classification algorithms (a_i) on two datasets (d_j) defined by estimates of the classification accuracy of those algorithms on those datasets.

Table 4. Accuracy of four learning algorithms on two classification problems

	a_1	a_2	a_3	a_4
d_1	90% (1)	61% (3)	82% (2)	55% (4)
d_2	84% (2)	86% (1)	60%(4)	79% (3)

Selecting the algorithm by trying out all alternatives is generally not a viable option. As explained in [15]:

> In many cases, running an algorithm on a given task can be time consuming, especially when complex tasks are involved. It is therefore desirable to be able to predict the performance of a given algorithm on a given task from description and without actually running the algorithm.

The learning approach to the problem of algorithm recommendation consists of using a learning algorithm to model the relation between the characteristics of learning problems (e.g., application domain, number of examples, proportion of symbolic attributes) and the relative performance (or ranking) of a set of algorithms [2]. We refer to this approach as *metalearning* because we are learning about the performance of learning algorithms.

Metalearning approaches commonly cast the algorithm recommendation problem as a classification task. Therefore, the recommendation provided to the user consists of a single algorithm. In this approach, the examples are datasets and the classes are algorithms. However, this is not the most suitable form of recommendation. Although the computational cost of executing all the algorithms is very high, it is often the case that it is possible to run a few of the available algorithms. Therefore, it makes more sense to provide recommendation in the form of a ranking, i.e. address the problem using a label ranking approach, where the labels are the algorithms. The user can then execute the algorithms in the suggested order, until no computational resources (or time) are available.

In the metalearning datasets, each example (x_i, y_i) represents a machine learning problem, referred to here as base-level dataset (BLD). The x_i is the set of metafeatures that represent characteristics of the BLD (e.g., mutual information between symbolic attributes and the target) and the y_i is the target ranking, representing the relative performance of a set of learning algorithms on the corresponding BLD. More details can be found in [2].

6   Experiment Results

We empirically tested the proposed adaptation of the naive Bayes algorithm for learning label rankings on some ranking problems obtained from metalearning applications. We start by describing the experimental setup and then we discuss the results.

6.1   Experimental Setup

We used the following metalearning datasets in our experiments:

- class: these data represent the performance of ten algorithms on a set of 57 classification BLD. The BLD are characterized by a set of metafeatures which obtained good results with the k-NN algorithm [2].
- regr: these data represent the performance of nine algorithms on a set of 42 regression BLD. The set of metafeatures used here has also obtained good results previously [14].
- svm-*: we have tried four different datasets describing the performance of different variants of the Support Vector Machines algorithm on the same 42 regression BLD as in the previous set and also using the same set of metafeatures [13]. The difference between the first three sets, svm-5, svm-eps01 and svm-21 is in the number of different values of the kernel parameter that were considered. The remaining dataset svm-eps01 uses the same 11 alternative kernel parameters as svm-11 but the value of the kernel parameter ϵ is 0.128 and not 0.001 as in the other sets.

Given that the attributes in the metalearning datasets are numerical and the NB algorithm is for symbolic attributes, they must be discretized. We used a simple equal-width binning method using 10 bins.

The baseline is a simple method based on the mean rank of each label (i.e., algorithm or parameter setting in these datasets) over all training examples (i.e., BLDs) [3].

$$\bar{\pi}_j^{-1} = \frac{\sum_{i=1}^{n} \pi_{i,j}^{-1}}{n} \tag{11}$$

where $\pi_{i,j}^{-1}$ is the rank of label λ_j on dataset i. The final ranking is obtained by ordering the mean ranks and assigning them to the labels accordingly. This ranking is usually called the *default ranking*, in parallel to the default class in classification.

The performance of the label ranking methods was estimated using a methodology that has been used previously for this purpose [2]. It is based on the leave-one-out performance estimation method because of the small size of the datasets. The accuracy of the rankings predicted by methods was evaluated by comparing them to the target rankings (i.e., the rankings based on the observed performance of the algorithms) using the Spearman's correlation coefficient (Eq. 1). The code for all the examples in this paper has been written in R [12].

6.2   Results

The results of the experiments are presented in Table 5. As the table shows, the algorithm is significantly better than the baseline on all datasets. In two datasets, with a 99% confidence level, in one with 95% and the remaining with 90% confidence. Despite the small size of the datasets (less than 60 examples), the algorithm is able to detect some patterns with predictive value.

Table 5. Experimental results of the adapted naive Bayes algorithm for label ranking compared to the baseline. Items with (*),(**), and (***) have statistical significance at 10% , 5% , and 1% confidence level respectively.

Dataset	NBr	Baseline	p-values
class	0.506	0.479	0.000***
regr	0.658	0.523	0.056*
svm-5	0.326	0.083	0.000***
svm-11	0.372	0.144	0.029**
svm-21	0.362	0.229	0.055*
svm-eps01	0.369	0.244	0.091*

7   Conclusion

In this paper we presented an adaptation of the naive Bayes algorithm for label ranking that is based on similarities of the rankings taking advantage of a parallel that can be established between the concepts of likelihood and distance. We tested the new algorithm on a number of metalearning datasets and conclude that it consistently outperforms a baseline method.

A number of issues remain open, which we plan to address in the future. Firstly, we are currently working on creating new datasets for ranking applications in different areas, including finance (e.g., predicting the rankings of the financial analysts based on their recommendations). These new datasets will enable us to better understand the behaviour of the proposed algorithm. In addition, we assume that target rankings are total orders. In practice, this is often not true [4,2]. We plan to address the problem of partial orders in the future. Finally, we plan to compare the new method with existing ones.

Acknowledgement

This work was partially supported by FCT project Rank! (PTDC/EIA/81178/ 2006). We thank the anonymous referees for useful comments.

References

1. Black, F., Litterman, R.: Global portfolio optimization. Financial Analysts Journal 48(5), 28–43 (1992)
2. Brazdil, P., Soares, C., Costa, J.: Ranking Learning Algorithms: Using IBL and Meta-Learning on Accuracy and Time Results. Machine Learning 50(3), 251–277 (2003)
3. Brazdil, P., Soares, C., Giraud-Carrier, C., Vilalta, R.: Metalearning Applications to Data Mining. Springer, Heidelberg (2009)
4. Cheng, W., Dembczynski, K., Hüllermeier, E.: Label Ranking Methods based on the Plackett-Luce Model. In: 27th International Conference on Machine Learning, Haifa, Israel (2010)
5. Cheng, W., Hühn, J., Hüllermeier, E.: Decision tree and instance-based learning for label ranking. In: ICML 2009: Proceedings of the 26th Annual International Conference on Machine Learning, pp. 161–168. ACM, New York (2009)
6. Dekel, O., Manning, C., Singer, Y.: Log-linear models for label ranking. Advances in Neural Information Processing Systems 16 (2003)
7. Domingos, P., Pazzani, M.: On the optimality of the simple bayesian classifier under zero-one loss. Machine learning 29(2), 103–130 (1997)
8. Har-Peled, S., Roth, D., Zimak, D.: Constraint Classification: A New Approach to Multiclass Classification. In: Cesa-Bianchi, N., Numao, M., Reischuk, R. (eds.) ALT 2002. LNCS (LNAI), vol. 2533, p. 365. Springer, Heidelberg (2002)
9. Hüllermeier, E., Fürnkranz, J., Cheng, W., Brinker, K.: Label ranking by learning pairwise preferences. Artificial Intelligence 172(2008), 1897–1916 (2008)
10. Lebanon, G., Lafferty, J.: Cranking: Combining Rankings Using Conditional Probability Models on Permutations. In: Proceedings of the Nineteenth International Conference on Machine Learning, p. 370. Morgan Kaufmann Publishers Inc., San Francisco (2002)
11. Mitchell, T.: Machine Learning. McGraw-Hill, New York (1997)

12. R Development Core Team: R: A Language and Environment for Statistical Computing. R Foundation for Statistical Computing, Vienna, Austria (2008) ISBN 3-900051-07-0
13. Soares, C., Brazdil, P., Kuba, P.: A meta-learning method to select the kernel width in support vector regression. Machine Learning 54, 195–209 (2004)
14. Soares, C.: Learning Rankings of Learning Algorithms. Ph.D. thesis, Department of Computer Science, Faculty of Sciences, University of Porto (2004); supervisors: Pavel Brazdil and Joaquim Pinto da Costa,
    http://www.liaad.up.pt/pub/2004/Soa04
15. Todorovski, L., Blockeel, H., Dzeroski, S.: Ranking with predictive clustering trees. In: Elomaa, T., Mannila, H., Toivonen, H. (eds.) ECML 2002. LNCS (LNAI), vol. 2430, pp. 123–137. Springer, Heidelberg (2002)
16. Vembu, S., Gärtner, T.: Preference Learning. Springer, Heidelberg (October 2010)
17. Vogt, M., Godden, J., Bajorath, J.: Bayesian interpretation of a distance function for navigating high-dimensional descriptor spaces. Journal of Chemical Information and Modeling 47(1), 39–46 (2007)

Topology Preserving SOM with Transductive Confidence Machine

Bin Tong[1,2], ZhiGuang Qin[1], and Einoshin Suzuki[2,3]

[1] CCSE, University of Electronic Science and Technology of China, China
[2] Graduate School of Systems Life Sciences, Kyushu University, Japan
[3] Department of Informatics, ISEE, Kyushu University, Japan

Abstract. We propose a novel topology preserving self-organized map (SOM) classifier with transductive confidence machine (TPSOM-TCM). Typically, SOM acts as a dimension reduction tool for mapping training samples from a high-dimensional input space onto a neuron grid. However, current SOM-based classifiers can not provide degrees of classification reliability for new unlabeled samples so that they are difficult to be used in risk-sensitive applications where incorrect predictions may result in serious consequences. Our method extends a typical SOM classifier to allow it to supply such reliability degrees. To achieve this objective, we define a nonconformity measurement with which a randomness test can predict how nonconforming a new unlabeled sample is with respect to the training samples. In addition, we notice that the definition of nonconformity measurement is more dependent on the quality of topology preservation than that of quantization error reduction. We thus incorporate the grey relation coefficient (GRC) into the calculation of neighborhood radii to improve the topology preservation without increasing the quantization error. Our method is able to improve the time efficiency of a previous method kNN-TCM, when the number of samples is large. Extensive experiments on both the UCI and KDDCUP 99 data sets show the effectiveness of our method.

1 Introduction

Self-organized map (SOM) [8] has been successfully used in a wide variety of applications, including image processing, intrusion detection, etc. SOM performs a mapping from a high-dimensional input space onto a neuron grid, such that it is capable of exhibiting a human-interpretable visualization for the data. A major characteristic of the SOM mapping is that training samples which are relatively close in the input space should be mapped to neuron nodes that are relatively close on the neuron grid. Although SOM is a specific type of clustering algorithm, its variants can serve as classifiers [6,13,9]. In general, a simple SOM classifier performs in the following way. In its training phase, each sample is assigned to the nearest weight vector. Note that each neuron node in the neuron grid is associated with a weight vector that has the same dimensionality with samples in the input space. Then, each assigned neuron node, which is also referred to as

B. Pfahringer, G. Holmes, and A. Hoffmann (Eds.): DS 2010, LNAI 6332, pp. 27–41, 2010.

winner neuron node, can be regarded as a representer of some samples, and its class label is voted by the samples. In the test phase, an unlabeled sample has the class label the same as the neuron node whose weight vector is nearest to it.

However, current SOM-based classifiers are usually difficult to be used in risk-sensitive applications, such as medical diagnosis and financial analysis, where incorrect predictions may result in serious consequences, because they require predictions to be qualified with degrees of reliability. The efforts to solve this problem for other kinds of classifiers by using transductive confidence machine (TCM) [4] can be traced back to kNN-TCM [11], where the kNN algorithm is embedded into the framework of TCM. A major advantage of using TCM to provide degrees of prediction reliability is that we do not rely on stronger assumptions than the i.i.d. one, which is very natural for most applications. In the kNN-TCM method, a nonconformity measure is defined to map each sample into a single real value with which a valid randomness test [15] is able to measure how nonconforming a new unlabeled sample is with respect to the training examples. kNN-TCM succeeds in obtaining a competitive prediction performance along with degrees of prediction reliability. However, kNN-TCM would suffer from its exhaustive computation, as handling a distance matrix is time consuming when either samples from the input space are of high-dimensionality or the number of them is large. Fortunately, SOM supplies a desirable dimension reduction tool such that the number of winner neuron nodes is much smaller than that of training samples. It implies a clue that integrating a SOM-based classifier with the TCM framework makes it possible that the computation time efficiency can be improved.

In this paper, a novel topology preserving SOM classifier with transductive confidence machine (TPSOM-TCM) is proposed. To achieve it, two significant issues should be addressed appropriately. The first issue is to design a nonconformity measurement for the winner neuron nodes, considering the distribution of the winner neuron nodes on the neuron grid. We also observe that the definition of the nonconformity measurement largely depends on the quality of topology preservation [7]. Thus, the second issue is aimed at obtaining a sophisticated topology preservation without increasing the quantization error. However, there is often a tradeoff between enhancing the quality of topology preservation and reducing the quantization error [7] when the dimensionality of the input space is higher than that of the neuron grid. To handle this challenging issue, we integrate the grey relation coefficient (GRC) [5] with the calculation of neighborhood radius for each neuron node.

2 Preliminaries

2.1 Transductive Confidence Machine

The framework of transductive confidence machine (TCM) allows to extend classifiers to produce predictions complemented with a confidence value that is able to provide an upper bound of the error rate [14]. The algorithmic randomness test [15] built in TCM is designed to find the regularities in a sample sequence.

For example, consider the following two sample sequences with the same length which are "0101010101...01010101" and "1001011110...01100000". It is obvious that the first sequence is more regular than the second one. One example of a valid randomness test [10] for real values produced under the $i.i.d.$ assumption is to measure how likely it is that a new real value is generated by the previous ones. In order to utilize the valid randomness test, a nonconformity measurement, which measures how nonconforming a sample is with respect to other available samples, should be designed to map each sample to a single real value.

A classifier with TCM generally works in the transductive way that it infers the class label for every new sample by utilizing the whole training samples. This process is formulated as follows. Denote $\mathcal{X}$ and $\mathcal{Y}$ as the input space and the corresponding label space, respectively. Suppose we are given a sequence of n training samples $X = \{\mathbf{x}_1, \mathbf{x}_2, \ldots, \mathbf{x}_n\}$ where $\mathbf{x}_i \in \mathcal{X}$, $i = 1, 2, \ldots, n$, and each $\mathbf{x}_i$ corresponds to a class label y_i, where $y_i \in \mathcal{Y}$. In addition, we suppose that every training sample is drawn from the same unknown probability distribution $\mathcal{P}$ that satisfies the $i.i.d.$ assumption. To predict the class label of a new unlabeled sample $\mathbf{x}_{n+1}$, we construct an extended sample sequence $X' = X \cup \mathbf{x}_{n+1}$, and each possible label $y_{n+1} \in \mathcal{Y}$ is tentatively labeled as the class label of $\mathbf{x}_{n+1}$. We then have a sequence of nonconformity scores denoted by $\alpha = \{\alpha_1, \alpha_2, \ldots, \alpha_n, \alpha_{n+1}\}$, where α_i, $i = 1, 2, \ldots, n$, represents the nonconformity score of $\mathbf{x}_i$, and α_{n+1} is for $\mathbf{x}_{n+1}$ given a specific class label y_{n+1}. In each tentative labeling, we perform a randomness test for the sequence X'. The randomness test function [10] is defined as follows.

$$p(\alpha_{n+1}) = \frac{card\{i \mid \alpha_i \geq \alpha_{n+1},\ 1 \leq i \leq n+1\}}{n+1} \tag{1}$$

where $card$ represents the cardinality of a set. The output of the randomness test function is called p-value. We can observe from Eq. (1) that if the p-value is close to its lower bound $\frac{1}{n+1}$, the new sample $\mathbf{x}_{n+1}$ with the specified label y_{n+1} is very nonconforming to the training sample sequence X. That is, the closer the p-value is to the upper bound 1, the more conforming the new sample $\mathbf{x}_{n+1}$ with the class label y_{n+1} is with respect to X. When a significance level ε is fixed, the classifier with TCM outputs a set of class labels $Y = \{y_{n+1} \in \mathcal{Y} \mid p(\alpha_{n+1}) > \varepsilon\}$ for the new sample $\mathbf{x}_{n+1}$. That is, we accept possible classifications, whose p-values are above the significance level ε, with a confidence at least $1 - \varepsilon$.

2.2 Self-Organized Map

In this subsection, we briefly review the self-organized map (SOM) [8], which serves as a fundamental of our method. The self-organized map (SOM) projects training samples onto neuron nodes in a low-dimensional neuron grid. In the following sections, we only consider the low-dimensional neuron grid as a two-dimensional one. Note that each neuron node u_j is associated with a weight vector $\mathbf{w}_j = [w_{j1}, w_{j2}, \ldots, w_{jd}]$, $j = 1, 2, \ldots, N$, where d is the dimensionality of the weigh vector which is the same as that of the training samples, and N

indicates the number of neuron nodes. For each neuron node u_j, we also specify its coordinate in the neuron grid as r_j. The basic learning process of SOM is illustrated as follows.

(a) Initialize randomly each weight vector $\mathbf{w}_j$, $j = 1, 2, \ldots, N$, and set the initial iteration epoch t to be 1;
(b) Present a training sample $\mathbf{x}_i$ at epoch t.
(c) Calculate the distances between the sample $\mathbf{x}_i$ and each neuron weight vector $\mathbf{w}_j$, $j = 1, 2, \ldots, N$, to identify the winner neuron u_c, $c \in N$, which is also called the Best-Matching Unit (BMU).

$$c = \arg\min_{j}\{\|\mathbf{x}_i - \mathbf{w}_j\|\} \qquad (2)$$

where $j = 1, 2, \ldots, N$ and $\|\cdot\|$ represents the Euclidean Norm.
(d) Adjust all the weight vectors in the neighborhood of the winner neuron u_c.

$$\mathbf{w}_j(t + 1) = \mathbf{w}_j(t) + \eta(t)h_{cj}(t)[\mathbf{x}_i - \mathbf{w}_j(t)] \qquad (3)$$

where $\eta(t)$ and $\sigma(t)$ are the learning rate and the neighborhood radius at epoch t, respectively. Note that both the learning rate and the neighborhood radius are decreasing over time. $h_{cj}(t) = \exp(\frac{-\|r_c - r_j\|^2}{\sigma(t)^2})$, where r_c and r_j are the coordinates of the winner neuron node u_c and the updated neuron node u_j, respectively.
(e) $t = t + 1$.
(f) Repeat the step (b)-(e) until a convergence condition is satisfied.

When a convergence is reached, the training samples from the high-dimensional input space are mapped to the winner neuron nodes on the neuron grid. In general, each winner neuron node can be regarded as a representer for the samples which are mapped to it. The winner neuron nodes preserve the topology and distribution of the training samples. Precisely speaking, the clusters of the samples with different class labels in the high-dimensional input space would be distributed in different areas on the grid. This characteristic is also pointed out in the SOM literature [7,12].

3   TPSOM-TCM

3.1   Nonconformity Measurement for SOM

In this subsection, the first issue of how to define a nonconformity measurement in the neuron grid is discussed. As mentioned in section 2.1, in order to integrate a SOM classifier with the TCM framework, a nonconformity measurement needs to be designed such that the randomness test can be performed. In this paper, in contrast to kNN-TCM [11] that designs the nonconformity measurement for each training or test samples, we design it for each winner neuron node in the neuron grid, since each winner neuron node is able to represent some of the training samples.

Given a sample $\mathbf{x}_i$, $i = 1, 2, \ldots, n$, the winner neuron node of the sample $\mathbf{x}_i$ is denoted by u_c and its class label is voted by y_i, $y_i \in \mathcal{Y}$. We define $\mathcal{N}_{u_c} = \{u_c^1, u_c^2, \ldots, u_c^K\}$ as a set of K-nearest winner neuron nodes according to the coordinate of the neuron grid. For each winner neuron node $u_c^j \in \mathcal{N}_{u_c}$, $j = 1, 2, \ldots, K$, we define $\nu_{u_c^j}$ as the number of samples which are mapped to the neuron node u_c^j. Among the $\nu_{u_c^j}$ samples, we further define $\nu_{u_c^j}^{y_i}$ as the number of samples whose class labels are equal to y_i, and $\nu_{u_c^j}^{-y_i}$ as the number of samples whose class labels are different from y_i, such that $\nu^{u_c^j} = \nu_{u_c^j}^{-y_i} + \nu_{u_c^j}^{y_i}$. Following the idea in kNN-TCM [11] that the nonconformity measurement is the ratio of the sum of k nearest distances from the same class to the sum of the k nearest distances from all other classes, the nonconformity score α_{u_c} for the winner neuron node u_c is defined as follows:

$$\alpha_{u_c} = \frac{\displaystyle\sum_{j=1:u_c^j \in \mathcal{N}_c}^{K} \nu_{u_c^j}^{-y_i}}{\displaystyle\sum_{j=1:u_c^j \in \mathcal{N}_c}^{K} \nu_{u_c^j}^{y_i}} \tag{4}$$

At the first glance, the form of Eq. (4) is different from that of the nonconformity measurement in kNN-TCM. However, they are conceptually the same. In SOM, samples close in the input space are mapped to neuron nodes that are close to each other in the neuron grid. It is natural to consider that the samples that are close in the input space are more likely to have the same class label, such that for a given sample, the sum of k nearest distances from the same class becomes small. From the viewpoint of the neuron grid, for a given neuron node, the density of samples with that class label becomes high. The intuition behind the nonconformity measurement is that, for each winner neuron node $u_c^j \in \mathcal{N}_{u_c}$, $j = 1, 2, \ldots, K$, the larger the value of $\nu_{u_c^j}^{y_i}$ is, the higher density the samples with class label y_i have. Therefore, it may lead to a smaller nonconformity score for the winner neuron node u_c. We observe from Eq. (4) that it is largely dependent on the topology structure of the neuron grid. That is, a sophisticated topology preservation for training samples is much more desirable for the nonconformity measurement.

3.2   Topology Preservation for SOM

In this subsection, how to improve the topology preservation for SOM is discussed. In SOM-based methods, the topographic error [7] and the quantization error [7] act as two criteria of estimating the quality of mapping. It was pointed out in [7,16] that improving the quality of topology preservation and reducing the quantization error always conflict when the dimension of training samples is larger than that of the neuron grid, hence there exists a tradeoff between the two criteria. Since the nonconformity measurement requires a high quality of the topology preservation, we are motivated to improve the quality of topology preservation without increasing the quantization error.

One method to improve the topology preservation is introduced in AdSOM [7]. The topographic error is proposed to measure the continuity for the mapping that reflects the probability distribution of the training samples. For a sample x_i, $i = 1, 2, \ldots, n$, assume that its nearest weight vector is $\mathbf{w}_c$ and the second nearest one is $\mathbf{w}_q$. If their corresponding neuron nodes, which are u_c and u_q respectively, are not adjacent according to their coordinates of the neuron grid, a local topographic error is generated. In AdSOM, the neighborhood radius for the neuron node u_j in the updating rule Eq. (3) is given as follows:

$$h_{cj}^*(t) = \begin{cases} \dfrac{\exp[-\frac{1}{2}(\frac{l}{\sigma_j})^2]-\exp(-\frac{1}{2})}{\sigma_j[1-\exp(-\frac{1}{2})]}, & l < \sigma_j \\ 0, & \text{otherwise.} \end{cases} \tag{5}$$

where $l = \|r_c - r_j\|$ and σ_j is defined as follows:

$$\sigma_j = \begin{cases} \|r_c - r_q\|, & \text{if } \mathbf{t} \le \|r_c - r_q\| \\ \|r_c - r_q\| - \mathbf{s}, & \text{otherwise when } \mathbf{s} < \|r_c - r_q\| \\ 1, & \text{otherwise} \end{cases} \tag{6}$$

where $\mathbf{t} = \max\{\|r_j - r_c\|, \|r_j - r_q\|\}$ and $\mathbf{s} = \min\{\|r_j - r_c\|, \|r_j - r_q\|\}$. For the sample x_i the neuron node u_j, σ_j is equal to the distance on the neuron grid between u_c and u_q if u_j is between them; if u_j is outside that area but not far from u_c and u_q, σ_j is equal to $\|r_c - r_q\| - \mathbf{s}$; otherwise, σ_j is set to be 1. From Eq. (6), we notice that, given the sample x_i, σ_j depends only on the distance between r_c and r_q, which means that the relationship between $\mathbf{w}_c$ and $\mathbf{w}_q$ might be neglected.

In order to improve the quality of topology preservation without increasing the quantization error, we firstly consider the relationships between the given sample x_i and each weight vector $\mathbf{w}_j$, $j = 1, 2, \ldots, N$, and exploit these relationships to improve the calculation of σ_j. Here, the grey relation coefficient (GRC) [5] is employed to discover the relationships between weight vectors. Given the sample $x_i = [x_{i1}, x_{i2}, \ldots, x_{id}]$ and the weight vector $\mathbf{w}_j = [w_{j1}, w_{j2}, \ldots, w_{jd}]$, $j = 1, 2, \ldots, N$, $k = 1, 2, \ldots, d$ with the normalized form, we define the grey relation coefficient between x_i and $\mathbf{w}_j$ as ξ_j, which is computed as follows.

$$\xi_{jk} = \frac{\Delta_{\min} + \rho\Delta_{\max}}{\Delta_{jk} + \rho\Delta_{\max}} \tag{7}$$

where ρ $(0 \le \rho \le 1)$ is a discriminative coefficient, and is usually set to be 0.5.

$$\Delta_{\min} = \min_j \min_k |x_{ik} - w_{jk}| \tag{8}$$

$$\Delta_{\max} = \max_j \max_k |x_{ik} - w_{jk}| \tag{9}$$

$$\Delta_{jk} = |x_{ik} - w_{jk}| \tag{10}$$

where $1 \leq j \leq N$, $1 \leq k \leq d$, and $|\cdot|$ represents the absolute value operator. We then derive ξ_j by $\xi_j = \frac{1}{d} \sum_{k=1}^{d} \xi_{jk}$. Given the example $\mathbf{x}_i$, we can associate ξ_j with each weight vector $\mathbf{w}_j$, $j = 1, 2, \ldots, N$.

To consider the relationships between weight vectors, we modify the calculation of σ_j as follows.

$$
\sigma_j = \begin{cases} (\xi_c + \xi_q) \| r_c - r_q \|, & \text{if } \mathbf{t}' \leq (\xi_c + \xi_q) \| r_c - r_q \| \\ (\xi_c + \xi_q) \| r_c - r_q \| - \mathbf{s}', & \text{otherwise when } \mathbf{s}' < (\xi_c + \xi_q) \| r_c - r_q \| \\ 1, & \text{otherwise.} \end{cases} \quad (11)
$$

where $\mathbf{t}' = \max\{(\xi_j + \xi_c) \| r_j - r_c \|, (\xi_j + \xi_q) \| r_j - r_q \|\}$ and $\mathbf{s}' = \min\{(\xi_j + \xi_c) \| r_j - r_c \|, (\xi_j + \xi_q) \| r_j - r_q \|\}$. The updating rule for the weight vector $\mathbf{w}_j$ in Eq. (6) is then recalculated as follows:

$$
w_{jk}(t + 1) = w_{jk}(t) + \eta(t)h_{cj}^*(t)\xi_{jk}[x_{ik} - w_{jk}(t)] \quad (12)
$$

where $j = 1, 2, \ldots, N$ and $k = 1, 2, \ldots, d$.

3.3   Framework of TPSOM-TCM

The topology preserving SOM with transductive confidence machine (TPSOM-TCM) mainly consists of two steps. The first step is to obtain a well-trained SOM map by using the training samples. Our objective in this step is to improve the quality of the topology preservation without increasing the quantization error. Our innovation is to integrate the grey relation coefficient (GRC) with the calculation of neighborhood radii. The second step is to perform the randomness test for the winner neuron nodes and the new unlabeled sample. Our innovation is to invent the nonconformity measurement such that the typical SOM-based classifier can be embedded into the TCM framework. The main steps of TPSOM-TCM are presented in Algorithm 1.

We now discuss the time complexity of TPSOM-TCM and kNN-TCM. Let the number of the unlabeled test samples be m. As pointed out in [2], kNN-TCM requires $O(n^2)$ distance computations when computing the nonconformity scores for the training samples. The complexity $O(nm)$ is required when computing the nonconformity scores of the test samples. If both m and n are very large, to compute the nonconformity scores for the training and test samples is extremely time consuming. However, in TPSOM-TCM, we only compute the nonconformity scores for the winner neuron nodes and the test samples. Suppose n' ($n' \ll n$) to be the number of winner neuron nodes in the neuron grid. The time complexity of TPSOM-TCM when computing the p-values for the test samples would be $O(n'm)$, hence the time computation efficiency can be improved. It explains why TPSOM-TCM is able to behave more appropriately than kNN-TCM when the number of samples is large.

Algorithm 1

Input: the training sample sequence X and corresponding label space $\mathcal{Y}$, K, the number of the neuron nodes in the neuron grid N, the unlabeled test sample $\mathbf{x}_{n+1}$

Output: the class label and the confidence for $\mathbf{x}_{n+1}$

1: Determine the size of the neuron grid according to N.
2: Present the training samples to train the SOM map by following the typical training steps of SOM, except that the Eq. (12) is employed to update the weight vectors.
3: Calculate the nonconformity score for each winner neuron node.
4: **for** $q = 1$ to $card(\mathcal{Y})$ **do**
5:    Associate $\mathbf{x}_{n+1}$ with its nearest neuron node in the neuron grid, and modify the statistical information of this neuron node.
6:    Recalculate the nonconformity scores for the neuron nodes if their calculations are involved with winner neuron nodes whose statistical information is modified.
7:    Calculate the nonconformity score of $\mathbf{x}_{n+1}$ with class label $\mathcal{Y}_q$ by using Eq. (4).
8:    Calculate the p-value of $\mathbf{x}_{n+1}$ with class label $\mathcal{Y}_q$ by using Eq.(1).
9: **end for**
10: Predict the class label of $\mathbf{x}_{n+1}$ with the largest p-value.
11: Calculate the confidence for $\mathbf{x}_{n+1}$ one minus the 2nd largest p-value .

---

4   Experiments on UCI Data Sets

All the data sets in this experiment come from the UCI benchmark repository[1]. Before the experiment, all the samples with missing feature values as well as duplicate were removed. Each data set was normalized to have zero mean and unit variance. The data sets used in the experiment are specified in Table 1.

Table 1. Summary of the benchmark data sets from UCI

Data set	Dimension	Instance	Class	Data set	Dimension	Instance	Class
heart	13	270	2	Balance	4	625	3
Ionosphere	34	351	2	Wdbc	30	569	2
Pima	8	768	2	Wpbc	33	194	2

As mentioned in [8], Kohonen suggested that the training phase of SOM could be divided into two steps which are the ordering step and the convergence step. Without a specific explanation, the experiment uses common parameters listed below. Since the numbers of instances in the data sets are not numerous, in order to obtain a stable result, the iteration number of the ordering step is set to be 25 times as large as the number of training instances, while the iteration number of the convergence step is 5 times as large as the number of training instances. The initial learning rate of the ordering step is 0.95 and the ending learning rate is 0.05, while the two learning rates in the convergence step are set to be 0.05 and 0.01, respectively. The configuration of the neuron grid follows the default setting of the *som_topol_struct* function in the SOM toolbox[2]. For more details about this function, please refer to the function specification[3]. Note

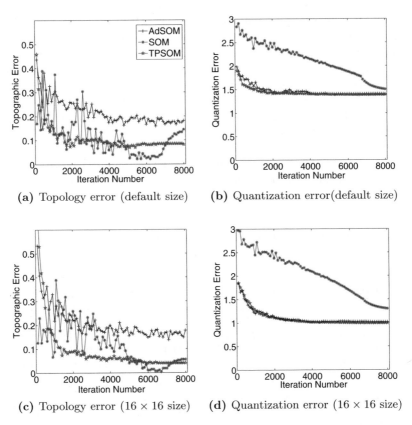

(a) Topology error (default size)   (b) Quantization error(default size)

(c) Topology error (16 × 16 size)   (d) Quantization error (16 × 16 size)

Fig. 1. The performance on the topology preservation and the quantization error reduction in two neuron grids with different sizes

that we denote our method without considering the topology preservation by SOM-TCM, and denote our method without considering TCM by TPSOM.

4.1 Analysis of Experiments

First, we examine various SOM methods, i.e., SOM, AdSOM and TPSOM, to evaluate the performances on the topology preservation and the quantization error reduction. As discussed in previous sections, the nonconformity measurement depends on the quality of the topology preservation. Thus, a high quality of the topology preservation without increasing the quantization error is desirable. Our TPSOM method considers the relationships between weight vectors associated with neuron nodes to improve the calculation of the neighborhood

---
[2] http://www.cis.hut.fi/somtoolbox/
[3] http://www.cis.hut.fi/somtoolbox/package/docs2/som_topol_struct.html

radii. We conduct experiments on two neuron grids with different sizes. The size of the first neuron gird follows the default setting of the *som_topol_struct* function in the SOM toolbox[2], and the other is set to be a square with 16×16 size. The experiments were conducted on the heart data set, and the results are shown in Fig. 1. From Fig. 1a and Fig. 1c, it is obvious that TPSOM has a stable change of the topographic error over the iteration period. In most cases, TPSOM keeps the smallest topographic error in the two neuron grids. In Fig. 1b and Fig. 1d, TPSOM and AdSOM have similar performances on the quantization error reduction, and outperform that of SOM apparently. From the viewpoint of the two quality criteria, TPSOM outperforms AdSOM and SOM. We attribute the fact to the reason that the grey relation coefficient (GRC) can help discover the relationships between weight vectors so that it is beneficial for the calculation of the neighborhood radii.

Table 2. Comparison of classification accuracies and degrees of classification reliability. A pair of values is shown in the form of classification accuracy/reliability degree.

Methods	Heart		Pima		Ionosphere	
	Average	Best	Average	Best	Average	Best
SOM	72.3	73.0	73.1	73.2	84.9	84.9
TPSOM	73.3	76.7	71.7	72.6	84.8	87.1
SOM-TCM	80.3/95.3	82.2/94.9	73.0/91.9	73.9/92.0	87.3/94.0	88.9/96.1
TPSOM-TCM	81.3/94.5	83.7/95.5	73.5/93.3	74.8/93.8	86.0/96.2	87.1/97.1
kNN-TCM	82.2/93.8	84.8/94.1	73.1/90.1	74.5/90.4	85.0/96.7	86.6/96.9

Methods	Balance		Wdbc		Wpbc	
	Average	Best	Average	Best	Average	Best
SOM	78.4	78.4	92.4	92.4	63.2	63.2
TPSOM	77.5	79.8	93.2	95.0	69.3	71.6
SOM-TCM	83.3/95.9	84.2/96.4	95.3/98.7	96.1/98.8	71.6/90.4	74.2/92.0
TPSOM-TCM	86.02/95.0	87.0/95.8	95.8/98.2	96.1/99.3	75.8/92.1	79.0/93.9
kNN-TCM	85.9/94.0	86.6/96.9	96.5/99.5	96.6/99.5	75.1/88.8	77.4/89.4

Second, we discuss the performance of TPSOM-TCM compared with SOM, TPSOM, TPSOM-TCM and kNN-TCM. Note that, for the two methods SOM and TPSOM, we only show the classification accuracy due to their lack of TCM. In this experiment, 10×5-fold cross-validation was utilized. The K values for searching neighbors used in all the methods are set to be 3. The average results and the best results are shown in Table 2. We can see that, in most cases, there is no significant improvement on the classification accuracy from SOM to TPSOM. The possible reason is that the purpose for designing TP-SOM is to improve the topology preservation for SOM without increasing the quantization error. Therefore, we believe that the improvement on the classification accuracy is more likely to be dependent on reducing the quantization error. It is worthy of noting that TPSOM-TCM and SOM-TCM outperform TPSOM and SOM. We attribute the fact to the reason that, in the SOM-based classifiers, the class label of a winner neuron node is voted by samples. Thus, the factor resulting from the samples with class labels, which are different from the voted one, is arbitrarily ignored in the prediction. However, in the nonconformity measurement for TPSOM-TCM and SOM-TCM, the density of samples with different class labels is taken into account such that it would be helpful for improving the performance of the classification accuracy. We believe that the nonconformity measurement for TPSOM-TCM and SOM-TCM is effective, and then TCM helps improving the SOM-based classifiers by using a transductive way. With the comparison between TPSOM-TCM and SOM-TCM, we notice that, although TPSOM-TCM has a similar performance on the classification accuracy with SOM-TCM, TPSOM-TCM outperforms SOM-TCM in terms of the degree of classification reliability. A possible reason is that TPSOM has a better topology preservation than SOM itself. Compared with kNN-TCM, we observe that TPSOM-TCM is competitive in terms of both the classification accuracy and the degree of classification reliability. We can see that Eq.(4) is effective in measuring the nonconformity degrees of samples.

5 Experiments on Intrusion Detection

In this section, we make use of the intrusion detection database KDDCUP 99[4] to examine the performance of TPSOM-TCM compared with other methods, i.e., SOM-TCM, kNN-TCM [11] and Multi-Class SVM (MC-SVM) [1]. The KDD-CUP 99 data set presents the network flow where the normal one and four types of attacks, i.e., DoS, Probe, R2L and U2R, are collected. Before the experiment, we reorganized the data set in the following way. We randomly extracted 5915 training data from the "10_percent_corrected" file and 9063 test data from the "correct" file. The specific details of the data are described in Table 3. We conducted the experiments on the Windows XP platform with Intel 3.16 GHz E8500 processor and 3G main memory. Note that we used LibSVM[5] library to simulate the MC-SVM method and the other methods were simulated by Matlab. The training iteration times of TPSOM-TCM and SOM-TCM are equivalent, each of which is 5 times the size of the training data. In addition, the neuron grid was set to be a square with 10×10 size.

Table 3. Data from KDDCUP 99

Class Label	Training Data	Test Data	Class Label	Training Data	Test Data
Normal	1946	3030	Dos	1764	2299
Probe	1027	1138	R2L	1126	2117
U2R	52	228	Total	5919	9063

---

[4] http://kdd.ics.uci.edu/databases/kddcup99/kddcup99.html
[5] http://www.csie.ntu.edu.tw/~cjlin/libsvm/

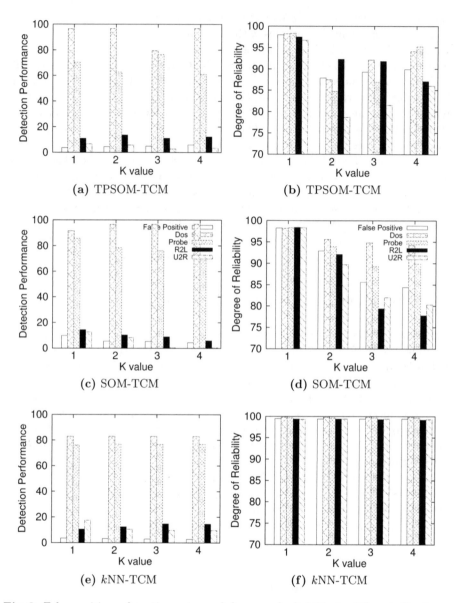

Fig. 2. False positives, detection rates and degrees of reliability in different K values

5.1   Analysis of Experiments

We firstly evaluate the performances of TPSOM-TCM, SOM-TCM and kNN-TCM on the false positives and the detection rates along with the degrees of classification reliability for four types of attacks. We vary the value K from 1 to 4 for searching neighbors when computing the nonconformity score. Although the training step for TPSOM-TCM and SOM-TCM is a random process, the results of the two methods are stable due to the well trained SOM map. Thus, for each chosen K, the results of TPSOM-TCM and SOM-TCM from only one run are reported. We can observe from Fig. 2 that, when the K value is set to be 1, the degrees of classification reliability in TPSOM-TCM and SOM-TCM keep the highest, while the reliability degrees of kNN-TCM are high and stable against the change of the K value. A possible reason to explain this fact is that the nonconformity scores of winner neuron nodes are only taken into account in TPSOM-TCM and SOM-TCM when computing p-values of the test data, while all the training data are utilized in kNN-TCM. Although each winner neuron node is able to represent some training data, the distribution of the winner neuron nodes is difficult to depict the accurate distribution of the training data, because the number of neuron nodes is much smaller than that of training data. Another reason is probably that the characteristics of various attacks are similar to those of the normal network flow. Hence, the neuron nodes with each possible class label are diversely distributed in the neuron grid. In this case, the K nearest neighbor set of a winner neuron node may contain neuron nodes with various different labels of training data, which would make the nonconformity measurement ineffective. It is worthy of noting from Fig. 2 that the detection rates of R2L and U2R are much lower than those of DoS and Probe. A possible reason is that the features of R2L and U2R network flow are very similar to those of the normal ones, which is also pointed out in the intrusion detection literature [3,6].

Table 4. The best results of various methods for false positives, detection accuracies and degrees of classification reliability. 'Rate' and 'Conf.' represent percentage of performance and degree of classification reliability, respectively.

Methods	False Positive		Dos		Probe		R2L		U2R	
	Rate	Conf.	Rate	Conf.	Rate	Conf.	Rate	Conf.	Rate	Conf.
TPSOM-TCM	3.7	98.0	96.6	98.2	70.4	98.3	11.1	97.5	6.6	96.7
SOM-TCM	9.9	98.2	91.6	98.2	85.7	98.3	14.5	98.4	12.7	98.3
kNN-TCM	3.6	99.5	82.9	99.8	76.1	99.7	10.6	99.4	17.5	99.3
MC-SVM	1.7	–	82.6	–	81.3	–	0.7	–	4.4	–

We summarize the best result of each method from different K value settings, as shown in Table 4. It can be seen that TPSOM-TCM is competitive to kNN-TCM, in the aspects of false positive, detection rate and degree of classification reliability. The Dos detection rate of TPSOM-TCM is much higher than that of kNN-TCM. Although SOM-TCM has a similar performance with TPSOM-TCM and SOM-TCM in terms of the detection rate and the degree of classification reliability, SOM-TCM is inferior to TPSOM-TCM and SOM-TCM due to the reason that the high false positive would place a negative effect on the reliability of an intrusion detection system. Table 4 also reports the result of MC-SVM. Although it is able to obtain the lowest false positive, TPSOM-TCM outperforms

MC-SVM in terms of Dos detection rate, R2L detection rate and U2R detection rate.

We then examine the computation time for TPSOM-TCM, SOM-TCM and kNN-TCM in the setting of different K values. In this experiment, we divide the computation time into two parts, which are modeling time and detection time. By observing the two parts, we are able to take a deep insight on the analysis of time complexity for each method. It is obvious from Table 5 that, when the K value changes from 1 to 4, the detection time of each method increases. We attribute this fact to the reason that, when computing the nonconformity scores for the test data, a larger value of K would lead to an increase of time by using Eq. (4). We can also see that kNN-TCM occupies the longest detection time while TPSOM-TCM and SOM-TCM take much shorter time on it, since the computation of the distance matrix involved with a large amount of training data and test data are extremely time consuming. We can draw a conclusion from Table 5 that TPSOM-TCM is superior to kNN-TCM in the aspect of the detection time, hence TPSOM-TCM is more likely to be adequate for real-time detection, especially when the network data is significantly huge. In addition, we notice that the modeling time of TPSOM-TCM is longer than that of SOM-TCM. Its reason is that the calculation of the grey relation coefficient for weight vectors and establishing the neighborhood radius for each neuron node take up extra time. Note that, in spite of this drawback, TPSOM-TCM almost has the same detection time with SOM-TCM.

Table 5. modeling time and detection time for various methods in the setting of different K values. All values are in seconds.

	TPSOM-TCM		SOM-TCM		kNN-TCM	
K Value	Modeling	Detection	Modeling	Detection	Modeling	Detection
1	49.4	9	6.0	9.1	11.6	134.6
2	49.3	9.7	6.5	10.3	12.5	147.3
3	47.9	11.9	7.9	12.2	11.9	150.0
4	50.0	12.5	6.5	14.3	12.3	157.2

6 Conclusion

In this paper, we proposed a novel topology preserving SOM classifier with transductive confidence machine (TPSOM-TCM) which is able to provide the degree of the classification reliability for new unlabeled samples. To achieve this objective, we firstly invented a nonconformity measurement for SOM, such that a typical SOM classifier can be easily embedded in the TCM framework. We then incorporated the grey relation coefficient (GRC) into the calculation of neighborhood radii to improve the topology preservation without increasing the quantization error. The experimental results on both the UCI and KDDCUP 99 data sets illustrate the effectiveness of our method.

Acknowledgments. This work is partially supported by the grant-in-aid for scientific research on fundamental research (B) 21300053 from the Japanese Ministry of Education, Culture, Sports, Science and Technology. Bin Tong is sponsored by the China Scholarship Council (CSC).

References

1. Ambwani, T.: Multi Class Support Vector Machine Implementation to Intrusion Detection. In: Proceedings of the International Joint Conference on Neural Networks (2003)
2. Barbará, D., Domeniconi, C., Rogers, J.P.: Detecting Outliers Using Transduction and Statistical Testing. In: KDD 2006: Proceedings of the Twelveth ACM SIGKDD International Conference on Knowledge Discovery and Data Mining (2006)
3. Cho, S.B.: Incorporating Soft Computing Techniques into A Probabilistic Intrusion Detection System. IEEE Transactions on Systems, Man, and Cybernetics, Part C 32(2), 154–160 (2002)
4. Gammerman, A., Vovk, V.: Prediction Algorithms and Confidence Measures Based on Algorithmic Randomness Theory. Theor. Comput. Sci. 287(1), 209–217 (2002)
5. Hu, Y.C., Chen, R.S., Hsu, Y.T., Tzeng, G.H.: Grey Self-organizing Feature Maps. Neurocomputing 48(1-4), 863–877 (2002)
6. Kayacik, H.G., Zincir-Heywood, A.N., Heywood, M.I.: A Hierarchical SOM-based Intrusion Detection System. Eng. Appl. of AI 20(4), 439–451 (2007)
7. Kiviluoto, K.: Topology Preservation in Self-organizing Maps. In: IEEE International Conference on Neural Networks (1996)
8. Kohonen, T., Schroeder, M.R., Huang, T.S. (eds.): Self-Organizing Maps. Springer, Heidelberg (2001)
9. Martin, C., Diaz, N.N., Ontrup, J., Nattkemper, T.W.: Hyperbolic SOM-based Clustering of DNA Fragment Features for Taxonomic Visualization and Classification. Bioinformatics 24(14), 1568–1574 (2008)
10. Melluish, T., Saunders, C., Nouretdinov, I., Vovk, V.: Comparing the Bayes and Typicalness Frameworks. In: EMCL 2001: Proceedings of the Twelfth European Conference on Machine Learning, pp. 360–371 (2001)
11. Proedrou, K., Nouretdinov, I., Vovk, V., Gammerman, A.: Transductive Confidence Machines for Pattern Recognition. In: Elomaa, T., Mannila, H., Toivonen, H. (eds.) ECML 2002. LNCS (LNAI), vol. 2430, pp. 381–390. Springer, Heidelberg (2002)
12. Su, M.C., Chang, H.T., Chou, C.H.: A Novel Measure for Quantifying the Topology Preservation of Self-Organizing Feature Maps. Neural Process. Lett. 15(2), 137–145 (2002)
13. Suganthan, P.N.: Hierarchical Overlapped SOM's for Pattern Classification. IEEE Transactions on Neural Networks 10(1), 193–196 (1999)
14. Vanderlooy, S., Maaten, L., Sprinkhuizen-Kuyper, I.: Off-Line Learning with Transductive Confidence Machines: An Empirical Evaluation. In: Perner, P. (ed.) MLDM 2007. LNCS (LNAI), vol. 4571, pp. 310–323. Springer, Heidelberg (2007)
15. Vanderlooy, S., Sprinkhuizen-Kuyper, I.: An Overview of Algorithmic Randomness and its Application to Reliable Instance Classification. Technical Report MICC-IKAT 07-02, Universiteit Maastricht (2007)
16. Villmann, T., Der, R., Herrmann, M., Martinetz, T.M.: Topology Preservation in Self-organizing Feature Map: Exact Definition and Measurement. IEEE Transactions on Neural Networks 8(2), 256–266 (1997)

An Artificial Experimenter for Enzymatic Response Characterisation

Chris Lovell, Gareth Jones, Steve R. Gunn, and Klaus-Peter Zauner

School of Electronics and Computer Science,
University of Southampton, UK, SO17 1BJ
{cjl07r,gj07r,srg,kpz}@ecs.soton.ac.uk

Abstract. Identifying the characteristics of biological systems through physical experimentation, is restricted by the resources available, which are limited in comparison to the size of the parameter spaces being investigated. New tools are required to assist scientists in the effective characterisation of such behaviours. By combining artificial intelligence techniques for active experiment selection, with a microfluidic experimentation platform that reduces the volumes of reactants required per experiment, a fully autonomous experimentation machine is in development to assist biological response characterisation. Part of this machine, an artificial experimenter, has been designed that automatically proposes hypotheses, then determines experiments to test those hypotheses and explore the parameter space. Using a multiple hypotheses approach that allows for representative models of response behaviours to be produced with few observations, the artificial experimenter has been employed in a laboratory setting, where it selected experiments for a human scientist to perform, to investigate the optical absorbance properties of NADH.

1 Introduction

Biological systems exhibit many complex behaviours, for which there are few models. Take for example the proteins known as enzymes, which are believed to act as biochemical computers [19]. Whilst much is understood within a physiological context, there exists a wide parameter space not yet investigated that may open up the development of biological computers. However, such investigation is restricted by the available resources, which require effective usage to explore the parameter spaces. Biological reactants add an additional problem, as they can undergo undetectable physical changes, which will alter the way they react, leading to observations not representative of the true underlying behaviours. There is therefore need for a new tool, which can aid the creation of response models of biological behaviours. Presented here are artificial intelligence techniques, designed to build models of response behaviours, investigated through an effective exploration of the parameter space. Key to this is the use of a multiple hypotheses technique, which helps manage the uncertainties present in experimentation with few, potentially erroneous, observations. These algorithms, or artificial experimenter, will in the future work with an automated lab-on-chip experiment platform, to provide a fully autonomous experimentation machine.

B. Pfahringer, G. Holmes, and A. Hoffmann (Eds.): DS 2010, LNAI 6332, pp. 42–56, 2010.

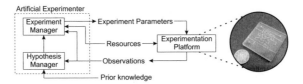

Fig. 1. Flow of experimentation between an artificial experimenter and an automated experimentation platform. A prototype of a lab-on-chip platform in development for conducting the experiments on is shown.

Autonomous experimentation is a closed-loop technique, which computationally builds hypotheses and determines experiments to perform, with chosen experiments being automatically performed by a physical experimentation platform, as shown in Fig.1. Currently few examples of such closed-loop experimentation systems exist in the literature [20,12,8], whilst another approach provided the artificial experimenter computational side of the system [9]. Of those that do exist, none consider learning from a small number of observations that could be erroneous. One approach works within a limited domain using extensive prior information to produce a set of hypotheses likely to contain the true hypothesis, allowing the experiment selection strategy to focus on identifying the true hypothesis from a set of hypotheses as cost effectively as possible [8]. However, as such prior knowledge does not exist in the domain of interest to us, our techniques must use also experiments to build a database of information to hypothesise from. Additionally, active learning considers algorithms that sequentially select the observations to learn from [11,5], however the current literature does not consider learning from small and potentially erroneous sets of observations.

Here we consider the development of an artificial experimenter, where in Section 2 we first consider the issues of building hypotheses in situations where observations are limited and potentially erroneous. Next in Section 3 we consider how such hypotheses can be separated, to efficiently identify the true hypothesis from a set of potential hypotheses, where we introduce a maximum discrepancy algorithm that is able to outperform a selection of existing active learning strategies. In Section 4 we present the design for an artificial experimenter, which is evaluated through simulation in Section 5 and in a laboratory setting in Section 6, to show proof-of-concept of the techniques developed.

2 Hypothesis Manager

The goal for the hypothesis manager is to develop accurate response predictions of the underlying behaviours being investigated, with as few experiments as possible. A key issue is dealing with erroneous observations, which are not representative of the true underlying behaviour being investigated. Whilst the validity of all observations could be determined through repeat experiments, doing so will cut into the resources available for investigating and identifying uncharacterised

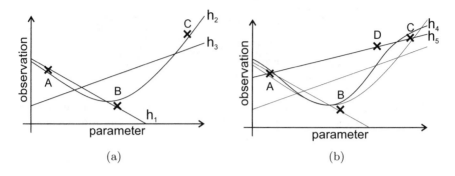

Fig. 2. Validity of observations affecting hypothesis proposal. Hypotheses (lines) are formed after observations (crosses) are obtained. In (a), h_1 formed after A and B are obtained questions the validity of C, whilst h_2 and h_3 consider all observations to be valid with differing levels of accuracy. In (b), D looks to confirm the validity of C, however now h_4 and h_5 differ in opinion about the validity of B.

behaviours. Therefore a hypothesis manager should employ computational methods to handle such uncertainty, built with the view that computation is cheap compared to the cost of experimentation, meaning that computational complexity is unimportant, so long as a solution is feasible.

In experimentation, all observations will be noisy, both in terms of the response value returned and also in the experiment parameter requested. Such noise can be thought of as being Gaussian, until a better noise model can be determined experimentally. As such, we consider a hypothesis as taking the form of a least squares based regression. In particular we use a spline based approach, since it is well defined, can be placed within a Bayesian framework to provide error bars and does not impose a particular spectral scale [18]. A hypothesis is built from a subset of the available observations, a smoothing parameter and a set of weights for the observations, which we will discuss more later.

Erroneous observations however, add a different type of noise, which can be considered as shock noise that provides an observation unrepresentative of the true underlying behaviour. The noise from an erroneous observation is likely to be greater than experimental Gaussian noise, meaning that potentially erroneous observations can be identified as observations that do not agree with the prediction of a hypothesis. The term potentially erroneous is important, as if an observation does not agree with a hypothesis, it may not be the observation that is incorrect, but rather the hypothesis that is failing to model an area of the experiment parameter space. In such limited resource scenarios, when presented with an observation that does not agree with a hypothesis, the hypothesis manager needs to determine whether it is the observation or the hypothesis, or both, which are erroneous.

A possible solution to this problem is to consider multiple hypotheses in parallel, each with a differing view of the observations. Such multiple hypotheses techniques are promoted in philosophy of science literature, as they can ensure alternate views are not disregarded without proper evaluation, making experimentation more complete [4]. Whilst there are multiple hypotheses based approaches in the literature that produce hypotheses from random subsets of the observations available [6,1], we believe additional more principled techniques can be applied to aid hypothesis creation. In particular, when a conflicting observation and hypothesis are identified, the hypothesis can be refined into 2 new hypotheses, one that considers the observation to be true, and one that considers the observation to be erroneous. To achieve this, the parameters of the hypothesis are copied into the new hypotheses, however one hypothesis is additionally trained with the potentially erroneous observation having a high weighting, whilst the other is additionally trained with that observation having a zero weighting. By giving the observation a higher weighting, the hypothesis considers the observation to be valid, by having its regression prediction forced closer to that observation. Whilst the zero weighting of the observation makes the hypothesis consider the observation erroneous and removes it from the regression calculation. The handling of potentially erroneous observations through multiple hypotheses, is illustrated in Fig. 2. Next we consider how these hypotheses can be used to guide experiment selection.

3   Effective Separation of the Hypotheses

With the hypothesis manager providing a set of competing hypotheses, there is now the problem of identifying the hypothesis that best represents the true underlying behaviour. To do this we consider methods of separating the hypotheses using experimental design and active learning techniques, evaluated on a simulated set of hypotheses. As the hypotheses will be built from the same small set of observations, their predictions are likely to be similar to each other, with some differences coming from potentially erroneous observations. Therefore, the metric we are interested in, is how well the separation methods perform when the hypotheses have different levels of similarity. To do this the techniques presented will be evaluated using abstract sets of hypotheses, which are described through a single parameter of similarity.

3.1   Techniques

Design of experiments, sequential learning and active learning techniques have considered this problem of hypothesis separation. In particular there is the experimental design technique of T-optimality [2]. However the authors suggest that such designs can perform poorly if the most likely hypothesis is similar to the alternate hypotheses or if there is experimental error [2], which is likely in the experimentation scenario we consider. Whilst many active learning techniques consider this problem in a classification scenario, where there are discrete predictions from the hypotheses [15], meaning that such techniques will require some alteration for a regression problem. The technique we apply to make this alteration, is to use the predictions of the hypotheses as the different classification labels.

In the following, an experiment parameter is represented as x, with its associated observation y. Hypotheses, $h_i(x)$, can provide predictions for experiment parameters through $\hat{h}_i(x)$. Each hypothesis can have its confidence calculated based on the existing observations as:

$$C(h) = \frac{1}{N} \sum_{n=1}^{N} \exp\left(\frac{-\left(\hat{h}(x_n) - y_n\right)^2}{2\sigma^2}\right) \tag{1}$$

where N is the number of observations available. A hypothesis calculates its belief that parameter x brings about observation y through:

$$P_{h_i}(y|x) = \exp\left(\frac{-\left(\hat{h}_i(x) - y\right)^2}{2\sigma_i^2}\right) \tag{2}$$

where σ_i^2 will be kept constant for the abstract hypotheses in the simulated evaluation presented in this section, but is substituted for the error bar of the hypothesis when applied to real hypotheses discussed in Section 5 and Section 6. Additionally, where observations are to be predicted, the hypotheses provide predictions through substituting y for $\hat{h}(x)$. Finally, the working set of hypotheses under consideration is defined as $\mathcal{H}$, which has a size of $|\mathcal{H}|$. We now consider different active learning techniques.

Variance. The difference amongst a group of hypotheses has been previously considered through looking at the variance of the hypotheses predictions [3]. Experiments are selected where the variance of the predictions is greatest. So as to allow for previous experiments to be taken into consideration on subsequent calls to the experiment selection method, the confidence of the hypothesis can be used to provide a weighted variance of the predictions, based on how well each hypothesis currently matches the available observations:

$$x_{\text{Var}}^* = \arg\max_x k \sum_{i=1}^{|\mathcal{H}|} C(h_i)\left(\hat{h}_i(x) - \mu^*\right)^2 \tag{3}$$

where

$$\mu^* = \frac{1}{\sum_{i=1}^{|\mathcal{H}|} C(h_i)} \sum_{i=1}^{|\mathcal{H}|} C(h_i)\hat{h}_i(x) \tag{4}$$

and k is a normalising constant for weighted variance.

KL Divergence. The Kullback-Liebler divergence [10], has been employed as a method for separating hypotheses where there are discrete known labels [13]:

$$x_{\text{KLM}}^* = \arg\max_x \frac{1}{|\mathcal{H}|} \sum_{i=1}^{|\mathcal{H}|} \sum_{j=1}^{|\mathcal{H}|} P_{h_i}\left(\hat{h}_j(x)|x\right) \log \frac{P_{h_i}\left(\hat{h}_j(x)|x\right)}{P_{\mathcal{H}}\left(\hat{h}_j(x)|x\right)} \tag{5}$$

where

$$P_{\mathcal{H}}\left(\hat{h}_j(x)|x\right) = \frac{1}{|\mathcal{H}|} \sum_{k=1}^{|\mathcal{H}|} P_{h_k}\left(\hat{h}_j(x)|x\right) \qquad (6)$$

which is the consensus probability between all hypotheses that the observation y_j will be obtained, within some margin of error, when experiment x is performed. This discrepancy measure selects the experiment that causes the largest mean difference between the individual hypotheses and the consensus over the observation distributions.

In its current form this approach requires hypotheses that do not match the observations to be removed. However, if $P_{h_i}(\hat{h}_j(x)|x)$ is multiplied by the confidence of the hypothesis, $C(h_i)$, and the normalising term $\frac{1}{|\mathcal{H}|}$ in (5) and (6) is replaced with the inverse of sum of the confidences, $\frac{1}{C}$, the impact a hypothesis has on the decision process can be scaled by its confidence.

Bayesian Surprise. The KL divergence has also been applied to formulate a notion of surprise, within a Bayesian framework [7]. The prior probability is determined from the available observations:

$$P_{h_i}(Y|X) = \frac{1}{n} \sum_{j=1}^{n} P_{h_i}(y_j|x_j) \qquad (7)$$

Whilst the predicted posterior probability also takes into consideration what the new probability of the hypothesis would be if a particular experiment x_p was performed that resulted in a specific y_p:

$$P_{h_i}(Y, y_p|X, x_p) = \frac{1}{n+1}\left(nP_{h_i}(Y|X) + P_{h_i}(y_p|x_p)\right) \qquad (8)$$

Using these distributions, we consider all predicted observations to determine a surprise term:

$$x^*_{\text{surprise}} = \arg\min_x \frac{1}{|\mathcal{H}|} \sum_{i=1}^{|\mathcal{H}|} \sum_{j=1}^{|\mathcal{H}|} K\left(h_i, \hat{h}_j(x)\right) \qquad (9)$$

where K is the KL divergence to provide Bayesian surprise [7]

$$K(h_i, y_j) = P_{h_i}(Y, y_j|X, x) \log \frac{P_{h_i}(Y, y_j|X, x)}{P_{h_i}(Y|X)} \qquad (10)$$

Importantly the experiment with the lowest KL divergence is selected, so as to find the experiment that weakens all hypotheses. If the maximum value were used, it would select the experiment that improves all hypotheses, which by definition will limit the difference between the hypotheses. It can be shown using the framework presented here, that using the minimum KL divergence value results in a better performing discrepancy technique than using the maximum KL divergence.

Maximum Discrepancy. Separating the hypotheses can be thought of as identifying experiments that maximise the disagreement between the predictions of hypotheses. Mathematically we consider maximising the integration of the differences between all of the hypotheses, over all possible experiment outcomes:

$$A = \sum_{i=1}^{|\mathcal{H}|} \sum_{j=1}^{|\mathcal{H}|} \int (h_i - h_j)^2 \, dy_t \tag{11}$$

where the likelihood function $P_h(y|x)$ can be used to determine the differences in the hypotheses:

$$A = \sum_{i=1}^{|\mathcal{H}|} \sum_{j=1}^{|\mathcal{H}|} \int \left(P_{h_i}(y|x) - P_{h_j}(y|x) \right)^2 dy \tag{12}$$

then as $P_{h_i}(y|x)$ is a Gaussian distribution, and distinct y can be taken from the predictions of the hypotheses, we can formulate a discrepancy measure:

$$x^*_{\text{discrepancy}} = \arg \max_x \sum_{i=1}^{|\mathcal{H}|} \sum_{i=j}^{|\mathcal{H}|} 1 - P_{h_i} \left(\hat{h}_j(x)|x \right) \tag{13}$$

where we look for the experiment parameter where the hypotheses disagree the most. Next a method of using the prior information is required. On subsequent runs, the discrepancy within the sets of currently agreeing hypotheses should be found, whilst also taking into consideration how well those hypotheses fit the observations. The disagreement term, $1 - P_{h_i}(y_j|x)$, can therefore be multiplied by $P(h_i, h_j|\mathcal{D})$, defined as:

$$P(h_i, h_j|\mathcal{D}) = C(h_i)C(h_j)S(h_i|h_j) \tag{14}$$

where

$$S(h_i, h_j) = \frac{1}{N} \sum_{n=1}^{N} \exp \left(\frac{- \left(\hat{h}_i(x_n) - \hat{h}_j(x_n) \right)^2}{2\sigma_i^2} \right) \tag{15}$$

is the similarity between two hypotheses predictions for the previously performed experiments, with σ_i coming from the error bar of h_i at x for real hypotheses, and is kept constant in the abstract trial discussed next.

3.2   Hypothesis Separation Results

To evaluate the experiment selection techniques, an arbitrary function is used to create a set of potential training observations. These observations are distorted from the function through Gaussian noise, where the amount of noise is the parameter that controls how different the hypotheses in the set are. Twenty hypotheses are then trained from random subsets of the training observations using an arbitrary regression technique. The hypotheses are then compared to

Table 1. Number of experiments until the hypothesis with the highest confidence is the true hypothesis. The similarity is shown as the Gaussian noise applied to the initial training data, where a noisier set of training data provides hypotheses less similar to each other. The best strategy in each case is highlighted in bold.

Hypothesis Similarity (increasing order)	Strategy				
	Random	Variance	Max Discrepancy	Surprise	KL Divergence
$N(0, 4^2)$	3	**2**	**2**	3	**2**
$N(0, 2^2)$	8	4	**3**	7	4
$N(0, 1^2)$	18	**7**	**7**	13	11

each other, with the hypothesis that is most similar to all other hypotheses being chosen to act as the true hypothesis. The training observations are then discarded. The true hypothesis provides the observations for the experiments that the active learning techniques request, distorted by Gaussian noise $N(0, 0.5^2)$. The goal is for the active learning techniques to provide evidence to make the true hypothesis have the sustained highest confidence of all the hypotheses in consideration, where the techniques do not know which is the true hypothesis.

Shown in Table 1 are the results for the average number of experiments, over 100 trials, required for the most confident hypothesis to be the true hypothesis, for sets of hypotheses of increasing similarity. As the similarity between the hypotheses increases, it is clear that the variance and maximum discrepancy experiment selection techniques provide the most efficient methods for selecting experiments to separate the hypotheses. However, the variance approach can suffer if there is a hypothesis that makes a prediction that is significantly different to the other hypotheses. As illustrated in Fig. 3(a), alongside an example set of hypotheses in (b), the variance approach can select an experiment where the majority of the hypotheses have the same view, which will likely result in no information gain from that observation. The maximum discrepancy approach however, provides a more robust approach at selecting experiments to separate hypotheses and as such, it will form the basis for the experiment selection strategy employed by the artificial experimenter. The design of which we discuss in the next section.

4   Artificial Experimenter

Building on the concepts discussed earlier of multiple hypotheses and maximum discrepancy experiment selection, we now discuss the design of the artificial experimenter. To begin a number of exploratory experiments are performed, positioned equidistant in the parameter space. In the simulated and laboratory evaluation, 5 experiments are initially performed. After these experiments are performed, an initial set of working hypotheses are created using random subsets of the available observations and randomly selected smoothing parameters. The smoothing parameter is chosen from a set of predetermined smoothing parameters that allow for a range of fits. Initially 200 hypotheses are created in this

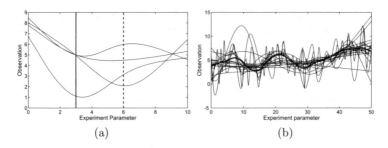

Fig. 3. In (a) is an illustration of where the variance approach can fail, where the solid line is the experiment parameter chosen by the variance approach and dashed is where the maximum discrepancy approach chooses, for the hypotheses shown as curved lines. The variance approach is mislead by a single hypothesis. In (b) is an example set of hypotheses used to test the different active learning techniques for separating a corpus of similar hypotheses, where the bold hypothesis is the true hypothesis.

manner. The observations are then compared against all of the hypotheses to find observations that do not agree with the hypotheses. An observation is determined to be in disagreement with a hypothesis, if that observation is outside the 95% error bar for the hypothesis. If a hypothesis and observation disagree, the parameters of the hypothesis are used to build 2 new hypotheses. These 2 new hypotheses are refinements, where one hypothesis will consider the observation as valid by applying a weight of 100 to the observation, whilst the other hypothesis considers the observation erroneous by applying a weight of 0 to the observation. All 3 hypotheses are then retained in the working set of hypotheses.

After this process of refinement, the hypotheses are evaluated against all available observations, using the confidence function in Eqn. 1. For computational efficiency, the worst performing hypotheses can at this stage be removed from the working set of hypothesis. Currently the best 20% hypotheses are kept into the next stage of experimentation, as initial tests have indicated that higher percentages provided little additional benefit and only increased the computational complexity.

Next a set of experiments to perform are determined by evaluating the hypotheses with the discrepancy equation:

$$D(x) = \sum_{i=1}^{|\mathcal{H}|} \sum_{j=1}^{|\mathcal{H}|} \left(1 - P_{h_i}\left(\hat{h}_j(x) | x \right) \right) C(h_i) C(h_j) S(h_i, h_j) \tag{16}$$

with the error bars of the hypotheses providing σ_i for Eqn. 2. Whilst the discrepancy approach has been shown to be efficient in identifying the best fitting hypothesis from a set of hypotheses, it is not designed to explore the parameter space to help build those hypotheses. Therefore, to promote exploration, the peaks of Eqn. 16 are used to determine the locations for the set of experiments to next perform. This allows experiments to be performed that investigate differences between the hypotheses in several areas of the parameter space. Additionally,

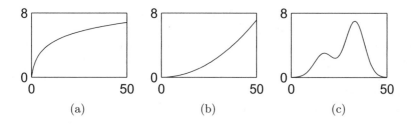

Fig. 4. Underlying behaviours used to evaluate the artificial experimenter, motivated from possible enzyme experiment responses

repeat experiments are not performed. The set of experiments are then performed sequentially, where after each experiment is performed, a new set of hypotheses are created, merged with the working hypotheses, which are refined, evaluated and reduced in the process described previously. Once all experiments in the set are performed, a new set of experiments are determined by evaluating the current working set of hypotheses with the discrepancy equation again.

5   Simulated Results

Evaluating the ability of the technique to build suitable models of biological response characteristics, requires underlying behaviours to compare the predictions against. Whilst documented models of the enzymatic behaviours to be investigated do not exist, there are some possible characteristics that may be observed defined in the literature. In Fig. 4 we consider three potential behaviours, motivated from the literature, where (a) is similar to Michaelis-Menton kinetics [14], (b) is similar to responses where there is a presence of cooperativity between substrates and enzymes [17], whilst (c) considers nonmonotonic behaviours that may exist in enzymatic responses [19].

To perform the simulation, we assume that a behaviour being investigated is captured by some function $f(x)$. Calls to this function produce an observation y, however, experimental noise in both the observations obtained (ϵ) and the experiment parameters (δ), deviate this observation from the true response. Additionally, erroneous observations can in some experiments occur through a form of shock noise (ϕ). Whilst ϵ and δ may occur in all experiments, represented through a Gaussian noise function, ϕ will only occur for a small proportion of experiments and will be in the form of a larger offset from the true observation. Therefore we use the following function to represent performing an experiment:

$$y = f(x + \delta) + \epsilon + \phi \tag{17}$$

with the goal of the artificial experimenter being to determine a function $g(x)$ that suitably represents the behaviours exhibited by $f(x)$.

In the simulation, $\epsilon = N(0, 0.5^2)$ for all experiments and $\phi = N(3, 1)$ for 20% of the experiments performed, with one of the first 5 being guaranteed to be erroneous. Shock noise δ is currently not used for clarity of results. In each trial, 5 initial experiments are performed, with a further 15 experiments being chosen through an active learning technique. In addition to the multiple hypotheses approach presented here, for comparison a single hypothesis approach is tested that is trained with all available observations, using cross-validation to determine the smoothing parameter. The single hypothesis approach is evaluated using two experiment selection methods, which are random selection and placing experiments where the error bar of the hypothesis is maximal. The multiple hypotheses approach is evaluated using three experiment selection methods, which are random selection, the multiple peaks of the discrepancy equation as presented previously, and choosing the single highest peak of the discrepancy equation for each experiment. For each technique and underlying behaviour, 100 trials are conducted, with the bias and variance of the most confident hypothesis of each trial compared to the true underlying behaviour, being used to evaluate the techniques:

$$E = \frac{1}{N} \sum_{n=1}^{N} \left(\left(\bar{b}(x_n) - f(x_n) \right)^2 + \frac{1}{M} \sum_{m=1}^{M} \left(\hat{b}_m(x_n) - \bar{b}(x_n) \right)^2 \right) \tag{18}$$

where $\bar{b}(x_n)$ is the mean of the predictions of the most confident hypotheses, $\hat{b}_m(x_n)$ is the prediction of the most confident hypothesis in trial m, M is the number of trials and N is the number of possible experiment parameters.

In Fig. 5, the performance of the different artificial experimenter techniques are shown. The single hypothesis approaches only perform well in the monotonic behaviours shown in (a) and (b), as the cross-validation allows for errors to be smoothed out quickly. However, in the nonmonotonic behaviour, the single hypothesis approaches perform worse, as the features of the behaviours are smoothed out by the cross-validation, as shown in (d), where the single hypothesis approach misses the majority of the features in the behaviour. On the other hand, the multiple hypotheses approach using the presented technique, fairs well in all behaviours. After 15 experiments it has the lowest prediction error of the techniques tested all three behaviours tested here. However, the multiple hypotheses approach using random experiment selection, is able to reduce the error at a faster rate in the nonmonotonic behaviour (c). Whilst as expected, choosing the single highest peak in the discrepancy equation after each experiment, performs the worst of the multiple hypotheses techniques as expected throughout, as that approach does not effectively explore the parameter space.

The difference between the random and multiple peaks experiment selection strategy, is due to the multiple peaks strategy initially finding the differences between hypotheses that poorly represent the underlying behaviour. These early experiments will investigate discrepancies that will return more general information about the behaviour, with it being possible for experiments within a particular set obtaining similar information. However, as the hypotheses better represent the underlying behaviour, the discrepancies between the hypotheses are more likely to indicate where more specific differences in the hypotheses exist, for example a smaller peak in the behaviour being investigated. This is why

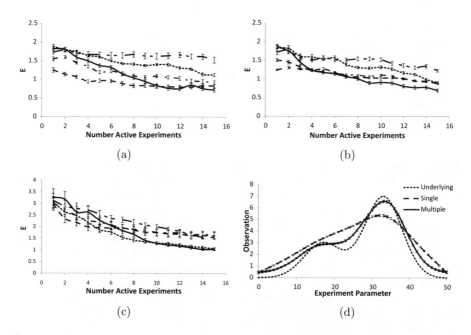

Fig. 5. Comparison of error over number of actively chosen experiments, where 20% of the observations are erroneous in (a-c), with comparison to true underlying for (c) shown in (d). Figures (a-c) correspond to the behaviours in Fig. 4. In (a-c) the lines represent: single hypothesis - variance (dashed), single hypothesis - random (dash dot), multiple hypotheses - discrepancy peaks (solid), multiple hypotheses - random (dots), multiple hypotheses - single max discrepancy (dash dot dot). The multiple hypotheses technique using the peaks of the discrepancy function provides the lowest error after 15 actively selected experiments consistently. The single hypothesis approach fails to identify features in nonmonotonic behaviours shown in (d).

in all three of the behaviours tested, the multiple peaks experiment selection strategy is initially one of the worst performing strategies, but then reduces its error at a faster rate than any of the other strategies. These results suggest that the multiple peaks experiment strategy may in some scenarios benefit from additional exploration, before the active strategy begins. Next we consider an evaluation of the technique within a laboratory setting.

6   Laboratory Evaluation

Further to the simulated evaluation, the artificial experimenter has been tested within a real laboratory setting. Here the artificial experimenter has guided a human scientist to characterise the optical absorbance profile of the coenzyme NADH, where the rate of change of absorbance can be compared to the Beer-Lambert law. NADH is commonly used for monitoring enzymatic catalytic activity.

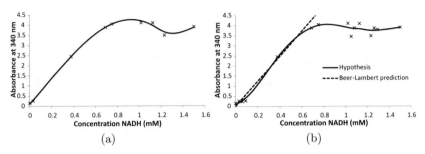

Fig. 6. Most confident hypothesis and experiments chosen for NADH absorbance characterisation. A stock solution of 5 mM NADH and a 10 mM Tris buffer at pH 8.5 were prepared. Dilutions of NADH requested by the artificial experimenter were produced by mixing volumes taken from the stock solution and the buffer. Measurements of optical absorbance at 340 nm were recorded with a PerkinElmer Lambda 650 UV-Vis Spectrophotometer to provide the observations. The photometric range of the spectrophotometer was 6 A. In (a) the most confident hypothesis after 4 active experiments is shown, where a slight dip in absorbance has been detected. Further experiments determine this dip does not exist, as shown in (b). The hypothesis identifies a linear region in good agreement with the Beer-Lambert law, whilst also identifying a nonlinear region that is likely caused by nonlinear optical effects, as all measurements are within the operational range of the spectrophotometer.

To perform the test, the artificial experimenter was first provided the boundary to which it could explore in the parameter space, 0.001–1.5 mM. The parameter space was coded to the parameter space used in the simulation, allowing for same set of smoothing parameters to be used ($\lambda = \{10, 50, 150, 100, 500, 1000\}$). The artificial experimenter requested an initial 5 experiments, placed equidistant within the parameter space. Using the procedure described in Fig.6, the human scientist performed the experiments as directed, providing the observations to the artificial experimenter. The artificial experimenter then presented a graph of the observations, along with the current best hypothesis, the alternate hypotheses and the discrepancy amongst them. The artificial experimenter was then allowed to select an additional 10 experiments using the multiple peaks active experiment selection technique described.

In Fig 6, the results of those experiments are shown. After the initial exploratory experiments, the artificial experimenter identifies the key feature that there is an increase in absorbance between 0.001 and 0.75 mM, that then begins to level off. The first active experiment looks at roughly where the increase in absorbance ends at 0.69 mM, with the observation agreeing with the initial trend of the data. The second active experiment at 1.23 mM, providing an observation lower than the initial prediction, makes the artificial experimenter consider the possibility that rather than a leveling off in absorbance, the absorbance lowers again with a similar rate to that which it increased. The remainder of experiments then look to investigate whether the absorbance lowers or remains largely flat, with a few additional experiments investigating where the rise in absorbance begins. The hypothesis after 15 experiments matches the expected Beer-Lambert

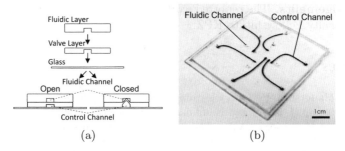

(a) (b)

Fig. 7. Microfluidic chip layered design (left) and photo of prototype chip (right). Reactants flow in channels between the fluidic and valve layers, whilst control channels exist between the valve and glass layers. Pressure on the control channels control whether fluidic channels are open or closed, to allow reactants to pass. On-chip absorbance measurement will allow for all experimentation to take place on chip.

law rate of change in absorbance prediction, using the indicated extinction coefficient of 6.22 at a wavelength of 340 nm [16], as shown in Fig. 6(b).

7 Conclusion

Presented here is an artificial experimenter that can direct experimentation in order to efficiently build response models of behaviours, where the number of experiments possible is limited and the observations are potentially erroneous. The domain of enzymatic experiments is used to motivate the approach, however the technique is designed to be general purpose and could be applied to other experimentation settings where there are similar limiting factors. The technique uses a multiple hypotheses approach, where different views of the observations are taken simultaneously, in order to deal with the uncertainty that comes from having potentially erroneous observations and limited resources to test them. A technique of experiment selection that places experiments in locations of the parameter space where the hypotheses disagree has been proposed. Whilst this approach appears to perform consistently across simulated behaviours, perhaps additional measures of exploration could be added to the technique, so as to better manage the exploration-exploitation trade-off. Additionally the approach should also consider when to terminate experimentation by monitoring the change in hypotheses over time, rather than using fixed numbers of experiments allowed.

The next stage is to couple the artificial experimenter with the lab-on-chip experiment platform in development, which is shown in Fig. 7. This autonomous experimentation machine, will allow the artificial experimenter to request experiments to be performed, which the hardware will automatically perform, returning the result of the experiment back to the computational system. As such, it will provide a tool for scientists, which will not only allow them to reduce experimentation costs, but will also allow them to redirect their time from monotonous characterisation experiments, to analysing the results, building theories and determining uses for those results.

Acknowledgements. The reported work was supported in part by a Microsoft Research Faculty Fellowship to KPZ.

References

1. Abe, N., Mamitsuka, H.: Query learning strategies using boosting and bagging. In: ICML 1998, pp. 1–9. Morgan Kauffmann, San Francisco (1998)
2. Atkinson, A.C., Fedorov, V.V.: The design of experiments for discriminating between several models. Biometrika 62(2), 289–303 (1975)
3. Burbidge, R., Rowland, J.J., King, R.D.: Active learning for regression based on query by committee. In: Yin, H., Tino, P., Corchado, E., Byrne, W., Yao, X. (eds.) IDEAL 2007. LNCS, vol. 4881, pp. 209–218. Springer, Heidelberg (2007)
4. Chamberlin, T.C.: The method of multiple working hypotheses. Science (old series) 15, 92–96 (1890); Reprinted in: Science, vol. 148, p. 754–759 (May 1965)
5. Cohn, D.A., Ghahramani, Z., Jordan, M.I.: Active learning with statistical models. Journal of Artificial Intelligence Research 4, 129–145 (1996)
6. Freund, Y., Seung, H.S., Shamir, E., Tishby, N.: Selective sampling using the query by committee algorithm. Machine Learning 28, 133–168 (1997)
7. Itti, L., Baldi, P.: Bayesian surprise attracts human attention. Vision Research 49, 1295–1306 (2009)
8. King, R.D., Whelan, K.E., Jones, F.M., Reiser, P.G.K., Bryant, C.H., Muggleton, S.H., Kell, D.B., Oliver, S.G.: Functional genomic hypothesis generation and experimentation by a robot scientist. Nature 427, 247–252 (2004)
9. Kulkarni, D., Simon, H.A.: Experimentation in machine discovery. In: Shrager, J., Langley, P. (eds.) Computational Models of Scientific Discovery and Theory Formation, pp. 255–273. Morgan Kaufmann Publishers, San Mateo (1990)
10. Kullback, S., Leibler, R.A.: On information and sufficiency. Annals of Mathematical Statistics 22, 79–86 (1951)
11. MacKay, D.J.C.: Information–based objective functions for active data selection. Neural Computation 4, 589–603 (1992)
12. Matsumaru, N., Colombano, S., Zauner, K.-P.: Scouting enzyme behavior. In: CEC, pp. 19–24. IEEE, Piscataway (2002)
13. McCallum, A.K., Nigam, K.: Employing em and pool-based active learning for text classification. In: ICML, pp. 584–591. Morgan Kaufmann, San Francisco (1998)
14. Nelson, D.L., Cox, M.M.: Lehninger Principles of Biochemistry, 5th edn. W. H. Freeman and Company, New York (2008)
15. Settles, B.: Active learning literature survey. Tech. rep., University of Wisconsin-Madison (2009)
16. Siegel, J.M., Montgomery, G.A., Bock, R.M.: Ultraviolet absroption spectra of dpn and analogs of dpn. Archives of Biochemistry and Biophysics 82(2), 288–299 (1959)
17. Tipton, K.F.: Enzyme Assays, 2nd edn., pp. 1–44. Oxford University Press, Oxford (2002)
18. Wahba, G.: Spline Models for Observational Data, CBMS-NSF Regional Conference series in applied mathematics, vol. 59. SIAM, Philadelphia (1990)
19. Zauner, K.-P., Conrad, M.: Enzymatic computing. Biotechnol. Prog. 17, 553–559 (2001)
20. Żytkow, J., Zhu, M.: Automated discovery in a chemistry laboratory. In: AAAI-90, pp. 889–894. AAAI Press / MIT Press (1990)

Subgroup Discovery for Election Analysis: A Case Study in Descriptive Data Mining

Henrik Grosskreutz, Mario Boley, and Maike Krause-Traudes

Fraunhofer IAIS, Schloss Birlinghoven, Sankt Augustin, Germany
{firstname.lastname}@iais.fraunhofer.de

Abstract. In this paper, we investigate the application of descriptive data mining techniques, namely subgroup discovery, for the purpose of the ad-hoc analysis of election results. Our inquiry is based on the 2009 German federal Bundestag election (restricted to the City of Cologne) and additional socio-economic information about Cologne's polling districts. The task is to describe relations between socio-economic variables and the votes in order to summarize interesting aspects of the voting behavior. Motivated by the specific challenges of election data analysis we propose novel quality functions and visualizations for subgroup discovery.

1 Introduction

After a major election of public interest is held, there is a large and diverse set of societal players that publishes a first analysis of the results within the first day after the ballots are closed. Examples include traditional mass media like newspapers and television, citizen media like political blogs, but also political parties and public agencies. An instance of the last type is the Office of City Development and Statistics of the City of Cologne. The morning after major elections that include Cologne's municipal area, the office publishes a first analysis report on the results within the city[1]. In this report, socio-economic variables (e.g., average income, age structure, and denomination) are related to the voting behavior on the level of polling districts. The Office of City Development and Statistics performs much of the analysis, such as selecting a few candidate hypotheses, beforehand, i.e., based on previous election results—a course of action that might neglect interesting emerging developments. However, due to the strict time limit involved, there appears to be no alternative as long as an analyst mainly relies on time-consuming manual data operations. This motivates the application of semi-automatized data analysis tools.

Therefore, in this academic study, we take on the perspective of an analyst who is involved in the publication of a short-term initial analysis of election

---

[1] The report on the 2009 Bundestag election can be found (in German language) at
http://www.stadt-koeln.de/mediaasset/content/pdf32/wahlen/
bundestags\wahl2009/kurzanalyse.pdf

B. Pfahringer, G. Holmes, and A. Hoffmann (Eds.): DS 2010, LNAI 6332, pp. 57–71, 2010.

Party	description	result 2009	change
■ SPD	social democrats	26%	-12.1
■ CDU	conservatives	26.9%	-0.3
▨ FDP	liberals	15.5%	+4
▨ GRUENE	greens	17.7%	+2.8
■ LINKE	dem. socialists	9.1%	+3.4

Fig. 1. Results of the 2009 Bundestag election in Cologne

results, and we investigate how data mining can support the corresponding ad-hoc data analysis. In order to narrow down the task, we focus on the following *analysis question*:

> *What socio-economic variables characterize a voting behavior that considerably differs from the global voting behavior?*

This question is of central interest because it asks for interesting phenomenons that are *not* captured by the global election result, which can be considered as base knowledge in our context. Thus, answers to this question have the potential to constitute novel, hence, particularly news-worthy, knowledge and hypotheses. This scenario is a prototypic example for *descriptive* knowledge discovery: instead of deducing a global data model from a limited data sample, we aim to discover, describe, and communicate interesting aspects of it.

(a)	(b)

Fig. 2. Spatial visualization of polling districts. Color indicates: (a) above average FDP votes; (b) high share of households with monthly income greater than 4500€

We base our study on the German 2009 federal Bundestag election restricted to the results of Cologne. For this election we analyzed the data during a corresponding project with Cologne's Office of City Development and Statistics. See Figure 1 for the list of participating parties, their 2009 election results, and the difference in percentage points to their 2005 results. The data describes the election results on the level of the 800 polling districts of the city, i.e., there is one data record for each district, each of which corresponds to exactly one polling place. Moreover, for each district it contains the values of 80 socio-economic variables (see Appendix A for more details). Figure 2 shows the geographical alignment of the districts. To illustrate that indeed there is a relation between voting behavior and socio-economic variables, the figure additionally shows the districts with a high share of households with high income (Fig. 2(b)). The high overlap between these districts and those with above-average votes for party FDP (Fig. 2(a)) is an indication that those two properties are correlated.

The semi-automatized data analysis we propose in this work, i.e., the descriptive pattern and hypotheses generation, is meant purely indicative: the evaluation of all discovered patterns with respect to plausibility (e.g., to avoid ecological inference fallacy [18]) and their interestingness are up to the analyst and her background knowledge (hence, *semi-automatized* data analysis). In fact, almost all of the analyst's limited time and attention has to be reserved for the manual preparation and creation of communicable content. Thus, in addition to fast execution times, a feasible tool has to meet the following requirements:

(R1) The tool has to discover findings that directly support answers to the analysis question above. In particular, it must suitably define how to assess "notably different voting behavior." That is, it must provide an *operationalization* that relates this notion to the measurement constituted by the poll.

(R2) Operating the tool has to be simple. In particular, it either has to avoid complicated iterative schemes and parameter specifications, or there must be clear guidelines for how to use all degrees of freedom.

(R3) The tool's output must be intuitively interpretable and communicable. This involves avoiding redundant or otherwise distracting output as well as providing a suitable visualization.

In the next section we show how *subgroup discovery* can be configured to meet these requirements. In particular, we propose new quality functions for subgroup discovery that are capable of handling *vector-valued target variables* as they are constituted by election results, and that *avoid the generation of redundant subgroups* without requiring the specification of any additional parameter. Moreover, we propose a *visualization technique* that is tailor-made for rendering subgroups with respect to election results.

2   Approach

There are existing approaches to election analysis (e.g., [8,14]) that are either based on a global regression model or on an unsupervised clustering of the population. In contrast, we approach the problem from a supervised local pattern discovery perspective [15]. As we show in this section, our analysis question naturally relates to the task of *subgroup discovery* [21]. After a brief introduction to this technique, we discuss how a subgroup discovery system can be configured and extended such that it satisfies the initially identified requirements (R1)-(R3).

2.1   Subgroup Discovery

Subgroup discovery is a descriptive data mining technique from the family of *supervised descriptive rule induction methods* (see [17,16]; other members of the family include *contrast set mining*, *emerging patterns* and *correlated itemset mining*). It aims to discover local sub-portions of a given population that are a) large enough to be relevant and that b) exhibit a substantially differing behavior from that of the global population. This difference is defined with respect to a designated target variable, which in our scenario is the election result. The data sub-portions are called *subgroups*, and they are sets of data records that can be described by a conjunction of required features (in our case the features are constraints on the values of the socio-economic variables).

For a formal definition of subgroup discovery, let DB denote the given database of N data records $d_1, \ldots, d_N$ described by a set of n (binary) features $(f_1(d_i), \ldots, f_n(d_i)) \in \{0,1\}^n$ for $i \leq N$. A *subgroup description* is a subset of the feature set $sd \subseteq \{f_1, \ldots, f_n\}$, and a data record d satisfies sd if $f(d) = 1$ for all $f \in sd$, i.e. a subgroup description is interpreted conjunctively. The *subgroup* described by sd in a database DB, denoted by $DB[sd]$, is the set of records $d \in DB$ that satisfy sd. Sometimes, $DB[sd]$ is also called the *extension* of sd in DB. The interestingness of a subgroup description sd in the context of a database DB is then measured by a *quality function* q that assigns a real-valued quality $q(sd, DB) \in \mathbb{R}$ to sd. This is usually a combination of the subgroup's size and its unusualness with respect to a designated target variable.

In case the target variable T is real-valued, that is, T is a mapping from the data records to the reals, the unusualness can for instance be defined as the deviation of the mean value of T within the subgroup from the global mean value of T. A common choice is the *mean test quality function* [9]:

$$q_{mt}(DB, sd) = \sqrt{\frac{|DB[sd]|}{|DB|}} \cdot (m(DB[sd]) - m(DB)) \qquad (1)$$

where $m(D)$ denotes the mean value of T among a set of data records D, i.e., $m(D) = 1/|D| \sum_{d \in D} T(d)$.

Generally, quality functions order the subgroup descriptions according to their interestingness (greater qualities correspond to more interesting subgroups). One is then usually interested in k highest quality subgroup descriptions of length at most l where the length of a subgroup description is defined as the number of features it contains.

2.2   Application to Election Analysis

Requirement (R1) of Section 1 includes the support of a suitable operationalization of "notably different voting behavior". If we choose to perform this operationalization on the level of individual parties, i.e., as a notably different share of votes of one specific party, the analysis question from Section 1 can directly be translated into subgroup discovery tasks: choose q_{mt} as quality function, create a set of features based on the socio-economic variables, and as target variable T choose either a) the 2009 election result of a particular party or b) the difference of the 2009 and the 2005 result. The latter option defines "voting behavior" with respect to the *change* in the share of votes. This is a common perspective that is usually used to interpret the outcome with respect to the success or failure of individual parties.

In order to answer the analysis question independently of a specific party, one can just run these subgroup discoveries once for each of the possible party targets and then choose the k best findings among all returned patterns. With this approach, the patterns for the overall voting behavior are chosen from the union of the most interesting patterns with respect to the individual parties.

There are, however, several important relations among the parties. For instance, they can be grouped according to their ideology (in our case, e.g., SPD, GRUENE, and LINKE as "center-left to left-wing"), or one can distinguish between major parties (SPD and CDU) and minor parties (FDP, GRUENE and LINKE). Voting behavior can alternatively be characterized with respect to such groups (e.g., "in districts with a high number of social security claimants, the major parties lost more than average.") This indicates that a subgroup may be interesting although no individual party has an interesting result deviation in it, but because the *total* share of two or more parties is notably different.

This is not reflected in the initial approach, hence it is desirable to extend the subgroup discovery approach such that it captures requirement (R1) more adequately. In particular, we need a quality function that does not only rely on a single target variable but instead on a set of k real-valued target variables $T_1, \ldots, T_k$. With this prerequisite we can define a new quality function analogously to the mean test quality by

$$q_{\mathrm{dst}}(DB, sd) = \sqrt{\frac{|DB[sd]|}{|DB|}} \, \|\mathbf{m}(DB[sd]) - \mathbf{m}(DB)\|_1 \qquad (2)$$

where $\mathbf{m}(D)$ denotes the mean vector of the $T_1, \ldots, T_k$ values among a set of data records D, i.e.,

$$\mathbf{m}(D) = 1/|D| \sum_{d \in D} (T_1(d), \ldots, T_k(d)) \ .$$

and $\|(x_1, \ldots, x_k)\|_1$ denotes the 1-norm, i.e., $\sum_{i=1}^{k} |x_i|$. Using this quality function we arrive at an alternative instantiation of subgroup discovery. We can choose q_{dst} as quality function and either a) T_i as the 2009 share of votes of party i or b) T_i as the gain (2009 result minus 2005 result) of that party.

2.3   Avoidance of Redundant Output

Requirement (R3) demands the avoidance of redundant output, but, unfortunately, a problem with the straightforward discovery of subgroups and other descriptive patterns is that a substantial part of the discovered patterns can be very similar. That is, many patterns tend to be only slight variations of each other, essentially describing the same data records. The reason for this is twofold. Firstly, there may be many highly correlated variables that provide interchangeable descriptions. We can get rid of these by performing a correlation analysis during preprocessing (see Section 3.1). In addition, for an interesting subgroup sd it is likely that there are some strict *specializations* $sd' \supset sd$ with an equal or slightly higher quality. Although the truly relevant and interesting portion of the subgroup may be described most adequately by sd, those specializations are at least equally likely to appear in the output, causing redundancy or—even worse—pushing sd out of the result set altogether.

We now present an approach that generalizes a common principle of some of the existing methods to address this problem [3,7,19,20], namely to discard subgroups sd that do not substantially improve their strict *generalization* $sd' \subset sd$. As captured in our requirement (R2) we want to avoid the introduction of additional parameters. Therefore, unlike the cited approaches, we do not introduce a minimum improvement threshold, but instead we *use the quality function itself* to measure the sufficiency of an improvement. That is, for some arbitrary *base* quality function q, we propose to assess the quality of a pattern sd as the minimum of the quality of sd with respect to the extension of all its generalizations. More precisely, we consider the quality function q^Δ that is defined as $q^\Delta(DB, \emptyset) = q(DB, \emptyset)$ for the empty subgroup description $\emptyset$ and

$$q^\Delta(DB, sd) = \min_{sd' \mid sd' \subset sd} q(DB[sd'], sd) \qquad (3)$$

otherwise. We call q^Δ the *incremental version* of q. After giving some additional definitions, we discuss in the remainder of this subsection that q^Δ has some desirable properties.

We call a subgroup description sd *tautological* with respect to a database DB if $DB[sd] = DB$, and we call sd *non-minimal* with respect to DB if there is a generalization $sd' \subset sd$ having the same extension, i.e., $DB[sd] = DB[sd']$. Moreover, we say that a quality function q is *reasonable* if $q(DB, sd) \leq 0$ whenever sd is tautological with respect to DB.

Proposition 1. *Let DB be a database, q a quality function, and q^Δ its incremental version. If q is reasonable, then q^Δ is non-positive for all non-minimal subgroup descriptions in DB.*

Proof. Note that, by definition, every non-minimal subgroup has a strict generalization sd' with identical extension. Therefore,

$$q^\Delta(DB, sd) \leq q(DB[sd'], sd) = q(DB[sd], sd) = 0$$

where the last equality follows from q being reasonable.   □

This property assures that non-minimal subgroup descriptions are filtered from the result set. Such descriptions are considered redundant respectively trivial [4,19]. For quality functions based on the mean deviation (e.g., Eq. 1) an even stronger statement holds: for such functions all descriptions are filtered that do not provide an improvement in the mean deviation. Thus, the incremental quality directly follows other filtering paradigms from descriptive rule induction; namely it eliminates patterns that do not provide a *confidence improvement* [3] respectively that are not *productive* [20]. Finally, we remark that the incremental quality is bounded by the base quality, i.e., for all subgroup descriptions sd it holds that $q^{\Delta}(DB, sd) \leq q(DB, sd)$.

2.4   Visualization

In order to completely meet the last requirement (R3), we need an appropriate visualization technique. Although there is existing work on subgroup visualization (e.g., [1,10]), we choose to design a new technique that is tailor-made for election analysis and allows for multiple target attributes. In fact we propose four visualizations, one for each possible subgroup discovery configuration discussed in Section 2.2. A common element is that every subgroup is visualized by a grey box having a color intensity that reflects the subgroup's quality. Higher qualities correspond to more intense grey shades. Every box shows the subgroup description and the size of its extension, plus additional information that depends on the quality function as well as on the operationalization of "voting behavior".

Fig. 3. Visualization for the different combinations of quality functions and operationalizations of election result: (a) single party result, (b) result distribution, (c) single party gain and (d) gain/loss vector. The result of the particular parties are plotted using their official colors, listed in Figure 1.

Figure 3 shows the four cases: (a) absolute results of a single party, (b) combined absolute results for all parties, (c) gain for a single party, and (d) combined gains and losses of all parties. In case the mean test is used as quality function, we show the mean value of the target variable, i.e. the result for a particular party, in the subgroup next to the extension size. For the vector-valued quality function this space is occupied by the 1-norm of the mean vector difference. Beside this figure, the boxes include a visualization of the election result of all parties. Depending on whether absolute 2009 results or the gains with respect to 2005 are considered, the results are rendered in a different fashion. The absolute 2009 results are represented by two bars (Figure 3(a) and 3(b)): the upper bar corresponds to the distribution over parties in the subgroup, while the lower bar visualizes the overall distribution. The different segments in the bars represent the share of votes for the different parties. They are visualized from left to right using the parties' official colors: red (SPD), black (CDU), yellow (FDP), green (GRUENE) and magenta (LINKE). If, instead, the gains respectively losses are considered (Figure 3(c) and 3(d)), the result is displayed as bar chart that is centered around a gain of 0. Gains are visualized by upward bars, while losses are visualized by downward bars. Again, the global gains are also plotted for easy comparison: For every party, a first bar shows the local gains in the subgroup, while a second bar on its right-hand shows the global gains. This second bar provides the context information required to interpret the gains in a particular subgroup.

3   Experiment

After the introduction of our tools we are now ready to describe our case study on the 2009 Bundestag election. Before we provide and discuss the results we briefly summarize our experimental setup.

3.1   Setup

From the raw input data to the final output we performed the following steps.

preprocessing. In order to avoid the occurrence of highly correlated features in the result, we performed a correlation analysis and removed one variable out of every pair of variables with a correlation of at least 0.85. The choice was based on background-knowledge and subjective preference. Moreover, we performed a 3-bin frequency discretization to all remaining numerical variables.

features. Based on the discretization, we defined the set of descriptive features as follows: for every variable and every bin, there is a binary feature that a data record possesses if and only if the variable value of this record lies in that bin. These features are denoted $V = h$, $V = m$, and $V = l$, respectively. There are, however, several exceptions. Some variables are part of a set of complementary variables that together describe a common underlying measurement. For instance, for the age structure there is one variable representing the number of inhabitants aged 16-24, the inhabitants aged 25-34, and so on, respectively. For such variables, we did not create features corresponding to the middle or lower bin because they would have only low descriptive potential. Altogether, there is a total of 64 descriptive features.

parameters. We used a length limit l of 3 for the subgroup descriptions, and a number of subgroups k of 10. These settings lead to a reasonably small set of results that can be manually inspected and that are short enough to be easily communicable.

targets. As stated in Section 2.2 there are several options for the operationalization of "voting behavior": one has to choose between individual parties and the combined results as well as between the absolute (2009) results and the difference between the 2009 and the previous (2005) results. This leaves us with four different configurations of quality functions and target variables.

C1. For absolute combined results, quality function $q = q^{\Delta}_{dst}$ (Eq. 2) with target variables $T_1, \ldots, T_5$ such that T_i is the 2009 share of votes of party i.

C2. As exemplary configuration for absolute results in the single party case, $q = q_{mt}^{\Delta}$ (Eq. 1) with the 2009 result of FDP as target T.

C3. For combined results measured by the difference to previous elections, $q = q_{dst}^{\Delta}$ with target variables $T_1, \ldots, T_5$ such that T_i as the gain (2009 result minus 2005 result) of party i.

C4. Again as exemplary configuration for differences in the single party case, $q = q_{mt}^{\Delta}$ with the difference between the 2009 and the 2005 result of FDP as target.

visualization. Finally, for each configuration the resulting subgroups are rendered using the appropriate visualization technique introduced in Section 2.4. Additionally, the boxes are joined by arrows corresponding to the transitive reduction of the specialization relation among the subgroups.

Some of the above steps are not fully consistent with our requirement (R2). In particular, in the preprocessing step the user is left with the decision which variables to keep. Moreover, the parameter settings (for the number of bins and the number of subgroups) are not the only viable option. However, they are a good starting point, given that a restriction to 3 bins results in bins with an easily communicable meaning ("low" and "high"), while 10 subgroups represent a manageable amount of patterns.

3.2 Results

After describing the setup of our experiments we now present the result it yielded. In order to put our findings into context, we first recap the most important aspects of the 2009 Bundestag election results: The parliamentary majority shifted from the so-called grand coalition (CDU and SPD) to a coalition of CDU and FDP. The expiration of the grand coalition was essentially caused by an all-time low result of the social-democratic SPD combined with substantial gains for the FDP. This development is also reflected in the local results of Cologne (see Figure 1).

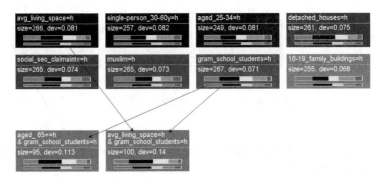

Fig. 4. Subgroups found using the distribution over parties as label

Absolute results of all parties. Figure 4 shows the subgroups obtained using Configuration C1, i.e., considering the combined absolute results of all parties in the 2009 election. There are several subgroups with a strong preference for the liberal-conservative election winners, FDP and CDU. These include the subgroup of districts with a "high average living space per accommodation," and the subgroup "high share of detached houses." The longer subgroup description "high average living space per accommodation and high share of grammar school students" is even more notable, as it has an extremely high share of FDP votes. While all other parties have lower results in these subgroups, the share of LINKE votes is particularly low. Another interesting subgroup is "high number of 30-60 year-old single-persons." This constraint is an indicator for a high share of GRUENE voters. All other parties obtained results below average in this subgroup, those of the CDU being particularly weak. There are also subgroups with a high share of SPD and LINKE votes, namely "high share of social security claimants" and "high share of muslims."

This first experiment shows that our tool reveals features that imply a strong voting preference for one particular party (e.g., GRUENE) as well as for political alliances or ideological blocks (e.g., CDU/FDP and SPD/LINKE). It is important to note that subgroups of the latter kind—although they have a clear interpretation and are easily communicable—can be missed if the analysis is performed using the single party operationalization: if one uses this option, for instance the "high share of social security claimants" subgroup is not among the top-10 subgroups.

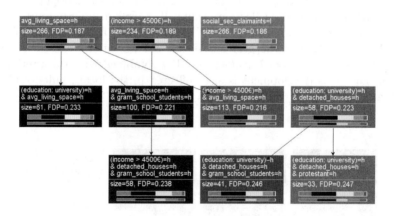

Fig. 5. Subgroups for target 'FDP'

Absolute result of FDP. Still, in case one is solely interested in one particular party, it is a reasonable choice to resort to the individual party configurations. Configuration C2 exemplary considers the FDP 2009 results—the party with the highest gain. Figure 5 shows corresponding subgroups.

While the figure shows some subgroups which are already identified using Configuration C1 (e.g., "high average living space" and the specialization with the additional constraint "high share of grammar school students"), it also contains additional results. For instance, districts with a high share of persons with "net income of more than 4500€" (see Figure 2(b) for a geographical visualization). This feature is confirmed by many other investigations to be an attribute associated with FDP voters.

Gains of all parties. We now move on to the alternative operationalization of voting behavior based on the gains respectively losses. Again, first we consider the combined gains and losses of all parties as specified in Configuration C3. Figure 6 shows the result.

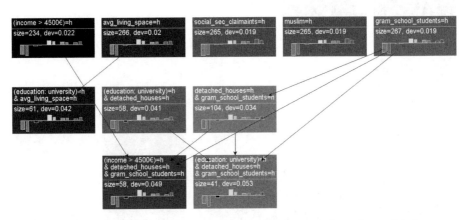

Fig. 6. Subgroups found using the distribution over the gains as label

The districts with a high share of persons with a "net income over 4500€" experienced over-average gains for the FDP, as well as (small) losses for the CDU. Such slightly over-average CDU losses can also be observed in the other subgroups with very strong FDP gains, like "average living space per accommodation" or the longer description "detached houses, grammar school students and high income". Another interesting observation is that these subgroups are also considered in the previous section, in which we considered subgroups with a high absolute share of FDP votes. This co-ocurrence indicates that FDP could achieve additional gains in its party stronghold. The inverse relation can also be observed for the SPD subgroups: the districts with a high share of "social security claimants"—which were observed to have high SPD results—actually witnessed above-average SPD losses. The same holds for the districts with a high share of muslims. This observation suggests that the SPD is losing popularity right in its party strongholds; an assumption shared by a broad range of media analysts.

One advantage of the visualization is that it not only allows identifying the winners in a subgroup, but that it also indicates where the votes could have come from. In the two subgroups above, which attract attention due to over-average SPD results *and* over-average SPD losses, the clear winner is the LINKE, while none of the other parties have above-average gains. This is a hint that a large part of a former SPD stronghold turned into LINKE voters.

Gains of FDP. Finally, it is possible to search for subgroups with particular gains or losses of a particular party. Using Configuration C4 we exemplarily do so again with the FDP gains. The result is shown in Figure 7. Beside confirming some results from the all parties configuration, it also reveals some additional observations. The most noteworthy is perhaps the subgroup with a high share of families having an upper middle-class monthly income (i.e. 3000-4500€). This group is not traditionally associated with the FDP, and can thus constitute a hypothetical part of an explanation of the FDP's success in this election.

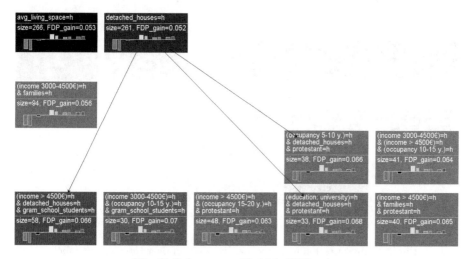

Fig. 7. Subgroups with high FDP gain

3.3   Comparison with the Traditional Approach

It is interesting to compare the results presented here with the findings reported in the Cologne report mentioned in the introduction. The main question considered there is the identification of party hot-spots and their characterization by socio-demographics attributes. This corresponds to our analysis configuration C2, which considers the absolute result of a particular party. If we compare our results with the report, we observe that our algorithm reveals the same socio-economic variables as those selected by the Cologne experts in a time-consuming manual investigation based on prior knowledge and experience. In the case of FDP, for example, the Cologne report also selects the proportion of persons with a high income and the grammar school students ratio to characterize polling districts with a high FDP support (see page 34 of the report).

3.4   Scalability

While the (manual) preprocessing steps can require some time depending on the complexity of the given data, the actual subgroup discovery is fast: for each of the four configurations, the computation takes less than 30 seconds on a standard Core 2 Duo E8400 PC. A detailed analysis of complexity issues is beyond the scope of this paper, but we not that subgroup discovery scales well in practice [2,6]—in particular with the numbers of polling districts, which is the quantity that is expected to vary the most in case the method is applied to other elections. Hence, given that the preprocessing is done in advance, the approach can be applied, e.g., during an election night.

4   Summary and Discussion

In this paper, we have demonstrated the application of a descriptive data mining technique, namely subgroup discovery, to ad-hoc election data analysis. This demonstration included a case study based on the 2009 Bundestag elections restricted to the data of Cologne. Besides presenting the results of this study, we formulated several requirements for data analysis software in this application context and discussed how subgroup discovery tools can be configured to meet these requirements. In particular, we proposed a new quality function and a novel filtering scheme for the avoidance of redundant output. The quality function is an extension of the mean test quality that is based on the combined mean deviation of several target variables. The generally applicable filtering scheme is an incremental, i.e., higher order, quality function that is defined with respect to some desired base quality function. Its idea is to reevaluate all subgroups based on their base quality in the databases defined by their generalizations.

The quality function with several target variables is motivated by the fact that an election result is constituted by the combined results of several parties rather than just one party. Our experiments demonstrate that the introduction of several target attributes is a valuable extension: otherwise important patterns that have an interestingness resulting from the total unusualness of the results of two ore more parties can be dominated by less interesting patterns. We remark that subgroup discovery on datasets involving more than one target attribute is also known as *exceptional model mining* [13], and that our approach could thus be considered as a form of exceptional model mining (based on a new quality function).

Our other technical addition, the incremental quality function, generalizes the well-known idea of evaluating patterns with respect to their generalizations. Following our earlier specified requirements this filtering technique is completely parameter-free. This feature distinguishes our method from the other improvement-based techniques [3,7,19,20] and others, like the weighted covering scheme [12] or approaches based on affinity [5]. Note that although subgroup filtering based on the theory of relevancy [11] is also parameter-free, it only applies to data with a *binary* target variable and thus is not applicable here.

Acknowledgments

Part of this work was supported by the German Science Foundation (DFG) under the reference number 'GA 1615/1-1'. We would like to thank the Office of City Development and Statistics for their kind cooperation.

References

1. Atzmüller, M., Puppe, F.: Semi-automatic visual subgroup mining using vikamine. J. UCS 11(11), 1752–1765 (2005)
2. Atzmüller, M., Puppe, F.: SD-map - a fast algorithm for exhaustive subgroup discovery. In: Fürnkranz, J., Scheffer, T., Spiliopoulou, M. (eds.) PKDD 2006. LNCS (LNAI), vol. 4213, pp. 6–17. Springer, Heidelberg (2006)
3. Bayardo, R.J., Agrawal, R., Gunopulos, D.: Constraint-based rule mining in large, dense databases. Data Min. Knowl. Discov. 4(2/3), 217–240 (2000)
4. Boley, M., Grosskreutz, H.: Non-redundant subgroup discovery using a closure system. In: ECML/PKDD, vol. (1), pp. 179–194 (2009)
5. Gebhardt, F.: Choosing among competing generalizations. Knowledge Acquisition 3, 361–380 (1991)
6. Grosskreutz, H., Rüping, S., Wrobel, S.: Tight optimistic estimates for fast subgroup discovery. In: ECML/PKDD, vol. (1), pp. 440–456 (2008)
7. Huang, S., Webb, G.I.: Discarding insignificant rules during impact rule discovery in large, dense databases. In: SDM (2005)
8. Johnston, R., Pattie, C.: Putting Voters in Their Place: Geography and Elections in Great Britain. Oxford Univ. Press, Oxford (2006)
9. Klösgen, W.: Explora: A multipattern and multistrategy discovery assistant. In: Advances in Knowledge Discovery and Data Mining, pp. 249–271 (1996)
10. Kralj, P., Lavrač, N., Zupan, B.: Subgroup visualization. In: Proc. 8th Int. Multiconf. Information Society, pp. 228–231 (2005)
11. Lavrac, N., Gamberger, D.: Relevancy in constraint-based subgroup discovery. In: Boulicaut, J.-F., De Raedt, L., Mannila, H. (eds.) Constraint-Based Mining and Inductive Databases. LNCS (LNAI), vol. 3848, pp. 243–266. Springer, Heidelberg (2006)
12. Lavrac, N., Kavsek, B., Flach, P., Todorovski, L.: Subgroup discovery with cn2-sd. J. Mach. Learn. Res. 5(February), 153–188 (2004)
13. Leman, D., Feelders, A., Knobbe, A.: Exceptional model mining. In: Daelemans, W., Goethals, B., Morik, K. (eds.) ECML PKDD 2008, Part II. LNCS (LNAI), vol. 5212, pp. 1–16. Springer, Heidelberg (2008)
14. Mochmann, I.C.: Lifestyles, social milieus and voting behaviour in Germany: A comparative analysis of the developments in eastern and western Germany. PhD thesis, Justus-Liebig-University Giessen (2002)
15. Morik, K., Boulicaut, J.-F., Siebes, A. (eds.): Local Pattern Detection. LNCS (LNAI), vol. 3539. Springer, Heidelberg (2005)
16. Nijssen, S., Guns, T., Raedt, L.D.: Correlated itemset mining in roc space: a constraint programming approach. In: KDD, pp. 647–656 (2009)
17. Novak, P.K., Lavrač, N., Webb, G.I.: Supervised descriptive rule discovery: A unifying survey of contrast set, emerging pattern and subgroup mining. J. Mach. Learn. Res. 10, 377–403 (2009)
18. Robinson, W.S.: Ecological correlations and the behavior of individuals. Am. Sociolog. Rev. (1950)
19. Webb, G., Zhang, S.: Removing trivial associations in association rule discovery. In: ICAIS (2002)
20. Webb, G.I.: Discovering significant patterns. Mach. Learn. 71(1), 131 (2008)
21. Wrobel, S.: An algorithm for multi-relational discovery of subgroups. In: PKDD 1997, pp. 78–87. Springer, Heidelberg (1997)

A   Description of the Data

The data used in this paper consists of 800 records, one for every polling district in the City of Cologne. Beside the number of votes obtained by the different parties in the 2009 and 2005 Bundestag elections, the records include more than 80 descriptive variables, which were assembled from different sources. The primary data source is the official city statistics, gathered and published by the Office of Statistics. Second, commercial data was used to obtain information about the type of buildings and the debt-ratio. Finally, information about the average income and the education level was taken from an anonymous citizen survey conducted by the Office of City Development and Statistics in 2008/09. The survey data is a random sample, stratified according to age, sex and urban district, which includes about 11200 responses. All variables occurring in at least one of the subgroup reported in this paper are listed in the following table. Beside the description of the variable, we also indicate the data source (OS - Official Statistics, CO - Commercial, SU - Survey).

Variable	Description
aged_16-24, aged_25-34, aged_35-64, aged_65+	age structure, i.e. the number of inhabitants aged 16-24, 25-45, etc. (OS)
avg_living_space	average living space per accommodation (CO)
catholic, muslim, protestant	number of persons with a particular religious denomination (OS)
detached_houses ... 16-19_fam._buildings 20+_fam._buildings	type of buildings: number of detached houses, number of apartment buildings of different size (CO)
education:elem._school education:secondary education:university	highest level of general education (SU)
families	number of families with children (OS)
gram_school_students	number of grammar school students (OS)
income < 1500€ income 1500-3000€ income 3000-4500€ income > 4500€	household net income per month (SU)
men, women	percentage of male resp. female inhabitants (OS)
occupancy 5-10 y. occupancy 10-15 y. occupancy 15-20 y. ...	duration of living in Cologne (SU)
single_parents	number of single parent households (OS)
single-person_<30y single-person_30-60y	number of one-person householders aged under 30, resp. aged 30-60 (OS)
social_sec_claimants	number of social security claimants (OS)

On Enumerating Frequent Closed Patterns with Key in Multi-relational Data

Hirohisa Seki, Yuya Honda, and Shinya Nagano

Nagoya Inst. of Technology, Gokiso-cho, Showa-ku, Nagoya 466-8555, Japan
seki@nitech.ac.jp

Abstract. We study the problem of mining closed patterns in multi-relational databases. Garriga et al. (IJCAI'07) proposed an algorithm RelLCM2 for mining closed patterns (i.e., conjunctions of literals) in multi-relational data, which is an extension of LCM, an efficient enumeration algorithm for frequent closed itemsets mining proposed in the seminal paper by Uno et al. (DS'04). We assume that a database considered contains a special predicate called *key* (or *target*), which determines the entities of interest and what is to be counted. We introduce a notion of closed patterns with key (*key-closedness* for short), where variables in a pattern other than the one in a key predicate are considered to be existentially quantified, and they are linked to a given target object. We then define a closure operation (key-closure) for computing key-closed patterns, and show that the difference between the semantics of key-closed patterns and that of the closed patterns in RelLCM2 implies different properties of the closure operations; in particular, the uniqueness of closure does not hold for key-closure. Nevertheless, we show that we can enumerate key-closed patterns using the technique of ppc-extensions à la LCM, thereby making the enumeration possible without storage space for previously generated patterns. We also propose a literal order designed for mining key-closed patterns, which will require less search space. The correctness of our algorithm is shown, and its computational complexity is discussed. Some preliminary experimental results are also given.

1 Introduction

Multi-relational data mining (MRDM) has been extensively studied for more than a decade (e.g., [7,8] and references therein). The research topics discussed in the conventional data mining (e.g., [11]) have been considered in this more expressive framework of MRDM, where data and patterns are represented in the form of logical formulae such as Datalog (a class of first order logic). The framework is therefore suitable to use the techniques developed in computational logic, including, among others, inductive logic programming (ILP).

In contrast to the traditional data mining dealing with rather simple patterns such as itemsets, the expressive formalism of MRDM allows us to use more complex and structured data in a uniform way, including trees and graphs in particular, and multi-relational patterns in general.

B. Pfahringer, G. Holmes, and A. Hoffmann (Eds.): DS 2010, LNAI 6332, pp. 72–86, 2010.

WARMR [5] is one of the earliest works on frequent Datalog pattern mining; it "upgrades" the conventional frequent itemset mining method based on Apriori [1] to frequent Datalog pattern mining. Since the number of frequent patterns becomes extremely large, it is common to focus on more compact representations such as the set of frequent *closed* sets (e.g., [11,3]), whose size is much smaller, while preserving the same information of the set of frequent patterns. Garriga et al. proposed an algorithm called RelLCM2 [10] for mining closed patterns (i.e., conjunctions of literals) in multi-relational data, which is an "upgrade" of LCM, an efficient enumeration algorithm for frequent closed itemsets mining proposed in the seminal paper by Uno et al. [18].

In this paper, we study the problem of mining closed patterns in multi-relational data based on these precursors. In a typical task of MRDM, a user is usually expected to specify a special predicate *key* (or *target*) (e.g., [5,6]). The key is an atom which determines the entities of interest and what is to be counted. In SPADA system for spatial pattern mining [2], for example, a distinction is made between reference (target) objects and other task-relevant (non-target) objects, which are relevant for the task in hand and related to the target objects. We introduce a notion of closed patterns with key (*key-closedness*), where variables in a pattern other than the one in a key predicate are considered to be existentially quantified, representing non-target objects *linked* to a given target object. We then define a closure operation (key-closure) for computing key-closed patterns. The difference between the semantics of key-closed patterns and that of the closed patterns in RelLCM2 implies different properties of the closure operations. In particular, the uniqueness of closure does not hold for key-closure. Nevertheless, we show that we can enumerate key-closed patterns using the technique of ppc-extensions à la LCM, thereby making the enumeration possible without memory space for storing previously generated patterns. We also propose a literal order designed for mining key-closed patterns, which will require less search space. The correctness of our algorithm is shown, and its computational complexity is discussed. Some preliminary experimental results are also given.

The organization of the rest of this paper is as follows. In Section 2, we give some preliminaries, notations and definition, and, in particular, we explain the notion of a closed pattern in multi-relational data. In Section 3, we introduce a notion of closed patterns with key (key-closedness), and define a closure operation computing key-closed patterns (key-closure). We then explain our algorithm called ffLCM which can enumerate key-closed patterns using ppc-extensions. We also show the correctness and the computational complexity of our algorithm. In Section 4, we give a literal order tailored for mining key-closed patterns, and show its effectiveness by some preliminary experimental results. Finally, we give a summary of this work in Section 5.[1]

2 Preliminaries

2.1 Multi-relational Data Mining

In the task of frequent pattern mining in multi-relational databases, we assume that we have a given database DB, a language of patterns, and a notion of frequency which

---

[1] Due to space constraints, we omit most proofs and some details, which will appear in the full paper.

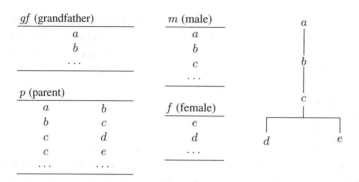

Fig. 1. A family example DB_0, including four relations. gf is a key atom (target).

measures how often a pattern occurs in the database. We use Datalog, or Prolog without function symbols other than constants, to represent data and patterns. We assume some familiarity with the notions of logic programming (e.g., [14]), although we introduce some notions and terminology in the following.

An *atom* (or *literal*) is an expression of the form $p(t_1, \ldots .t_n)$, where p is a *predicate* (or *relation*) of arity n, denoted by p/n, and each t_i is a *term*, i.e., a constant or a variable.

A substitution $\theta = \{X_1/t_1, \ldots, X_n/t_n\}$ is an assignment of terms to variables. The result of applying a substitution θ to an expression E is the expression $E\theta$, where all occurrences of variables V_i have been simultaneously replaced by the corresponding terms t_i in θ. The set of variables occurring in E is denoted by $Var(E)$.

A *pattern* is expressed as a conjunction of atoms (literals) $l_1 \wedge \cdots \wedge l_n$, denoted simply by $l_1, \ldots, l_n$. Let C be a pattern (i.e., a conjunction) and θ a substitution of $Var(C)$. When $C\theta$ is logically entailed by a database DB, we write it by $DB \models C\theta$. We will represent conjunctions in list notation, i.e., $[l_1, \ldots, l_n]$. For a conjunction C and an atom p, we denote by $[C, p]$ the conjunction that results from adding p after the last element of C.

In multi-relational data mining, one of predicates is often specified as a *key* (or *target*), which determines the entities of interest and what is to be counted.

Example 1 (Multi-relational Database). (Adapted from [15]) Consider a database DB_0 (Fig. 1), including relations, $p(X, Y)$ meaning that X is a parent of Y, $f(X)$ for female X, $m(X)$ for male X, and $gf(X)$ meaning that X is someone's grandfather. Let gf be a key (or target).

Consider a pattern C of the form: $gf(X), m(X), p(X, Y), m(Y)$. For a substitution $\theta = \{X/a, Y/b\}$, we have that $DB_0 \models C\theta$, since each literal in $C\theta = [gf(a), m(a), p(a, b), m(b)]$ is in DB_0.   □

2.2   Mining Closed Patterns in Multi-relational Data

Since the number of frequent patterns is huge and it is expensive to compute all frequent patterns, it is usual to consider the problem of mining *closed* patterns (e.g., [17,3] and

$\mathcal{T}$	
trans_ID	list of item_IDs
t100	1, 2, 5
t200	2, 4
t300	2, 3
t400	1, 2, 4, 5

key (trans_ID)
t100
t200
t300
t400

buys	
t100	1
t100	2
t100	5
t200	2
t200	4
t300	2
t300	3
t400	1
t400	2
t400	4
t400	5

	itemset case	multi-relational data
pattern	$\{1, 2\}$	$C_0 = [key(X), buys(X, 1), buys(X, 2)]$
occurrence set	$\{t100, t400\}$	$\{\{X/t100\}, \{X/t400\}\}$
closure	$\{1, 2, 5\}$	$[C_0, buys(X, 5)]$

Fig. 2. An Example of Transaction Database $\mathcal{T}$ and its Relational Data Representation

references therein). In frequent itemset mining, a notion of *closed* patterns is defined in terms of *occurrence sets*.

Example 2 (Multi-Relational DM includes Frequent Itemsets Mining). Following [18], we first recall some notions in frequent itemset mining (FIM). Consider a transaction database $\mathcal{T}$ in Fig. 2. Each element $t_i \in \mathcal{T}$ ($1 \leq i \leq |\mathcal{T}|$) is called a *transaction*. Let $\mathcal{I}$ be the set of *items*. A subset $P \subseteq \mathcal{I}$ is called a *pattern*. For a pattern P, a transaction including P is called an *occurrence* of P. The *denotation* of P, denoted by $\mathcal{T}(P)$, is the set of the occurrences of P.

A pattern P is said to be *closed*, if P is maximal w.r.t. set inclusion in the set of patterns with the same denotation as that of P, i.e., $\{Q \mid \mathcal{T}(P) = \mathcal{T}(Q)\}$. In the case of FIM, the closed pattern of P is computed by $\cap_{t \in \mathcal{T}(P)} t$, i.e., the set of items common to all transactions in $\mathcal{T}(P)$, which is called the *closure* of P.

Let $P_0 = \{1, 2\}$ be a pattern. P_0 will be represented by a conjunction C_0 of the form: $key(X), buys(X, 1), buys(X, 2)$. The set $\mathcal{T}(P_0)$ of occurrences of P_0 is $\{t100, t400\}$. As a counterpart of $\mathcal{T}(P_0)$, it will be appropriate to consider the set of substitutions $\{\{X/t100\}, \{X/t400\}\}$. The closure of P_0 is $\cap_{t \in \mathcal{T}(P_0)} t = \{1, 2, 5\}$, which will correspond to a conjunction $[C_0, buys(X, 5)]$. □

Definition 1 (Occurrence Set). [10] Let DB be a given database and C a conjunction. An *occurrence* of C in DB is a substitution θ of $Var(C)$ such that $DB \models C\theta$.

The *denotation* of C, denoted by $\mathcal{O}(C)$ is the set of the occurrences of C in DB. $|\mathcal{O}(C)|$ is called the *frequency* of C, denoted by $freq(C)$. A pattern (conjunction) C is said to be *frequent* if $freq(C) \geq min\_sup$, where min_sup is a given constant (threshold). □

Let DB be a given database. For a pair of patterns P and Q, we say that P and Q are equivalent to each other if $\mathcal{O}(P) = \mathcal{O}(Q)$. This relationship induces equivalence classes on patterns. In the case of conjunctions, it does not make sense to consider the case where P contains a variable not appearing in Q, because the resultant substitutions are not comparable. Therefore, we impose the following condition which requires that $Var(P) = Var(Q)$.

Definition 2 (Closed Conjunction). Let C be a conjunction, and DB a given database. C is said to be *closed*, if there exists no conjunction $C' \supset C$ such that (i) $Var(C') = Var(C)$, and (ii) the occurrence set of C' is the same as that of C, i.e., $\mathcal{O}(C') = \mathcal{O}(C)$. □

The *closure* of a set of items P is defined as the intersection of all the transactions in $\mathcal{T}(P)$ as explained in Example 2. On the other hand, the above notion of closedness in MRDM leads to the procedure for computing a closure of a given conjunction, which is shown in Fig. 3 [10].

Algorithm $Clo(C)$
input : conjunction C
output: closed conjunction C'

1   $C' \leftarrow C$;
2   **repeat**
3   Find an atom $p \in \rho_{RR}(C')$ s.t. $\mathcal{O}(C') = \mathcal{O}([C',p])$;
4      $C' \leftarrow [C',p]$
5   **until** *no such atom p is found*;
6   **return** C'

Fig. 3. Computing Range-Restricted Closed Conjunctions Clo [10]

In Fig. 3, a *refinement operator* ρ_{RR} is employed, which computes an atom p to be added into C' such that $p \notin C'$ and it satisfies the *range-restricted* condition (e.g., [6]), i.e., $Var(p) \subseteq Var(C')$.

We note that there is a caveat to the closure procedure in Fig. 3. The procedure Clo computes a closure in a non-deterministic way due to ρ_{RR}, which might depend on the order of additions of atom p. However, thanks to the definition of the occurrence set $\mathcal{O}$, $Clo(C)$ is well behaved, as the following lemma [10] shows:

Lemma 1 ([10]). Let C be a conjunction and p, q atoms such that $p \in \rho_{RR}(C)$. If $\mathcal{O}(C) = \mathcal{O}([C,p])$ for a given database DB, then $\mathcal{O}([C,q]) = \mathcal{O}([C,q,p])$. □

The procedure Clo therefore defines the unique closure for a given conjunction C, which is in fact the closed conjunction containing C. The following lemma is obvious from Def. 2 and Lemma 1.

Lemma 2. A conjunction C is closed if and only if $C = Clo(C)$. □

3 Mining Closed Patterns with Key

3.1 Key and Language Bias of Patterns

As explained in Sect. 1, in a typical task of MRDM, a user is usually expected to specify a special predicate *key* (or *target*) (e.g., [5,6]). The key is an atom which determines the entities of interest and what is to be counted. The key (target) is thus to be present in all patterns considered. In Example 1, the key is predicate gf. We assume henceforth that the arity of a key atom is 1 for simplifying the explanation.

A pattern containing a key is not always be meaningful to be mined. For example, let $C = [gf(X), f(Y), m(Z)]$ be a conjunction in Example 1. Variables Y and Z in C are not *linked* to variable X in key atom $gf(X)$; the objects represented by Y and Z will have nothing to do with key object X. In ILP, the following notion of *linked literals* [12] is a standard one to specify the so-called language bias, which is similar to a *first-order feature* in [13].

Definition 3 (Linked Literal). [12] Let $key(X)$ be a key atom and l a literal. l is said to be *linked* to $key(X)$, denoted by $key(X) \sim l$, if either $X \in Var(l)$Cor there exists a literal l_1 such that $key(X) \sim l_1$ and $Var(l_1) \cap Var(l) \neq \emptyset$.   □

Definition 4 (Bias Condition). Given a database DB and a key atom $key(X_1)$, we assume that there are predefined finite sets of predicate (resp. variables; resp. constant symbols), denoted by $\mathcal{P}$ (resp. $\mathcal{V}$; resp. $\mathcal{C}$), and that, for each literal l in a conjunction C, the predicate symbol (resp. variables; resp. constants) of l is in $\mathcal{P}$ (resp. $\mathcal{V}$; resp. $\mathcal{C}$).

Moreover, each pattern C of conjunctions to be mined satisfies the following conditions: $key(X_1) \in C$ and, for each $l \in C, l \sim key(X_1)$.   □

We denote by $\mathcal{L}$ the set of literals constructed using predicate symbols in $\mathcal{P}$ and variables (constants) in $\mathcal{V}$ ($\mathcal{C}$), respectively. We also denote by $\mathbb{P}$ the set of conjunctions (patterns) consisting of literals in $\mathcal{L}$ such that they satisfy the bias condition.

When a conjunction has a key $key(X_1)$, the notion of occurrence set is modified. Let θ be a substitution of the form: $\theta = \{X_1/t_1, X_2/t_2, \dots\}$. Then, the *restriction* of θ w. r. t. $key(X_1)$, denoted by θ_{key}, is defined by $\theta_{key} = X_1/t_1.$[2]

Definition 5 (Key-Occurrence Set). Let C be a conjunction with $key(X_1)$, and DB a database. Then, the *key-restricted occurrence set* (*key-occurrence set*, for short), denoted by $\mathcal{O}_{key}(C)$, is defined by $\mathcal{O}_{key}(C) = \{\theta_{key} | \theta \in \mathcal{O}(C)\}$.

$|\mathcal{O}_{key}(C)|$ is called the *frequency* of C, denoted by $freq_{key}(C)$.   □

Definition 6 (Key-closed Conjunction). Let C be a conjunction with $key(X_1)$, and DB a database. C is said to be *closed w.r.t. key-restriction* (or *key-closed*), if there exists no conjunction $C' \supset C$ such that (i) $Var(C') = Var(C)$, and (ii) the key-occurrence set of C' is the same as that of C, i.e., $\mathcal{O}_{key}(C') = \mathcal{O}_{key}(C)$.   □

In a conjunction C with $key(X_1)$, X_1 is a free variable, while the other variables in C are considered to be *existentially quantified*. Since $key(X_1)$ is specified as a target, we are only interested in the values substituted for X_1 to be counted; we do not care about the instantiations of those existentially quantified variables, as far as they exist in a given database. The purpose of considering key-closedness is the same as the case of closed itemsets: the number of frequent key-closed conjunctions will be less than

---

[2] We denote θ_{key} simply by $\theta_{key} = t_1$, when a variable X_1 in key is apparent. When there is no confusion, we also use a notation such as $X, Y, Z, \dots$ instead of $X_1, X_2, X_3, \dots$ for simplicity.

that of frequent (closed) conjunctions, while they have the same information w.r.t. the key-occurrence sets.

Due to Lemma 1, the closure of a given conjunction C is uniquely determined, which is computed by $Clo(C)$. On the other hand, a key-closed conjunction containing C is not unique as shown in Example 3.

Example 3 (Key-Closed Pattern is not unique). Consider database DB in Fig. 4.

In the figure, $C = [key(X), mem(Y, X)]$ is closed, but it is not key-closed, since there exists a conjunction $C_1 = [C, anc(Y, X)]$ which satisfies that $C_1 \supset C$ and $\mathcal{O}_{key}(C) = \mathcal{O}_{key}(C_1)$. On the other hand, C_1 is key-closed.

There exists another key-closed conjunction $C_2 = [C, des(Y, X)]$ which also satis-fies that $C_2 \supset C$ and $\mathcal{O}_{key}(C) = \mathcal{O}_{key}(C_2)$.

We note that $\mathcal{O}([anc(Y, X), des(Y, X)]) = \emptyset$.   □

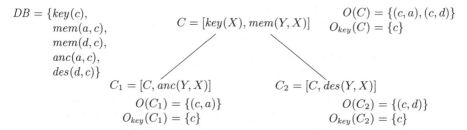

Fig. 4. A Key-Closed Conjunction is not Unique: C, C_1, C_2: closed C C_1, C_2: key-closed. In the figure, a substitution $\theta = \{X/t_1, Y/t_2\}$ in an occurrence set is denoted simply by (t_1, t_2).

We consider a procedure $Clo_{key}(C)$ defined in Fig. 5, which computes, in a non-deterministic way, a conjunction C' containing C with the same key-occurrence set of C. We call C' a *key-closure* of C.

Example 4 (Key-Closed Conjunction). Continued from Example 3. For conjunction C, there are two outputs computed by $Clo_{key}(C)$; one is C_1, and the other is C_2. Since $Clo_{key}(C_1) = C_1$, C_1 is the unique key-closure of itself. Similarly, we have that $Clo_{key}(C_2) = C_2$, thus C_2 is also the key-closure of itself.   □

The following lemma is immediate from the definitions of key-closedness and Clo_{key}, so the proof is omitted.

Lemma 3. *Let C be a conjunction and DB a given database. Then, C is closed if C is key-closed. Moreover, an output computed by $Clo_{key}(C)$ is key-closed, and it is thus closed.*   □

3.2   Enumerating Key-Closed Patterns Using PPC-Extensions

Since a key-closed conjunction containing C is not unique (Example 3), computing the key-closed conjunctions would entail non-deterministic choices in literals added to

---

Algorithm $Clo_{key}(C)$
input : conjunction C
output: a key-closed conjunction C'

1   $C' \leftarrow C$;
2   **repeat**
3   Find an atom $p \in \rho_{RR}(C')$ s.t. $\mathcal{O}_{key}(C') = \mathcal{O}_{key}([C', p])$;
4      $C' \leftarrow [C', p]$
5   **until** *no such atom p is found*;
6   **return** C'

Fig. 5. Computing Closed Conjunctions w.r.t. Key-Restriction Clo_{key}

C. However, because a key-closed conjunction is closed (Lemma 3), we can adopt the following simple method for enumerating key-closed patterns: (i) we first enumerate all the closed conjunctions for a given database DB, (ii) then, for each closed conjunction, we check whether it is key-closed.

In RelLCM2 [10], Garriga et al. upgrade LCM [18] to frequent closed pattern mining in multi-relational data. The key property of a closed pattern in their algorithms is that any closed pattern $P \neq \emptyset$ is a closure extension of other closed patterns, where P is a *closure extension* of Q if $P = Clo([Q, l])$ for some $l \notin P$. Uno et al. elaborated closure extension by introducing the notion of *prefix-preserving closure extension* (*ppc-extension*), which enables us to avoid duplicated enumeration of closed patterns without storing previously enumerated patterns.

Following [18,10], we explain some definitions necessary for ppc-extension. We first assume that each variable in $\mathcal{V}$ (Def. 4) has its index, denoted by X_i $((|\mathcal{V}| \geq i \geq 1)$, where X_1 is assumed to the variable appearing in key atom $key(X_1)$. We also assume that there exists a total order $\preceq$ on the set $\mathcal{L}$ of literals appearing in conjunctions (patterns) in $\mathbb{P}$. Then, there exists a mapping ι from $\mathcal{L}$ to the set of natural numbers $\mathcal{N} = \{0, 1, 2, \ldots\}$, where $\iota(key(X_1)) = 0$. We further assume that an order on $\mathbb{P}$ is induced based on the total order $\preceq$ (e.g., lexicographically).

Each literal in a conjunction $C = [l_1, \ldots, l_n]$ is supposed to be ordered in ascending order according to the total order $\preceq$. The *i-th prefix* of C, denoted by $C[i]$, is the prefix of C such that it consists only of literals whose indices are no greater than i, i.e., $\iota(l) \leq i$ for every $l \in C[i]$. Let C be a conjunction and pr the least prefix of C such that $\mathcal{O}(pr) = \mathcal{O}(C)$. Then, the *core index* of C, denoted by $core\_i(C)$, is the maximal (i.e., last) literal of pr.

Using these definitions, a ppc-extension is defined in an analogous way to LCM as follows:

Definition 7 (Prefix-preserving closure extension). [10] Let $C = [q_1, \ldots, q_n]$ be a closed conjunction. A conjunction C' is called a *prefix-preserving closure extension* (*ppc-extension*) of C, if the following conditions are satisfied:

1. $C' = Clo([C, p])$ for some literal $p \notin C$,
2. $p \succ core\_i(C)$, and
3. $C[\iota(p) - 1] = C'[\iota(p) - 1]$, that is, the $(\iota(p) - 1)$-prefix of C is preserved.   □

Algorithm ffLCM(C)
input : closed conjunction C, minimum support min_sup

1 **if** C *is not frequent* **then return**;
2 **for** *all refinements* $[C, p]$ *with* $p \in \rho_G(C)$ *such that* $p \succ core\_i(C)$ **do**
3   **if** $\mathcal{O}([C, p]) = \emptyset$ **then** skip refinement;
4   **else** $C' \leftarrow Clo([C, p])$;
5   **if** $C[\iota(p) - 1] = C'[\iota(p) - 1]$ **then** $//$ C' is a ppc-extension of C
6   **if** C' *is key-closed,* $C' \in \mathbb{P}$ *and frequent* **then** output C';
7   **call** ffLCM (C')
8 **end**

Fig. 6. Algorithm ffLCM

We are now in position to give our algorithm ffLCM, which is shown in Fig. 6. It is started by calling ffLCM($Clo(key(X_1))$). In the algorithm, a *refinement operator* ρ_G is employed (line 2), which computes an atom $p \in \mathcal{L}$ to be added into C in such a way that it only requires that $p \notin C$.[3]

When C' is a ppc-extension of C (line 5), ffLCM(C') is called recursively (line 7). The enumeration of closed patterns by the algorithm is therefore done in depth-first search.

The key-closedness of C' is checked (line 6) by calling procedure $Clo_{key}(C')$ (Fig. 5), i.e., C' is key-closed if and only if $C' = Clo_{key}(C')$. When C' is key-closed, it is an output of the algorithm (line 6), provided that it satisfies the bias condition (i.e., $C' \in \mathbb{P}$) and its frequency is no less than a given minimum support.

We now show the properties of the algorithm ffLCM. We introduce some notations. Let DB be a given database and p a predicate defined in DB. We denote the relation (i.e., the set of tuples) p by DB_p. We also denote the size of DB, the number of tuples in DB, by $|DB|$. Let C be a pattern (a conjunction of literals) with variables $X_{i_1}, \ldots, X_{i_j}$ $(1 \le i_1 < \cdots < i_j \le |\mathcal{V}|)$, and $\mathcal{O}(C)$ its occurrence set. We consider $\mathcal{O}(C)$ as a relation[4] (a set of tuples) with the set of its attributes $Var(C)$ which is defined by $\{(t_{i_1}, \ldots, t_{i_j}) \mid \theta = \{X_{i_1}/t_{i_1}, \ldots, X_{i_j}/t_{i_j}\} \in \mathcal{O}(C)\}$. In the following, we can thus use relational algebra expressions such as $\mathcal{O}(C) \bowtie DB_p$, a natural join of $\mathcal{O}(C)$ and DB_p, for example.

Theorem 1 (Algorithm ffLCM)

[Correctness]. Let C is a key-closed conjunction. Then, there exists a single closed conjunction C_0 such that C is a ppc-extension of C_0.

[Complexity]. Let DB be a given database. Let C_0 be a closed conjunction with its occurrence set $\mathcal{O}(C_0)$. Then, the algorithm ffLCM enumerates all frequent key-closed patterns in $O(|\mathcal{O}(C_0)| \times |DB|^2 \times |\mathcal{L}|^2)$ time for each closed pattern C_0 with $O(|\mathcal{O}(C_0)| \times |DB|^2)$ memory space.

---

[3] Note that, unlike ρ_{RR}, ρ_G allows p to have new variables not occurring in C.
[4] In other words, we regard C as a *view* defined by relations corresponding to literals in C.

Proof. *[Correctness]* The correctness of the algorithm ffLCM is straightforward from the properties of ppc-extension [18,10] and the fact that a key-closed conjunction is closed (Lemma 3).

[Complexity] Let $[C_0, p]$ be a refinement of C_0 with atom $p \in \rho_G(C_0)$ such that $p \succ core\_i(C_0)$. To compute $C = Clo([C_0, p])$, we first compute $\mathcal{O}([C_0, p])$. This is done by $\mathcal{O}(C_0) \bowtie DB_p$, taking $O(|\mathcal{O}(C_0)| \times |DB_p|)$ time.

The next step to compute closure $C = Clo([C_0, p])$ is that, for each literal $l \in \rho_{RR}([C_0, p]) \subseteq \mathcal{L}$, we compute $\mathcal{O}(C_0) \bowtie DB_p \bowtie DB_l$, and we then check to see whether the occurrence set of $[C_0, p, l]$ is the same as that of $[C_0, p]$, which is done by using difference operation $\mathcal{O}(C_0) \bowtie DB_p \bowtie DB_l - (\mathcal{O}(C_0) \bowtie DB_p)$. In the computation of $C = Clo([C_0, p])$, we thus need $O(|\mathcal{O}(C_0)| \times |DB_p| \times |DB_l|)$ time for each $l \in \mathcal{L}$, hence it takes $O(|\mathcal{O}(C_0)| \times |DB_p| \times |DB| \times |\mathcal{L}|)$ in total.

On the other hand, the computation of key-closure $Clo_{key}(C)$ is similarly done, except that checking the equivalence of the key-occurrence sets, $\mathcal{O}_{key}(C) = \mathcal{O}_{key}([C, l])$, is performed by using difference operation between the projected relations, i.e., the relations projected on variable X_1 occurring in a key atom, $\pi_{X_1}(\mathcal{O}(C) \bowtie DB_l) - \pi_{X_1}\mathcal{O}(C)$. From the fact that $\mathcal{O}(C) = \mathcal{O}([C_0, p])$, the time complexity is thus the same as that of computing closure $C = Clo([C_0, p])$.

It therefore follows that the enumeration of key-closed patterns takes $O(|\mathcal{O}(C_0)| \times |DB|^2 \times |\mathcal{L}|^2)$ time for each closed pattern C_0, while it requires $O(|\mathcal{O}(C_0)| \times |DB|^2)$ memory space.   □

Remark 1. For the readers familiar with LCM [18], we recall the complexity of LCM algorithm. For a given transaction database $\mathcal{T}$, it enumerates all frequent closed patterns in $O(||\mathcal{T}(P)|| \times |\mathcal{I}|)$ time for each pattern P with $O(||\mathcal{T}||)$ memory space, where $||\mathcal{T}(P)||$ is the total size of the occurrence set $\mathcal{T}(P)$, defined by $||\mathcal{T}(P)|| = \sum_{t \in \mathcal{T}(P)} |t|$. Therefore, LCM achieves polynomial-delay and polynomial-space complexity.

In LCM, to derive the closure of $P \cup \{i\}$ for an item i, it takes *only* $O(||\mathcal{T}(P)||)$ time. This is because the closure of $P \cup \{i\}$ is computed by $\cap_{t \in \mathcal{T}(P \cup \{i\})} t$, which is done by $O(||\mathcal{T}(P)||)$ time. On the other hand, to compute $C = Clo([C_0, p])$ in MRDM, it takes $O(|\mathcal{O}(C_0)| \times |DB_p| \times |DB| \times |\mathcal{L}|)$ time.

One way to see these differences is that a transaction database $\mathcal{T}$ can be regarded as a result of joining[5] relations $buys(X, i)$ (see Fig 2) up to $|\mathcal{I}|$ times, where $\mathcal{I}$ is the set of items. On the contrary, to compute the occurrence set, say, $\mathcal{O}([C_0, p])$ of $[C_0, p]$ in MRDM, we should perform a natural join $\mathcal{O}(C_0) \bowtie DB_p$; computing a counterpart of $\mathcal{T}$ thus requires the operations of natural joins *dynamically* when needed. The time complexity required for computing closure in MRDM would be therefore a price intrinsic to the representation of multi-relational data.   □

4   Reducing Search Space by a Literal Order w.r.t. Key

4.1   Literal Order w.r.t. Key

As explained in the previous section, the search of algorithm ffLCM is done by first enumerating all the closed conjunctions for a given database DB, and then checking

---

[5] To be precise, we shall use *outerjoin* (e.g., [9]), since the size of each transaction $t \in \mathcal{T}$ will be different from each other.

whether it is key-closed and satisfies the bias condition. Those closed conjunctions intermediately generated do not necessarily satisfy the bias condition.

Since we regard only key-closed patterns satisfying the bias condition as meaningful, it will be preferable to make the search of the algorithm ffLCM more limited so that those intermediately generated closed conjunctions should satisfy the bias condition, while preserving the completeness of the algorithm. To do that, we introduce the following order in the set $\mathcal{L}$ of literals.

Definition 8 (literal order w.r.t. key). Let $\mathcal{L}$ be the set of literals which satisfy the bias condition, and $X_i \in \mathcal{V}$ a variable with its index i ($1 \leq i \leq |\mathcal{V}|$). Let $\preceq$ be a total order on $\mathcal{L}$ such that $key(X_1)$ is the minimum of $\mathcal{L}$ w.r.t. $\preceq$.

Suppose that $\mathcal{L}$ is a disjoint union of $\mathcal{L}_i \subseteq \mathcal{L}$ ($0 \leq i \leq |\mathcal{V}|$), where $\mathcal{L}_i$ is defined as follows:

$$\mathcal{L}_0 = \{key(X_1)\}$$
$$\mathcal{L}_1 = \{p \mid X_1 \in Var(p), p \neq key(X_1)\}$$
$$\mathcal{L}_i = \{p \mid X_i \in Var(p), \text{ and } j \geq i \text{ for each } X_j \in Var(p)\} \quad (|\mathcal{V}| \geq i \geq 2)$$

Then, an order over $\mathcal{L}$ is said to be an *order w.r.t. key* (or *key-order*), denoted by $\preceq^\star$, if it satisfies the following conditions:

1. the restriction of $\preceq^\star$ to $\mathcal{L}_i$ ($|\mathcal{V}| \geq i \geq 0$) is the same as $\preceq$, and
2. $p \prec^\star q$, if $p \in \mathcal{L}_i, q \in \mathcal{L}_j$ and $i < j$.   □

In the above definition, $\mathcal{L}_1$ is the set of literals other than $key(X_1)$ which contain variable X_1. On the other hand, a literal p is in $\mathcal{L}_i$ ($2 \leq i \leq |\mathcal{V}|$), if variable X_i occurs in p and each variable in p has an index not less than i. We note that $\preceq^\star$ is a total order in $\mathcal{L}$.

For a conjunction C, we define the *normal form* of C, denote by $\mathrm{nf}(C)$, by the conjunction of literals in C such that each literal in C is in ascending order w.r.t. $\preceq^\star$. We first give the following technical lemma, which shows some useful properties of the order $\preceq^\star$.

Lemma 4. Let C be a conjunction which satisfies the bias condition (i.e., $l \sim key(X_1)$ for every $l \in C$). Then, there exists a conjunction C' such that (i) it is a renaming of C and (ii) every prefix of $\mathrm{nf}(C')$ satisfies the bias condition.   □

Example 5. Let C be a conjunction of the form: $C = key(X_1), p(X_1, X_3), m(X_2), p(X_2, X_3), m(X_3)$, where each literal in C is in ascending order w.r.t. $\preceq^\star$. C satisfies the bias condition; $key(X_1) \sim l$ for $\forall l \in C$. Its prefix: $key(X_1), p(X_1, X_3), m(X_2)$, however, does not satisfy the bias condition; $key(X_1) \not\sim m(X_2)$.

On the other hand, consider a renaming C' of C of the form: $C' = key(X_1), p(X_1, X_2), m(X_3), p(X_3, X_2), m(X_2)$. Then, for $\mathrm{nf}(C') = key(X_1), p(X_1, X_2), m(X_2), p(X_3, X_2), m(X_3)$, it holds that every prefix of $\mathrm{nf}(C')$ satisfies the bias condition.   □

Let $\rho_G^\star$ a refinement operator which, like ρ_G, computes an atom $p \in \mathcal{L}$ to be added into C in such a way that $p \notin C$ and, moreover, $[C, p]$ satisfies the bias condition. Let ffLCM* be the algorithm ffLCM defined in Fig. 6 which, instead of ρ_G, employs $\rho_G^\star$ and uses a total order $\preceq^\star$. The following theorem shows that the algorithm so defined still satisfies the completeness.

Theorem 2 (Correctness of ffLCM with literal order $\preceq^\star$). Let C be a frequent closed conjunction which satisfies the bias condition. Then, there exists a a renaming C' of C such that (i) every prefix of $\mathrm{nf}(C')$ satisfies the bias condition and (ii) $\mathrm{nf}(C')$ is computed in $\mathrm{ffLCM}^\star$.

Moreover, if C is a frequent key-closed conjunction which satisfies the bias condition, then $\mathrm{nf}(C')$ is an output of $\mathrm{ffLCM}^\star$.   □

Proof. (Sketch) The proof is by induction on the length of C. Due to Lemma 4, there exists a renaming C' of C such that every prefix of $\mathrm{nf}(C')$ satisfies the bias condition. Suppose that $\mathrm{nf}(C')$ is of the form: $Clo([C_0, p])$, i.e., $\mathrm{nf}(C')$ is the closure of conjunction C_0, added p at the last of it. Then, $\mathrm{nf}(C')$ is a ppc-extension of $Clo(C_0)$, and $Clo(C_0)$ satisfies the bias condition. It follows from the induction hypothesis that $Clo(C_0)$ is computed in $\mathrm{ffLCM}^\star$. Since renaming does not change key-closedness, $\mathrm{nf}(C')$ is key-closed if C is key-closed. Therefore, the proposition follows.   □

4.2   Experimental Results

To see the effectiveness of the literal order $\preceq^\star$ (key-order), we now present some results of our experiments of $\mathrm{ffLCM}^\star$ on two datasets. $\mathrm{ffLCM}^\star$ is implemented using SWI-Prolog (32 bits, Version 5.6.64), and the experiments were run on a PC with Pentium Core2Duo 1.8GHz, 2GB memory under Windows XP [16]. Table 1 summarizes the results for $\mathrm{ffLCM}^\star$ and ffLCM with lexicographical order, denoted by ffLCM^{lex}. The lexicographical order is employed in RelLCM2 [10].

The first dataset is the mutagenicity prediction,[6] containing 30 chemical compounds. Each compound is represented by a set of facts using predicates such as *atom*, *bond*, for example. The size of the set $\mathcal{P}$ of predicate symbols is 12. The size of key atom ($active(X_1)$) is 230, and minimum support $min\_sup = 0.05 \times 230$. We study the effects of using key-order for the mutagenesis dataset where patterns contain at most 4 variables (i.e., $1 \le |\mathcal{V}| \le 4$) and they contain no constant symbols (i.e., $|\mathcal{C}| = \emptyset$). Table 1 reports the numbers of frequent key-closures generated, as well as those of frequent closed conjunctions for $\mathrm{ffLCM}^\star$ and ffLCM^{lex}. In Table 1 (above), we show the rate of reduction to see the effects of literal orders employed. The rate of reduction is calculated by $(1 - \#\,\mathrm{closed}^\star / \#\,\mathrm{closed}^{lex}) \times 100$, where $\#\,\mathrm{closed}^\star$ ($\#\,\mathrm{closed}^{lex}$) is the number of frequent closed conjunctions generated in $\mathrm{ffLCM}^\star$ (ffLCM^{lex}), respectively. According to the increase of the number (# vars) of variables occurring in conjunctions, the use of key-order significantly decreases the number of closed conjunctions generated; the rate of reduction is more than 40% for $|\mathcal{V}| = 4$.

Next, we note that Table 1 (above) also show the numbers of frequent closed conjunctions generated in ffLCM^{lex} which satisfy the bias condition (the figures in parentheses). One can see that those numbers are still greater than those in $\mathrm{ffLCM}^\star$ for $|\mathcal{V}| \ge 3$. The reason is that, in $\mathrm{ffLCM}^\star$, it only generates conjunctions such that every prefix of a conjunction satisfies the bias condition (see Example 5). This means that the use of key-order is in fact effective for making the enumeration of closed conjunctions

---

[6] http://www.comlab.ox.ac.uk/activities/machinelearning/
mutagenesis.html

Table 1. Effects of Literal Orders in ffLCM for Datasets Mutagenesis (above) and English Corpora DB (below) with minimum support $min\_sup = 0.05 \times |key|$: # key-closed (# closed) is the number of frequent key-closures (frequent closed conjunctions) generated, respectively. The figures in parentheses mean the numbers of generated conjunctions which satisfy the bias condition. $-$: time out.

	lexicographical order				key-oder		reduction
# vars	# key-closed		# closed		# key-closed	# closed	rate %
1	1	(1)	1	(1)	1	1	0
2	16	(16)	16	(16)	16	16	0
3	206	(194)	486	(436)	149	320	34.2
4	2016	(1836)	8817	(7664)	1204	4920	44.2

	lexicographical order				key-oder	
# vars	# key-closed		# closed		# key-closed	# closed
1	1	(1)	1	(1)	1	1
2	42	(8)	59	(8)	8	8
3	–		–		40	48
4	–		–		150	248

restrictive, thereby reducing the number of patterns generated without losing the search completeness.

The second dataset is from the Penn Treebank Project[7], an annotated corpus of English, where each sentence is annotated with its part-of-speech (POS) tags. The tagset contains 36 POS tags, such as cc (coordinating conjunction), dt (determiner), and so on. Each tag is represented by a predicate here, and the size of the set of predicate symbols $\mathcal{P}$ is thus 37. The size of key atom ($sentence(X_1)$) is 300, and the other conditions are the same as before. Table 1 (below) shows that in this dataset, the effectiveness of the key-order is more prominent; the runtime of ffLCM^{lex} is over the prescribed time (48 hours) for $|\mathcal{V}| \geq 3$.

5   Concluding Remarks

We have studied the problem of mining closed patterns in multi-relational data. Unlike RelLCM2 by Garriga et al., a database is assumed to contain a special predicate called *key* (*target*), by which one can specify the entities of interest. We have then introduced a notion of closed patterns with key (key-closedness) which satisfy some bias conditions, requiring that each object represented by a variable be linked to a given target object. Therefore, a pattern satisfying the bias conditions can be regarded as a first-order representation of some *features* (or attributes) of the target object. We have also defined a closure operation computing key-closed patterns (key-closure), and shown that the uniqueness of closure does not hold for key-closure.

---

[7] http://www.cis.upenn.edu/~treebank/

We have proposed an algorithm, called ffLCM, which enumerates key-closed patterns using ppc-extensions à la LCM, thereby making the enumeration possible without storage space for previously generated patterns. We have also proposed a literal order designed for mining key-closed patterns, which will require less search space, while preserving the search completeness. We have discussed its computational complexity, and have compared our algorithm with the case of LCM. The effects of the proposed literal order have been exemplified by some experimental results.

As pointed out in [4], efficiency and scalability have been major concerns in multi-relational data mining. Research in this direction is found, for example, in a recent work by Appice et al.[2]. Since ffLCM is descended from LCM by Uno et al., it will be expected to be efficient at least in theory; it will inherit the advantages of polynomial-delay and polynomial-space algorithm of LCM. Moreover, since ffLCM (Fig. 6) consists of a simple for-loop, it would be amenable to data-parallelism. As future work, our plan is to implement our algorithm by exploiting the parallelism as much as possible.

Acknowledgement. The authors would like to thank anonymous reviewers for their constructive and useful comments on the previous version of the paper. The authors are also grateful to Nobuhiro Inuzuka for providing us the datasets for the experiments in this paper.

References

1. Agrawal, R., Srikant, R.: Fast Algorithms for Mining Association Rules. In: VLDB, pp. 487–499 (1994)
2. Appice, A., Ceci, M., Turi, A., Malerba, D.: A Parallel Distributed Algorithm for Relational Frequent Pattern Discovery from Very Large Data Sets. Intell. Data Anal. (2009) (to appear)
3. Arimura, H., Uno, T.: Polynomial-Delay and Polynomial-Space Algorithms for Mining Closed Sequences, Graphs, and Pictures in Accessible Set Systems. In: SIAM Int'l. Conf. on Data Mining, pp. 1087–1098 (2009)
4. Blockeel, H., Sebag, M.: Scalability and efficiency in multi-relational data mining. SIGKDD Explorations Newsletter 2003 4(2), 1–14 (2003)
5. Dehaspe, L.: Frequent pattern discovery in first-order logic, PhD thesis, Dept. Computer Science, Katholieke Universiteit Leuven (1998)
6. De Raedt, L., Ramon, J.: Condensed representations for Inductive Logic Programming. In: Proc. KR 2004, pp. 438–446 (2004)
7. Dzeroski, S.: Multi-Relational Data Mining: An Introduction. SIGKDD Explorations Newsletter 5(1), 1–16 (2003)
8. Dzeroski, S., Lavrač, N. (eds.): Relational Data Mining. Springer, Heidelberg (2001)
9. Garcia-Molina, H., Widom, J., Ullman, J.D.: Database System Implementation. Prentice-Hall, Inc., Englewood Cliffs (1999)
10. Garriga, G.C., Khardon, R., De Raedt, L.: On Mining Closed Sets in Multi-Relational Data. In: IJCAI 2007, pp.804–809 (2007)
11. Han, J., Kamber, M.: Data Mining: Concepts and Techniques, 2nd edn. Morgan Kaufmann Publishers Inc., San Francisco (2005)
12. Helft, N.: Induction as nonmonotonic inference. In: Proc. KR 1989, pp. 149–156 (1989)
13. Lavrač, N., Flach, P.A.: An Extended Transformation Approach to Inductive Logic Programming. ACM Trans. Computational Logic 2(4), 458–494 (2001)

14. Lloyd, J.W.: Foundations of Logic Programming, 2nd edn. Springer, Heidelberg (1987)
15. Motoyama, J., Urazawa, S., Nakano, T., Inuzuka, N.: A Mining Algorithm using Property Items Extracted from Sampled Examples. In: Muggleton, S.H., Otero, R., Tamaddoni-Nezhad, A. (eds.) ILP 2006. LNCS (LNAI), vol. 4455, pp. 335–350. Springer, Heidelberg (2007)
16. Nagano, S., Honda, Y., Seki, H.: On Enumerating Frequent Closed Patterns in Multi-Relational Data Mining. In: Proc. WiNF 2009, pp. 89–94 (2009) (in Japanese)
17. Pasquier, N., Bastide, Y., Taouil, R., Lakhal, L.: Discovering Frequent Closed Itemsets for Association Rules. In: Beeri, C., Bruneman, P. (eds.) ICDT 1999. LNCS, vol. 1540, pp. 398–416. Springer, Heidelberg (1998)
18. Uno, T., Asai, T., Uchida, Y., Arimura, H.: An Efficient Algorithm for Enumerating Closed Patterns in Transaction Databases. In: Suzuki, E., Arikawa, S. (eds.) DS 2004. LNCS (LNAI), vol. 3245, pp. 16–31. Springer, Heidelberg (2004)

Why Text Segment Classification Based on Part of Speech Feature Selection

Iulia Nagy*, Katsuyuki Tanaka, and Yasuo Ariki

Kobe University
1-1 Rokkodai-cho, Nada-ku, Kobe 657-8501, Japan
{nagy,katsutanaka}@me.cs.scitec.kobe-u.ac.jp, ariki@kobe-u.ac.jp

Abstract. The aim of our research is to develop a scalable automatic why question answering system for English based on supervised method that uses part of speech analysis. The prior approach consisted in building a why-classifier using function words. This paper investigates the performance of combining supervised data mining methods with various feature selection strategies in order to obtain a more accurate why classifier.Feature selection was performed a priori on the dataset to extract representative verbs and/or nouns and avoid the dimensionality curse. LogitBoost and SVM were used for the classification process. Three methods of extending the initial "function words only" approach, to handle context-dependent features, are proposed and experimentally evaluated on various datasets. The first considers function words and context-independent adverbs; the second incorporates selected lemmatized verbs; the third contains selected lemmatized verbs & nouns. Experiments on web-extracted datasets showed that all methods performed better than the baseline, with slightly more reliable results for the third one.

Keywords: Question-answering, supervised learning, feature selection.

1 Introduction

In the past years Internet has become a major source of information, many people relying on it to find the answers to their questions. Although very popular, search engines do not provide the user with a direct answer to his or her query but with a number of web pages the user has to browse manually to obtain the information he or she is looking for. A crucial step for the next generation search engines is to integrate a system allowing the user to obtain a straightforward and concise answer to his or her question. Such systems are known as question-answering (QA) systems and have undergone significant progress during past years. Two main types of question-answering systems can be distinguished : factoid, which address questions requiring simple answers such as person name, organization name, numeric expression, and non-factoid dealing with questions that require a more complex answer.

---

* Exchange student from INSA de Lyon, Computer Science Department.

B. Pfahringer, G. Holmes, and A. Hoffmann (Eds.): DS 2010, LNAI 6332, pp. 87–101, 2010.
© Springer-Verlag Berlin Heidelberg 2010

Our work focuses on creating a QA system for non factoid questions, more precisely a why-type QA system. While many such systems are presented in the QA literature, some of them suffer from domain dependency, since they address a specific domain such as medicine, or may prove difficult to build due to hand-crafted patterns and the considerable grammar expert knowledge needed. In the attempt to overcome these flaws, we adopted a machine learning approach for building our why-type QA system. The main purpose of our research is to build an effective QA system able to detect why text segments from arbitrarily built corpora and scalable to different languages.

More specifically, the task we address is building a classifier for QA-system able to identify the answers that actually respond to a why question. By applying this classifier in a preprocessing step we should be able to reduce the amount of data to analyze, by eliminating all text segment not answering a why question, and therefore facilitate the work of the answer extraction module of the QA system. Previous work focused on adapting to English an approach described in the Japanese literature [10] and evaluating its performance. In this method only function words are extracted from pre-labeled text segments, and then used to train a why-classifier. Considering the overall satisfying results of this experiment, we have decided to seek for methods to improve the performance of the existent classifier.

In this paper, we present the different techniques we applied in order to improve the initial classifier's performance. In order to achieve our goal we decided to enrich the initial feature space with other valuable features. Initially we added context-independent[1] adverbs to the feature space that contained only function words. Afterwards, using a priori feature selection techniques, lemmatized verbs and lemmatized verbs & nouns were also added to the feature space.

In order to evaluate how well our 3 methods work, we trained classifiers using both LogitBoost and SVM with a Pearson VII function based kernel. Moreover, in order to ensure the validity of our experience, we used various training and evaluation datasets, composed of web-extracted text segments.

This article is organized as follows: Section 2 describes the related work on why-type QA, Section 3 describes the previous work along with the method that initially inspired us. Section 4 presents the feature selection algorithms and the classifier algorithms proposed while Section 5 describes the experimental preparation and the results. Finally Section 6 presents the conclusion and the description of future works.

2   Related Work

With the continuous growth of the information base available on Internet, the importance of effective question-answering tools to facilitate the search process continues to increase. While research in building factoid QA systems has a long

---

[1] Context-independent words refer to words that have no intrinsic meaning; on the contrary, context-dependent words describe an action, a feeling or an object.

history, it is only recently that studies have started to focus also on the creation and development of QA systems for answering why-type questions.

One of the best known figures in the domain is Verberne [13–16] whose initial work consisted in retrieving why-answers with the use of Rhetorical Structure Theory. In [15] she presented a re-ranking method where the score assigned to a QA-pair by QAP ranking algorithm[2] is weighted by taking into consideration a number of syntactic features. In her latest work [16] Verberne implements a fully functional why-QA system by integrating the re-ranking algorithm described in paper number and also makes a throughout analysis of the advantages and disadvantages of the BOW model in a why-QA context. This system obtains a 20% improvement in terms of MRR. Though efficient this method is labor intensive: the values produced by the 2 parsers used, the Pelican (constituency parser) and the EP4IR parser (statistical parser), have to be extracted manually and assigned to the selected features. Moreover, this method requires advanced language processing skills that only an expert in language syntax and semantics would possess.

A slightly different approach encountered in scientific literature is to derive causal expression patterns by extracting causal expressions from corpora. More clearly, these methods extract why-answers based on the presence of certain causal verbs [4] or relators [2] in the text analyzed. Although they are simple to implement and effective, these methods have the disadvantage of a low domain coverage: they do not address all why-type QA but only those that fulfill a certain pattern.

A more general approach, where causal expressions are acquired automatically with the aid of the Japanese EDR[3] dictionary, is described by Higashinaka and Isozaki [5]. The EDR dictionary contains phrases gathered from heterogeneous sources thus a good coverage of causal expressions is ensured. In this approach each phrase of the EDR dictionary is processed and context-independent words that express cause are extracted. All other words are replaced with a "*" to maintain the structure of the phrase. The structures obtained, combined with manually extracted causality indicative rules, are used to train a ranker. While known to be the best-performing fully implemented why-QA system for Japanese, Higashinaka and Isozaki's system relies on information extracted from a hand-crafted resource and therefore is not fully automated. Moreover the EDR dictionary is a rather high-priced resource only available for a limited number of languages.

To overcome the disadvantages of the former method, Tanaka [9, 10] built a fully automated classifier using bag-of-words features. Although the classifier performed well on small datasets, it failed on very large ones. In order to improve the performance of his initial method, Tanaka removed all context-dependent terms (e.g. nouns, verbs, adjectives etc.) and only included in the analysis a small group of words: the function words. Since the dimension of the new feature space

---

[2] QAP is a scoring algorithm for passages developed for question answering tasks. For further detail refer to [15].

[3] Electronic Dictionary Research.

was rather small the dimensionality problem was corrected, while all the initial qualities of the system were preserved. The latter method has the advantage of being easy to implement, scalable and effective. Moreover, it proves that feature selection is a promising technique in classifying text samples. Therefore our previous work was dedicated to testing and adapting it to English.

3   Previous Work

In this section we document our efforts [7] to extend Tanaka's [10] method to English. A detailed description is needed because this paper presents our attempts to improve this method.

3.1   Terminology

A content word refers to a word that has a meaning, and usually serves to describe an action, a feeling, an object (e.g. verb, noun, adjective etc.).

A function word is defined as a word that holds no meaning in itself, its sole purpose being to connect and create relations between content words.

A text segment is a group of sentences that are an eligible candidate for answering a why-question.

Tanaka's [10] method will be referred to as "Bag of function words" henceforth.

Text segments that are eligible why-answers will be referred as why-TS while those that do not as other-TS.

3.2   Bag of Function Words - Method Outline

The fundamental quality of this method is its ability to build domain independent fully automated classifiers. In his work Tanaka argues that 3 conditions are primordial to obtaining the domain independence of a classifier:

- convergence and reasonable size of feature space
- generality of features in the feature space
- ability of the feature to discriminate between encoding or not encoding causation text segments.

After analyzing vocabulary syntax, Tanaka concluded that function words fulfill all three conditions stated beforehand: their number is limited contrary to words like nouns; they have no intrinsic meaning therefore they ensure generality of features; and, last but not least, each one of them can be used to express a specific context(definition, cause, explanation etc.).

In order to identify function words in corpora, Tanaka used syntactic parser for Japanese on each text segment. The words that fulfilled the conditions stated above were selected and included in the feature space; the subset obtained contained mainly Japanese particles (e.g. ga, wa, kara etc.). Subsequently these words were mapped in a training dataset, composed of both why-TS and other-TS. $Tf - idf$ was calculated for each function word and feature vectors were built for each text segment. A classification model was built using LogitBoost and tested on various datasets.

3.3   LogitBoost

LogitBoost is a boosting algorithm with a binomial log-likelihood loss function and is part of the ensemble learning methods. The principle that governs ensemble learning is that combining several models produced by a classification algorithm into an ensemble might guarantee better accuracy than a single classifier, under the condition that the models are different enough to avoid making similar errors. In other words, boosting works by combining weak or base learners into a more accurate ensemble classifier. During the boosting process a number of base classifiers are fitted iteratively to re-weighted data in order to build a strong classifier. With each iteration the weight of the misclassified data points is increased while decreasing that of the correctly classified. Therefore, at each next iteration, the base learner will concentrate on the misclassified samples, working on correctly classifying it. Any algorithm normally used for classification can be employed as the base learner, provided it allows weighting of samples.

In Tanaka's study, decision stumps were used as a base learners since they are was easy to use and gave promising results.

3.4   Adaptation of the Bag of Function Words Method to English

Since the "Bag of function words" method was originaly designed only for Japanese, our previous work was dedicated to implementing this method for English. First and foremost, we had to replace the Japanese part-of-speech tagger with one suited to English. We selected the Standford tagger due to its high accuracy (over 95%). This tagger uses the well known Penn Treebank style containing a total of 36 part-of-speech labels. Following the principle of Tanaka's method, we selected 12 part-of-speech labels that we considered labeled words that fulfilled the three conditions described previously. These parts of speech mainly consist in coordinations, conjunctions, prepositions, modal verbs, pronouns, particles and determiners.

Feature Extraction. The Stanford Tagger [11] is run on all the text segments from the training dataset and the function words are extracted. Afterwards every text segment is mapped in the feature space using $tf - idf$ where the term frequency equals the number of times a function word appears in a text segment, and the document frequency measures in how many different text segments the function word is present. After feature extraction the dataset is thus:

$$\{(\boldsymbol{x_i},\ y_i)\}, \quad i = 1, 2, \ldots N \qquad y_i \epsilon \ \{true,\ false\} \tag{1}$$

where x_i is the feature vector for a given text segment i, N is the total number of text segments and y_i indicates if the i-th text segment encodes (*true*) or does not encode causation (*false*).

Experimental Results. The preprocessed training dataset is used to build a classifier by using LogitBoost with decision stumps. The performance of the output classifier was evaluated using 10-fold cross-validation and measuring precision, recall, and F-measure of all the classifiers produced.

Our experiment concluded that the classifier was successful, yielding an average precision of 76.1%, and average recall of 70.6% for text segments encoding causality, respectively 72.6% and 77.9% for text segments that do not encode causality.

Although preliminary results were promising, we think the small datasets used for training and testing might affect the validity of our study. Moreover we want to investigate the potential of other words in the why-classification process.

4   Proposed Method

After further analysis of English syntax we concluded that other parts-of-speech hold precious information for why-type classification: along with the parts-of-speech that we considered as labeling function words, some adverbs also fulfilled the conditions to be considered function words. Adverbs such as "before", "less" or "only" are frequently present in any kind of text corpora and therefore they are not context-dependent. Moreover, since their number is limited, they successfully satisfy the reasonable feature space condition (section 3.2, 1st condition). Considering the properties of these words, we have decided to add them to our initial feature space. The extraction procedure is detailed in subsection 4.1 . This method will be considered as the first method for our tests.

An analysis on the Second Edition of the Oxford English dictionary [1] shows that, out of the 171476 words, over half of the words are nouns, while about a quarter are adjectives, and about a seventh are verbs. In this respect, we assume that nouns and verbs play an important part when it comes to expressing causality. In contrast, we consider adjectives only bring supplementary descriptive information but do not hold notable causality discrimination properties. Hence including verbs and nouns to our feature space might boost the classifier's performance providing their number remains limited.

On a first approach we considered including only verbs to our analysis since their number is rather limited. We noticed that for 1000 text segments approximately the same number of distinctive verbs were extracted. Therefore including all verbs will almost triple the dimension of the initial feature space. Moreover, only a small amount of these verbs are eligible candidates for causal expression. Given these results two options presented to us: use a predefined dictionary of causal verbs or attempt to automatically extract significant verbs from the set of verbs present in our training dataset. Although the first option is appealing, it implies using a resource build with the help of a linguist expert. Besides, there exists no record of an exhaustive list of causal verbs, most of them being the fruit of scientific papers that deal with a precise subject [6].

For these reasons, we selected the second option: acquiring causal verbs automatically from corpora. To avoid the dimensionality curse we opted for an a priori feature selection technique. With this technique, we are able to extract verbs that discriminate well between why-TS and other-TS. We believe this list also incorporates a fair amount of causal verbs. A full description of this method can be found in subsection 4.2. Due to the importance of nouns in the English language we decided to implement this method for nouns as well (see subsection 4.2).

4.1   Adverb Extraction and Selection

In order to extract the context-independent adverbs from the corpora, we use WordNet [3] as an external resource that will help identify the eligibility of an adverb. With the help of the Stanford Tagger we gather all adverbs in our corpora and select only those whose root does not correspond to content word. WordNet is only used to verify whether the root is identical to a lemma of a verb, noun or adjective, and exclude the adverb if that is the case. We decide to reject these adverbs because we believe they only have a descriptive role in the sentence, with little or no causality information. Moreover, most of them derive from adjectives (by adding the "-ly" suffix) that we have already excluded from analysis. The entire procedure is easy to implement and fully automated.

The WordNet dictionary is a resource broadly used for research purposes displaying a vast lexical database that can guarantee a good coverage of the English vocabulary. Moreover this dictionary is or will be available for many languages, thus guaranteeing the scalability of the present method.

4.2   Verb and Verb & Noun Extraction and Selection

The extraction process is identical for both verbs and nouns. All existing verbs are selected from corpora and lemmatized using the lemmatizer supplied by MorphAdorner [8]. The initial feature vectors, used only for feature selection purposes, are created by following the same procedure we used in our previous word (see section 3.4) by keeping only verbs. These feature vectors are fed to several a priori feature selection algorithms and the representative lemmas are selected. The lemmas extracted are added to the initial feature space, that contained only function words and selected adverbs. Finally, the final feature vectors, used for classification, are generated with the same method. In this feature vectors all features are represented (function words, adverbs and selected lemmas). We chose to perform the feature selection on lemmas only, because function words and adverbs seem to represent well each text segment due to their redundancy in text. Performing a feature selection on all feature will lead to the elimination of these words and therefore a poorer representation of each text segment.

In our last experiment we follow this procedure for both verbs and nouns. We chose to make the selection on both nouns and verbs at the same time because some of these parts-of-speech share the same lemma (e.g. cause, suggest-suggestion etc.); therefore instead of obtaining two different $tf - idf$ calculations for the same lemma, we obtain only one where the $tf - idf$ value reflects the presence of the lemma in the text and not of the verb or noun individually.

4.3   Feature Selection Algorithms

Feature selection is a data mining technique which consist in choosing representative input features and removing irrelevant and redundant ones. This method is used in supervised learning to find feature subsets that will boost the classification accuracy. Moreover, with fewer features to analyze the classification algorithm will operate faster and more effectively.

For our study we investigated the performance of Correlation based Feature Selection (CFS) and χ^2. The 2 methods differ by the fact that CFS uses one-sided metrics while χ^2 uses two-sided ones. Feature selection algorithms using two-sided metrics select features most indicative of both membership (positive features) and non-membership (negative feature), while feature selection using one-sided metrics only extracts features most indicative of membership.

Correlation Based Feature Selection. CFS uses a heuristic to measure the usefulness of each feature in predicting the class label by considering their average correlation to the class against the average inter-correlation. In other words, a feature has increased importance if it has high average correlation with the class and low inter-correlation with other features. The formula of the heuristic is:

$$G_s = \frac{k\,\overline{r_{ci}}}{\sqrt{k + k\,(k - 1)\,\overline{r_{ii}}}} \qquad (2)$$

where k is the number of features in the subset, $\overline{r_{ci}}$ the mean feature correlation with the class, and $\overline{r_{ii}}$ is the average feature-feature inter-correlation.

To determine which features are included in the output subset the heuristics is combined with a search strategy.

χ^2 Based Feature Selection. The χ^2 statistic measures the lack of independence between a word, w, and a given category, c_k. $\chi^2(w, c_k)$ has a natural value of zero if word w and category c_k are independent. Since $\chi^2(w, c_k)$ is per-class, the average is used to combine the scores and select the k most representative features.

This method outputs a ranked list of all the variables in the dataset with their respective score. The number of features to include in the final subset is determined empirically.

4.4   Classification Algorithms

To evaluate the performance of the different proposed methods we consider two classification algorithms : LogitBoost and Support Vector Machine (SVM) with a Pearson VII function based kernel (Puk) [12]. LogitBoost has already been used for classification purpose in previous work, a full description being available in section 3.3. Support Vector Machine is a very promising machine learning tool due to its generalization ability and robust behavior over a variety of different learning tasks. However, SVM can perform effectively only if a suitable kernel function is applied. Usually the latter is determined experimentally by applying various kernel functions and selecting the best performing.

In this paper we used Puk function because of its ability to behave as a generic kernel. The Puk function can be varied gradually from a Gaussian bell to a Lorentzian line shape just by changing its input parameters, σ and ω. The Puk kernel function is:

$$K(x_i, x_j) = \frac{1}{\left[1 + \left(\frac{2\sqrt{\|x_i - x_j\|^2}\sqrt{2^{1/\omega} - 1}}{\sigma} \right)^2 \right]^\omega} \tag{3}$$

In Eq. (4) the parameter σ determines the width (sharpness) of the Pearson VII function. The parameter ω controls the actual shape (tailing) of the function. The Euclidean distance between the two vector arguments is normalized ensuring that all distances between the input objects and the map weights are in the range [0-1]. Due to this uniform rescaling we can easily optimize the kernel function just by modifying the values of σ and ω.

5   Experimental Settings and Results

5.1   Datasets

The data used for the experiment came from three main sources : Yahoo!Answers, Wikipedia and the Why-TS made available by Verbene on her website. From Yahoo!Answers we have randomly extracted text segments that were the answer to a why-question, for the positive data, and also those that were the answer to other types of questions (e.g. when, what, who), for negative data. Only the answers from the best-answer category were selected. From Wikipedia we randomly extracted definitions to serve as negative data in our experiment, and also content-related passages to each why-TS from Verberne's dataset. The latter were extracted manually and served as negative examples that possessed similar word content as the text-segments from the Verberne's dataset.

From the data collection, we constructed the three training datasets displayed in Table 1. For each set the origin of negative/positive data is indicated with the mention whether the data was automatically extracted (A) or manually (M). The data used for training is balanced (same number of why-TS and other-TS). The TS column indicates the total number of text segments used for training.

Table 1. Training datasets

Name	TS	Negative Data	Positive Data
TR-V	432	Verberne Dataset	Wikipedia (M)
TR-Y	2000	Yahoo!Answers (A)	Yahoo!Answers (A)
TR-YW	2000	Yahoo!Answers (A)	Wikipedia (A)

Table 2. Test datasets

Name	Used with	Negative Data	Positive Data
Test-V	TR-V	Yahoo!Answers	Wikipedia
Test-Y	TR-Y	Yahoo!Answers	Yahoo!Answers
Test-YW	TR-YW	Yahoo!Answers	Wikipedia

For testing purposes we constructed incrementally several datasets in order to evaluate the performance of the algorithms with the increase of data. We created test sets of 2000, 4000, 6000, 8000 and 10000 samples. The origin of the data used to test each training dataset is displayed in Table 2. All data was gathered automatically.

5.2   Feature Extraction

The features were extracted from the datasets described in section 5.1. using Stanford Tagger for part-of-speech labeling and MorphAdorner Lemmatizer for extracting the lemma for verbs and nouns. A simple spell corrector algorithm was also used to correct recurrent spelling mistakes. Following the three methods described in section 4. we experimented with six possible feature vectors (see Fig. 1). There are twelve scenarios of the experiments in which three scenarios do not incorporate a feature selection step. The description of each is shown in Table 3.

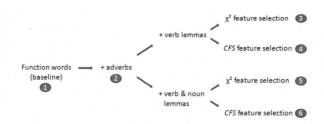

Fig. 1. Possible feature configurations vectors compared in the experiment

5.3   Parameter Optimization

In order to obtain maximum accuracy for the classification models we have to determine the optimal parameter setting for both classifiers. The optimization parameters were: the number of iterations, i, for the LogitBoost algorithm and σ, ω and the complexity parameter, c, for SVM-Puk. We evaluate the parameter setting performance over a 10-fold cross-validation performed on the training datasets; thus, the data used for parameter tunning is independent from the test sets. Table 4 contains the optimal parameter setting we have found.

Table 3. Description of scenarios

Features	Feature Selection	Classifier used	Scenario
Function words (F)	None	SVM - Puk	**F1**
		LogitBoost	**F2**
F + adverbs (FA)	None	SVM - Puk	**FA1**
		LogitBoost	**FA2**
FA + verbs	χ^2	SVM - Puk	**FV1**
		LogitBoost	**FV2**
	CFS	SVM - Puk	**FV3**
		LogitBoost	**FV4**
FA + verbs & nouns	χ^2	SVM - Puk	**FN1**
		LogitBoost	**FN2**
	CFS	SVM - Puk	**FN3**
		LogitBoost	**FN4**

Table 4. Optimal parameter setting

Parameters	TR/Test-V				TR/Test-YW				TR/Test-Y			
	i	c	ω	σ	i	c	ω	σ	i	c	ω	σ
F1	-	1.4	0.9	1.2	-	1.4	0.9	1.2	-	1.0	1.6	1.6
F2	50	-	-	-	110	-	-	-	200	-	-	-
FA1	-	1.4	2.0	2.3	-	1.4	1.5	1.9	-	0.8	1.1	1.1
FA2	80	-	-	-	110	-	-	-	80	-	-	-
FV1	-	1.4	2.0	2.4	-	1.3	4.0	4.0	-	1.3	2.2	2.8
FV2	90	-	-	-	200	-	-	-	200	-	-	-
FV3	-	1.4	2.5	2.5	-	1.0	1.6	2.0	-	1.1	0.9	1.1
FV4	100	-	-	-	200	-	-	-	200	-	-	-
FN1	-	1.2	3.0	3.0	-	1.2	2.0	2.2	-	1.2	2.0	2.1
FN2	100	-	-	-	200	-	-	-	300	-	-	-
FN3	-	1.5	4.0	4.0	-	1.1	2.5	2.5	-	1.4	1.6	1.5
FN4	95	-	-	-	200	-	-	-	300	-	-	-

5.4   Results

All twelve scenarios were executed on each of the three training databases. To estimate the performance of the model built with each scenario we use a 10-fold

Table 5. Results obtained using the SVM classifier. Percent improvement, as well as statistical significance is with respect to the SVM baseline (F1).

Scenario	TR/Test-YW	TR/Test-V	TR/Test-Y
F1 (baseline)	0.9108	0.8101	0.6418
FA1	0.9178 (0.70%)	**0.8318 (2.17%)**	0.6467 (0.49%)
FV1	0.9126 (0.18%)	0.8196 (0.95%)	0.6602 (1.84 %)
FV3	0.9158 (0.50%)	*0.8082 (-0.19%)* †	0.6514 (0.96%)
FN1	**0.9252 (1.44%)**	*0.7700 (-4.01%)*	**0.6654 (2.36%)**
FN3	0.9198 (0.90%)	*0.7992(-1.09%)*	**0.6654 (2.36%)**

Table 6. Results obtained using the LogitBoost classifier. Percent improvement, as well as statistical significance is with respect to the LogitBoost baseline (F2).

Scenario	TR/Test-YW	TR/Test-V	TR/Test-Y
F2 (baseline)	0.9356	0.5344	0.6326
FA2	0.9381 (0.25%)†	0.6490 (11.46%)	0.6410 (0.84%)
FV2	0.9432 (0.76%)	**0.6722 (13.78%)**	0.6432 (1.06%)
FV4	0.9432 (0.76%)	0.5758 (4.14%)	0.6440 (1.14%)
FN2	**0.9496 (1.40%)**	0.6300 (9.56%)	**0.6556 (2.30%)**
FN4	0.9428 (0.72%)	0.6434 (10.9%)	**0.6556 (2.30%)**

cross-validation. Once each model has been optimized over cross-validation, we perform the evaluation tests on test datasets.

Tables 5 and 6 contain the results of our findings. The displayed value represents an average of the 5 F-measures (for 2000, 4000, 6000, 8000 and 10000 text segments) we measured for each scenario during our experiment. A significance paired t-test was performed on the 5 F-measure scores measured for each scenario, and succeeded on almost all at a $p < 0.05$ level; the scenarios that passed the test only at the $p < 0.1$ level are denoted with a †. In order to determine the most significant features for the CFS method we used a hill climbing search algorithm; for the χ^2 selection process we selected all features that had a score superior to zero.

Results show that all 3 methods over-perform baseline, with one slight exception for the $TR/Test - V$ with SVM classifier group (refer to the results in italic from table 5). In this case both function words and function words plus adverbs yield better results than the methods that integrate verbs or verbs & nouns. We believe this is a consequence of the fact that negative data was built with similar content words that existed in the positive data. Therefore verbs and nouns have lost their discriminative power when integrated in the SVM classification model. On the contrary, the LogitBoost models built for this set (FV2, FV4, FN2, FN4) are less affected by the content similarity and perform better than baseline.

Both LogitBoost and SVM are successful classification models on all data, yielding similar performance, except for the $TR/Test - V$ data where SVM classification outperforms LogitBoost with over 20% (refer to second column of tables 5 and 6). In terms of feature selection χ^2 and CFS give similar results. While χ^2 is faster in ranking the results, CFS is easier to manipulate since we are not required to determine the cut-off value that would produce the best results. Globally we notice the verbs & nouns methods (FN) are the best performing ones except for the TR-V Test-V data. Results show that all methods discriminate very well between random definitions and why-TS (up to 94%) while applied to a more heterogeneous database the accuracy of classification falls down to 65%.

In terms of execution time we notice that the average speed decreases with the number of features that are included in the analysis, but also with the number of validation and training instances. Therefore the time to build the model varies from 2.7 seconds, on TR-V, to 115.5 seconds, on TR-Y, for the LogitBoost classifier and from 190 milliseconds, on TR-V, to 28.5 seconds, on TR-Y, for the SVM classifier. The worst execution time with respect to testing is obtained when performing the test on the 10000 instances on the Test-Y dataset; the time is of 1.97 seconds with a LogitBoost classification model and of 70.38 seconds with a SVM-Puk classification model.

We show the progression of our two classification models with the increase of test data in Fig. 2 for SVM-Puk and respectively Fig. 3 for LogitBoost; the dataset used in both figures is TR/Test-Y. We have excluded the FV1-2 and FN1-2 because we stated before that the χ^2 feature selection performance is similar to CFS while CFS is easier to manipulate.

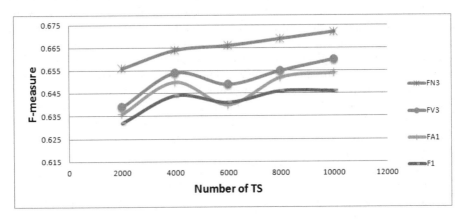

Fig. 2. F-measure value at various test dataset sizes for SVM-Puk Classifier

This graphics prove that the FN scenario is the best performing with both SVM and LogitBoost. We note that the SVM-Puk classfier is very sensitive

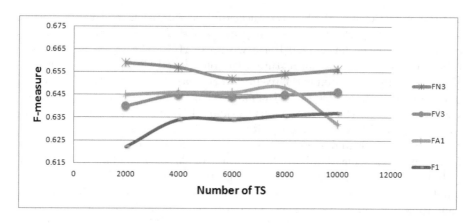

Fig. 3. F-measure value at various test dataset sizes for LogitBoost Classifier

to the quality of the test data, while the LogitBoost classifier suffers very little from it.

5.5   Conclusion and Future Works

In this paper we investigated several methods to improve the performance of the "Bag of function words" on English. Through our work we have shown the importance of adding new features (adverbs, verb lemmas and verb & noun lemmas) in boosting the classification of why-text segments. Initially, context-independent adverbs were added to the features showing small but valuable improvement of classification accuracy on all test datasets. Taking into account the amount of nouns and verbs in the English language we assumed they held significant information in terms of expressing causality and hence considered integrating them in the analysis. Confronted with their large number, we have added a feature selection step to our method to avoid the dimensionality curse. Adding the features selected by the feature selection algorithm has proven successful improving the classification performance with approximatively 2.5% for nouns & verb lemmas and 1% for verb lemmas.

We are tempted to think SVM with a Puk kernel might be a more appropriate classifier than LogitBoost since it can be parameterized to adapt to any kind of data and the results show that SVM slightly outperforms LogitBoost for most of the validation tests performed, but we believe this matter requires further investigation. However, during these experiments the optimum configuration parameters were determined by a local search performed manually. Therefore the accuracy of this classification model can be further improved by applying an automatic extensive search for the configuration parameters.

Future work will be dedicated to making our approach more robust to answers that contain noise (spelling mistakes, emoticons) and also handling answers that do not contain direct answers but an url to further resources.

References

1. AskOxford. How many words are there in the english language?,
   http://www.askoxford.com
2. Blanco, N., Castell, E., Moldovan, D.: Causal relation extraction. In: Proceedings
   of the Sixth International Language Resources and Evaluation, LREC 2008 (2008)
3. Fellbaum, C.: WordNet: An Electronic Lexical Database. Bradford Books (1998)
4. Girju, R.: Automatic detection of causal relations for question answering. In: Proceedings of the ACL 2003 workshop on Multilingual summarization and question
   answering, pp. 76–83 (2003)
5. Higashinaka, R., Isozaki, H.: Automatically acquiring causal expression patterns
   from relation-annotated corpora to improve question answering for why-questions.
   ACM Transactions on Asian Language Information Processing (TALIP) 7(2), 1–29
   (2008)
6. Khoo, C., Chan, S., Niu, Y.: Extracting causal knowledge from a medical database
   using graphical patterns. In: In Proceedings of 38th Annual Meeting of the ACL,
   Hong Kong, pp. 336–343 (2000)
7. Nagy, I., Tanaka, K., Takiguchi, T., Ariki, Y.: Extracting why text segment from
   web based on grammar-gram. In: Proceedings of the Fouth Spoken Document
   Processing Workshop (2010)
8. Philip, R.: "Pib" Burns of Academic and Northwestern University Research Technologies. English lemmatizer,
   http://morphadorner.northwestern.edu/morphadorner/lemmatizer/
9. Tanaka, T., Takiguchi, K., Ariki, Y.: Automatic why text segment classification
   and answer extraction by machine learning (japanese). Journal of Information Processing Society 49(6), 2234–2242 (2008)
10. Tanaka, T., Takiguchi, K., Ariki, Y.: Domain independent why text segment
    classification and answer extraction by grammar-gram and grammarverb-gram
    (japanese). WI2, pages pp. 89–94 (2009)
11. Toutanova, K., Christopher, D.: Manning. Enriching the knowledge sources used
    in a maximum entropy part-of-speech tagger. In: Proceedings of the 2000 Joint
    SIGDAT conference on Empirical methods in natural language processing and
    very large corpora, pp. 63–70 (2000)
12. Ustun, W.J., Melssen, B., Buydens, L.M.C.: Facilitating the application of support
    vector regression by using a universal pearson vii function based kernel. Chemo-
    metrics and Intelligent Laboratory Systems 81, 29–40 (2006)
13. Verberne, S.: Developing an approach for why-question answering. In: EACL 2006:
    Proceedings of the Eleventh Conference of the European Chapter of the Association
    for Computational Linguistics: Student Research Workshop, pp. 39–46 (2006)
14. Verberne, S., Boves, L., Oostdijk, N., Coppen, P.-A.: Evaluating discourse-based
    answer extraction for why-question answering. In: SIGIR 2007: Proceedings of the
    30th annual international ACM SIGIR conference on Research and development
    in information retrieval, pp. 735–736 (2007)
15. Verberne, S., Boves, L., Oostdijk, N., Coppen, P.-A.: Using syntactic information
    for improving why-question answering. In: COLING 2008: Proceedings of the 22nd
    International Conference on Computational Linguistics, pp. 953–960 (2008)
16. Verberne, S., Boves, L., Oostdijk, N., Coppen, P.-A.: What is not in the bag of
    words for why-qa? Comput. Linguist. 36(2), 229–245 (2010)

Speeding Up and Boosting Diverse Density Learning

James R. Foulds and Eibe Frank

Department of Computer Science, University of Waikato, New Zealand
{jf47,eibe}@cs.waikato.ac.nz

Abstract. In multi-instance learning, each example is described by a *bag* of instances instead of a single feature vector. In this paper, we revisit the idea of performing multi-instance classification based on a point-and-scaling concept by searching for the point in instance space with the highest diverse density. This is a computationally expensive process, and we describe several heuristics designed to improve runtime. Our results show that simple variants of existing algorithms can be used to find diverse density maxima more efficiently. We also show how significant increases in accuracy can be obtained by applying a boosting algorithm with a modified version of the diverse density algorithm as the weak learner.

1 Introduction

Multi-instance (MI) learning [7] is a variation of traditional supervised learning with applications in areas such as drug activity prediction [7], content-based image retrieval [26], stock market prediction [13] and text categorization [1].

In standard supervised learning, each example is represented by a single feature vector. In MI learning, examples are collections of feature vectors, called *bags*. The feature vectors within the bags are known as *instances*. Each instance is a vector of (typically real-valued) attribute values. Each bag has a class label, but labels are not given for the instances. The task is to learn a model from a set of training bags to predict the class labels of unseen future bags.

The original motivating application for MI learning was the *musk* drug activity prediction problem [7]. The task is to predict whether a given molecule will bind to a target "binding site" on another molecule, and hence emit a "musky" odor. A molecule may take on several different conformations (shapes) by rotating its internal bonds. If a single conformation can bind to the target binding site, the molecule is considered to be active. This is a difficult learning problem because it is not always clear which conformation is responsible for activity. Dieterich *et al.* represented each molecule as a bag containing the different conformations that the molecule can adopt.

Much of the work on MI learning, including all early work and notably including [7] and [14], makes a specific assumption regarding the relationship between the instances within a bag and its class label. We will follow [21], and refer to this assumption as the *standard MI assumption*. It states that each instance has a hidden class label $c \in \Omega = \{+, -\}$. Here, '+' is the *positive* class, and '−' is the *negative* class. The set of class labels for bags is also Ω. Under this assumption, a bag is positive if and only if at least

B. Pfahringer, G. Holmes, and A. Hoffmann (Eds.): DS 2010, LNAI 6332, pp. 102–116, 2010.
© Springer-Verlag Berlin Heidelberg 2010

one of its instances is positive. (i.e. belong to the '+' class). Thus, the bag-level class label is determined by the disjunction of the instance-level class labels.

A number of algorithms for MI learning can be found in the literature. Dieterich *et al.* [7] presented the first algorithms for MI learning, which use axis-parallel hyper-rectangles (APR) to solve the *musk* problem. The algorithms build a single APR that identifies the "positive" region of instance space. At classification time, any bag that contains an instance within the APR is labeled as positive, as per the standard MI assumption.

A different method for tackling MI learning problems is to transform the data so that unmodified single-instance learner can be applied, e.g. by computing summary statistics as in the relational learning system RELAGGS [12], or by labelling each instance with its bag's label and combining predictions [16].

A common approach to MI learning is to "upgrade" standard supervised algorithms to handle the MI scenario by modifying their internals. Such algorithms include k-nearest neighbours [20], support vector machines [1,11], decision trees and rules [6], logistic regression [24] and boosting [2,24,19]. In [3], a different approach is used: a *standard* boosting algorithm is applied in conjunction with an MI base learner—a special-purpose one that induces hyper-balls or hyper-rectangles. We pursue the same basic approach in this paper, and show that boosting can be applied successfully in conjunction with a modified version of an existing, established MI algorithm, the diverse density method proposed by Maron [14].

The diverse density method is a statistical approach to MI learning under the standard assumption. However, the original learning techniques within the framework are computationally expensive. In this paper we first investigate heuristics that are designed to improve computational efficiency. We show that runtime can be improved without loss of accuracy. We then show how to adapt the basic algorithm so that it can be boosted to increase predictive performance, yielding multi-instance classifiers that are competitive with the state-of-the-art.

The remainder of the paper is structured as follows: Sections 2 and 3 detail the diverse density framework and its associated algorithms respectively. Section 4 describes the heuristic variants of these algorithms we consider and Section 5 has experimental results. Section 6 shows how to adapt the basic method to perform boosting and presents the improvements in accuracy obtained. We conclude in Section 7.

2   Diverse Density

Diverse density $DD(h) : h \in H \to \mathbb{R}^+$, as defined in [13,14], is an objective function for determining the best hypothesis $\hat{h}$ in a certain class of probabilistic multi-instance classifiers H. The objective function is designed to model the intuition that under the standard MI assumption, the positive region of the instance space $\chi = \mathbb{R}^d$ is most likely to be close to instances from many different (i.e. "diverse") positive bags. In [13,14], diverse density is assumed to be proportional to the posterior density for the model parameters, so under this interpretation maximizing diverse density corresponds to finding a maximum a-posteriori estimate $\hat{h}_{MAP}$.

As this posterior density is not defined in terms of a conditional likelihood in [13,14], it is not immediately obvious that it is an appropriate objective function for classification learning. The "diverse"-ness property is only a heuristic motivation for $DD(h)$. However, diverse density can instead be understood as a conditional likelihood function under a different interpretation of some terms[1] (see also [23]). This interpretation is what we use here because it is more in line with work in probabilistic machine learning, and it motivates diverse density learning as a principled maximum-likelihood estimate of a discriminative model. Additionally, under this interpretation, we do not have to resort to cross-validation on the training set in order to choose a decision boundary[2].

Consider the task of learning a discriminative model $Pr(+|B_i, h)$ for predicting the probability that a bag B_i is positive, given a hypothesis h. Let $Pr(+|B_{ij}, h)$ be the probability that instance j of bag B_i is positive. Assuming independence, $Pr(+|B_i, h) = 1 - \prod_j(1 - Pr(+|B_{ij}, h))$, the probability that at least one instance is positive, in line with the standard MI assumption. Further assuming that bags are iid, the conditional likelihood function for h given a dataset D is

$$
\begin{aligned}
L(h) &= Pr(Y|D, h) \\
&= \prod_i Pr(+|B_i^+, h) \prod_i Pr(-|B_i^-, h) \\
&= \prod_i \left(1 - \prod_j(1 - Pr(+|B_{ij}^+, h))\right) \prod_i \left(\prod_j(1 - Pr(+|B_{ij}^-, h))\right),
\end{aligned}
$$

where Y is the set of labels for the bags in D, and the B_i^+s and B_i^-s are the positive and negative training bags, respectively.

The model parameters to be learnt are $h = \{x_1, \ldots x_d, s_1, \ldots, s_d\}$, where $x \in \chi$ is the location of the "target point" identifying the positive region of the instance space, and s is the feature scaling vector. To complete the model, we still need to specify $Pr(+|B_{ij}, h)$. Maron and Lozano-Perez use a radial "Gaussian-like" function $Pr(+|B_{ij}, h) = \exp(-||B_{ij} - x||^2)$, with $||B_{ij} - x||^2 = \sum_k s_k^2(B_{ijk} - x_k)^2$. In other words, it is assumed that the probability that instance j of bag B_i is positive drops exponentially with distance from point x, with the scaling of each dimension k in feature space determined by s_k.

Using this form for $Pr(+|B_{ij}, h)$, the conditional likelihood function is identical to Maron and Lozano Perez' diverse density function $DD(h)$ under what they call the "noisy-or model" (so named because the $Pr(+|B_i^+, h)$ term corresponds to a probabilistic version of a logical "or"). Hence, maximizing the (log) likelihood of this model is equivalent to maximizing diverse density. Maron also formulated the "most-likely-cause" model for diverse density learning. With this model, the likelihood needs to be modified so that $Pr(+|B_i, h) = \max_j Pr(+|B_{ij})$. The max operator can be viewed as an approximation to a logical "or", so the most-likely-cause model is also consistent with the standard MI assumption.

---

[1] Specifically, Maron and Lozano Perez' "$Pr(h|B)$" and "$Pr(h|B_{ij})$" are interpreted as $Pr(+|B, h)$ and $Pr(+|B_{ij}, h)$ respectively.

[2] This fact was also exploited implicitly in EM-DD [25].

3   Existing Diverse Density Algorithms

In this section we discuss existing MI learning algorithms that are based on the diverse density framework. In the next section we introduce new variants of these methods that are based on simple heuristics designed to reduce training time while maintaining classification accuracy on new data.

The original maxDD diverse density algorithm attempts to find the target concept by maximizing diverse density (i.e. conditional likelihood) over the instance space, using gradient ascent with multiple starting points. Because the feature scaling vector is also unknown, it is optimized simultaneously (and initialized with all values set to 1.0). Restarts are performed at every instance from each positive bag, as the target point is necessarily close to some of those instances.

Maron [13] also proposed an alternative DD maximization algorithm, Pointwise Diverse Density (PWDD), which was designed to speed up the training process. However, he did not evaluate PWDD, except on a simple artificial dataset that is used to illustrate the underlying idea. One contribution of this paper is that we compare PWDD to maxDD (and variants thereof) on a collection of datasets and thus attempt to close this gap in the literature.

While maxDD uses a gradient ascent method to search for the point of maximum diverse density—with different starting points—PWDD only computes diverse density at exactly the points corresponding to instances in positive bags in the training data. For each positive bag, the algorithm selects the instance with the highest diverse density, and the output hypothesis is the average point of these selected instances.

This version of the DD algorithm performs no gradient optimization and is therefore extremely fast; however it requires the feature scaling vector to be known. In practice, this is typically not the case, so the method must be extended to find the best scaling vector. Maron proposes several alternatives (but does not name them; the names are ours):

- *Scaling First* For each instance in a positive bag, perform gradient ascent to optimize the feature scaling vector for the highest diverse density at that point. Select the vector that produced the highest diverse density, and use this for PWDD.
- *Iterative* Pick an initial scaling, then use PWDD to find the best point (or points) with that scaling. Then use gradient ascent to optimize the scaling vector at the best point(s). Repeat using the new scaling.

Maron also mentions potential variants where the selection of the best instance is incorporated into the gradient ascent optimization routine by replacing the max operator with the differentiable $softmax$ function. However, we do not consider these softmax-based variants in this paper.

Zhang and Goldman [25] later formulated the EM-DD algorithm, a variant of the maxDD algorithm that is based on the expectation-maximization (EM) approach. The algorithm starts with an initial guess h of the target concept, obtained by selecting a point from a positive bag. It then performs an iterative procedure consisting of an *expectation* step followed by a *maximization* step.

The expectation step selects the instance from each bag that is most likely to be the cause of that bag's label given the current hypothesis h, using the most-likely-cause estimator. Then, the maximization step performs a gradient ascent search based on the selected instances to find a new h' that maximizes $DD(h')$. The current hypothesis h is reinitialized to h'. The EM loop is repeated until convergence.

Considering computational efficiency, EM-DD has an advantage over maxDD, as it selects only a single instance from each positive bag during the expectation step, which reduces the computational difficulty of the maximization step. Hence, EM-DD scales well with increasing bag size.

The authors of EM-DD initially reported improved performance over maxDD on the *musk* data, but Andrews et al. [1] pointed out that the original formulation of EM-DD selected the best hypothesis based on error rate on the test data. If this is corrected, the algorithm's accuracy is not generally superior to that of maxDD. However, it is still notable for its improved computational efficiency and this is why it is included in the experimental comparison presented in this paper.

Note that there are also algorithms related to the diverse density framework that are not based on direct optimization. They include DD-SVM [5] and its successor MILES [4], which create an instance-based feature space mapping by using diverse density to compute a similarity function between a bag and an instance (which corresponds to a feature in a new instance space), and build a support vector machine on the transformed dataset. The boosted diverse density approach presented in Section 6 generates a similar classifier because it also produces a linear combination of contributions from diverse density target points, where each training instance can potentially become a target point. However, in contrast to MILES, it allows the scaling of features in the original feature space to be adapted to the data at hand—automatically and individually for each target point. In Section 5, we compare the empirical performance of our approach with that of MILES.

4   A New Approach: QuickDD

In this section, we describe some simple variants of existing diverse density maximization algorithms that are designed to improve the computational efficiency of these techniques.

As Maron and Lozano-Pérez observed, the point of maximum diverse density is by definition close to instances from positive bags. A simple heuristic when searching for this point is therefore to only consider the exact locations of instances in positive bags. This heuristic is also used in PWDD, but then the best points from the different positive bags are averaged to form a hypothesis. We propose an even simpler approach, where the merging step is omitted, and we simply pick the best point from all positive bags as our target point.

We will refer to this heuristic as *QuickDD*, as we expect quicker execution over the standard gradient ascent approaches because the search space is greatly reduced in this method. We hypothesized that *(a)* merging of candidate points as in PWDD can produce undesirable target points and *(b)* gradient ascent search over instance space as in maxDD is unnecessary, as an instance from a positive training bag is a sufficient approximation of the target point in real-world problems. Regarding hypothesis *(a)*, the average of the highest diverse density points from each positive bag is not guaranteed to be a high diverse density point. If the best points for the positive bags belong to different local diverse density maxima, the average point may be in the trough between the maxima, and may not be the best hypothesis.

If the optimal feature scaling is already known, the execution of the maxDD algorithm with the QuickDD heuristic is very efficient: we merely compute the diverse density at each instance from each positive bag, and select the location of the instance with the highest diverse density, without performing any gradient optimization. Again, this is simpler than the PWDD approach, as we do not find the best instance from each positive bag and average the results, but merely select the best instance and use this as the target point.

If the optimal feature scaling is not known in advance, we must incorporate some method to compute it. The methods proposed by Maron for PWDD can easily be adapted for QuickDD:

- *Scaling Only.* For each instance in a positive bag, perform gradient ascent to optimize the feature scaling vector at that point. The hypothesis is the point and associated scaling vector that produces the highest diverse density. Here, although the gradient ascent optimization must be performed as many times as for maxDD, the number of parameters is halved because the coordinates of the target point are not optimized.
- *Iterative.* Pick an initial scaling, then compute the diverse density of the points in all positive bags with respect to that scaling, and select the point (or points) with the highest diverse density. Then use gradient ascent to optimize the scaling vector at the best point(s). Repeat using the new scaling.

A motivation for using QuickDD *Iterative* over PWDD *Iterative* is that the former monotonically increases the diverse density of the current hypothesis in each iteration—since it will not pick a scaling vector or target point location with a lower DD value than that of the current hypothesis—and thus is guaranteed to converge to a local DD maximum (or saddle point), while the averaging step in PWDD means that there are no such guarantees for that algorithm.

Note that the *Scaling Only* method is also applicable to EM-DD, by restricting the gradient search performed in each iteration to only optimize the scaling parameters for a fixed location in instance space.

We also consider the following simplifications of the above approaches:

- *No Scaling.* Initialize all entries of the feature scaling vector to the same value (i.e. 1.0 if we follow [14]). Compute the diverse density for each instance in each positive bag, and select the point with the highest diverse density. Do not perform any gradient ascent optimization.
- *Scaling Once.* Initialize the scaling vector as above. Compute the diverse density of the points in all positive bags with respect to that scaling, and select the point with the highest diverse density. Then use gradient ascent to optimize the feature scaling vector at that point. This is equivalent to the QuickDD *Iterative* method with the maximum number of iterations set to one.
- *Scaling Last.* This is an adaption of the *Scaling Once* method to PWDD. Execute PWDD to find a point with high diverse density with respect to an initial feature

Table 1. Datasets used in the Experiments

Name	Number of Bags	Number of Attributes	Avg. Number of Instances per Bag	Min. Number of Instances per Bag	Max. Number of Instances per Bag
musk1	92	166	5.2	2	40
musk2	102	166	64.7	1	1044
muta-atoms	188	10	8.6	5	15
muta-bonds	188	16	21.3	8	40
muta-chains	188	24	28.5	8	52
elephant	200	230	7.0	2	13
fox	200	230	6.6	2	13
tiger	200	230	6.1	1	13
maron	50	2	50.0	50	50

Table 2. Percentage Accuracy for EM-DD, maxDD and PWDD

Dataset	EM-DD	maxDD	PWDD Scaling First	PWDD Iterative	PWDD Scaling Last	PWDD No Scaling
musk1	85.4±11.6	86.8±11.5	86.9±10.2	86.5±11.5	85.5±11.5	48.9±4.9 ●
musk2	85.6±9.8	85.7±9.6	86.3±10.6	85.1±10.0	84.9±10.2	62.7±9.9 ●
muta-atoms	72.2±8.4	72.2±10.1	36.1±7.3 ●	42.9±15.0 ●	64.1±6.5 ●	66.5±2.3
muta-bonds	73.0±10.1	73.9±9.5	52.1±17.0 ●	70.4±8.9	73.2±8.5	66.5±2.3
muta-chains	73.5±11.1	79.1±7.6	67.8±7.0	65.2±11.1	80.3±7.8	66.5±2.3
elephant	75.9±10.3	81.9±8.9	78.5±6.3	10 82.6±8.7 ○	82.6±8.8 ○	56.4±5.9 ●
fox	60.3±8.5	61.3±10.8	59.7±8.9	62.3±10.7	60.4±10.9	55.4±5.3
tiger	71.9±10.3	75.4±9.7	72.0±9.6	70.5±11.0	71.4±9.9	49.9±6.8 ●
maron	93.4±11.0	96.4±7.7	94.2±10.0	94.4±9.9	94.4±9.9	61.6±14.4 ●

○, ●: significant increase or decrease vs EM-DD; number in small font: completed runs.

scaling vector, then perform a gradient ascent search to optimize the feature scaling vector at that point, i.e. perform a single iteration of PWDD *Iterative*.

5   Experimental Results

In this section we present the results of an empirical study of the classification performance and training time of the algorithms discussed above. The algorithms were evaluated on a variety of two-class datasets by averaging the results of ten repeats of ten-fold cross-validation, measuring both classification accuracy and training time. The datasets used were:

- *elephant, fox, tiger.* Content-based image retrieval datasets, originally provided by [1]. The MI bags represent photographs of animals, and the task is to predict whether an image contains the target animal (elephants, foxes and tigers, respectively).
- *musk1, musk2.* The musk data used in [7]. Each bag represents a molecule, and the task is to predict whether the molecule emits a musky odour. The *musk2* dataset is larger, both in terms of the number of molecules, and the number of instances per molecule.
- *mutagenesis.* The mutagenicity prediction problem [18], widely used as a benchmark for ILP algorithms. The learning problem is to identify mutagenic molecules. Three representations of molecules were used [17]: *muta-atoms, muta-bonds* and *muta-chains*.

Table 3. Training Times in CPU seconds (s), minutes (m) or hours (h) for EM-DD, maxDD and PWDD

Dataset	EM-DD	maxDD	PWDD Scaling First	PWDD Iterative	PWDD Scaling Last	PWDD No Scaling
musk1	2.7m±35.2s	22.7m±2.6m o	9.1m±43.2s o	6.6s±1.5s •	2.8s±0.6s •	0.4s±0.0s •
musk2	29.6m±27.7m	23.6h±9.3h o	24.5h±7.1h o	4.9m±2.1m •	1.6m±45.8s •	36.1s±4.9s •
muta-a	0.9s±0.4s	6.4m±2.6m o	27.4s±1.4s o	1.9s±0.8s o	0.8s±0.1s	0.9s±0.0s
muta-b	6.4s±2.2s	49.8m±2.4m o	7.1m±19.0s o	14.5s±4.1s o	9.4s±0.5s o	7.7s±0.1s
muta-c	16.9s±3.7s	6.1h±2.5h o	32.8m±50.2s o	27.0s±7.1s o	21.3s±0.8s o	18.9s±0.3s
elephant	20.1m±3.9m	12.7h±4.2h o	165.8h±47.1h 10	28.5m±33.5m	19.3m±15.9m	6.1s±0.1s •
fox	14.9m±3.5m	13.7h±3.6h o	19.7h±19.0h o	1.9m±2.1m •	2.3m±3.7m •	4.9s±0.1s •
tiger	9.5m±2.4m	6.2h±2.2h o	11.2h±7.4h o	1.9m±1.4m •	2.2m±1.8m •	3.8s±0.1s •
maron	2.5s±0.1s	4.2m±12.5s o	1.3m±2.2s o	4.0s±1.4s o	1.2s±0.0s •	1.1s±0.0s •

o, •: statistically significant increase or decrease vs EM-DD; number in small font: completed runs.

- *maron.* An artificial dataset based on one used in [14]. For each bag, 50 instances were sampled from a uniform distribution in $[0, 100] \times [0, 100] \subseteq \mathbb{R}^2$. Instances were *positive* if and only if they were within a 5×5 square in the middle of the domain, thus implementing the standard MI assumption. We generated 25 positive and 25 negative bags.

Key statistics of these datasets are summarized in Table 1.

The experiments were performed using WEKA [22], on 3.00 GHz Intel Pentium 4 CPU machines. All implementations were based on those of maxDD and EM-DD in WEKA, which use a quasi-Newton method with BFGS updates rather than plain gradient search. The details of this method can be found in Appendix B of [23]. For numeric stability, the negative logarithm of DD is minimized instead of maximizing DD directly.

The default behavior of the WEKA implementation of maxDD is to only consider instances from the *largest* positive bag as starting points for the optimization; we modified this to consider instances from *all* positive bags, to be consistent with the original description of maxDD. EM-DD was executed using instances from three random positive bags as starting points [25].

All iterative algorithms were restricted to a maximum of 10 iterations. We also applied normalization of attributes to the $[0, 1]$ interval to all datasets except for *maron*, as all of the algorithms typically performed poorly without this. We tested for significant differences between algorithms using the corrected resampled t-test [15] with significance level $\alpha = 0.05$.

Tables 2 and 3 display the accuracy and training time results for the three pre-existing algorithms EM-DD, maxDD and PWDD. Note that some of the 100 runs for the more expensive methods did not complete in time for submission. In those cases, no significance test was performed and the number of completed runs is given in small font next to the corresponding entry.

The results are consistent with the observations in [1], who disputed the superior classification performance of EM-DD over maxDD that was reported in earlier work: EM-DD performed similarly to maxDD on all datasets. However, its training time was several orders of magnitude lower in all cases. Hence, EM-DD is a worthwhile candidate in practical applications of MI learning.

Table 4. Percentage Accuracy for EM-DD and maxDD, Using 3 Random Positive Bags for Starting Points

Dataset	EM-DD	maxDD
musk1	85.4±11.6	87.0±11.4
musk2	85.6±9.8	85.8±10.1
muta-atoms	72.2±8.4	71.5±8.8
muta-bonds	73.0±10.1	74.1±9.5
muta-chains	73.5±11.1	79.2±7.7
elephant	75.9±10.3	81.9±8.5 ∘
fox	60.3±8.5	60.8±10.8
tiger	71.9±10.3	75.6±9.4
maron	93.4±11.0	96.4±7.7

∘, ● statistically significant increase
or decrease vs EM-DD.

Table 5. Training Times in CPU seconds (s), minutes (m) or hours (h) for EM-DD and maxDD, Using 3 Random Positive Bags for Starting Points

Dataset	EM-DD	maxDD
musk1	2.7m±35.2s	1.8m±27.0s ●
musk2	29.6m±27.7m	4.2h±3.6h ∘
muta-atoms	0.9s±0.4s	13.1s±2.8s ∘
muta-bonds	6.4s±2.2s	3.2m±41.6s ∘
muta-chains	16.9s±3.7s	12.7m±2.1m ∘
elephant	20.1m±3.9m	19.5m±5.2m
fox	14.9m±3.5m	15.6m±5.3m
tiger	9.5m±2.4m	8.8m±2.9m
maron	2.5s±0.1s	24.3s±1.9s ∘

∘, ● statistically significant increase
or decrease vs EM-DD.

When interpreting this result, it is important to remember that maxDD was executed using all instances from positive training bags as starting points for the gradient search, while EM-DD only used instances from three random positive bags. To isolate the effect of this modification, we performed a separate experiment where we used the same heuristic to reduce the number of starting points for the search in maxDD. Even with this modification maxDD frequently remains orders of magnitude slower than EM-DD. This can be seen from the results shown in Tables 4 and 5. Note that with this change to maxDD, the difference in accuracy on the elephant dataset becomes statistically significant.

Table 2 shows that all variants of PWDD suffered at least one significant loss in classification accuracy against EM-DD. In particular, PWDD struggled on the mutagenesis datasets, where most variants of the algorithm failed to improve on the 66.5% accuracy rate obtained by predicting the majority class. PWDD *No Scaling*, the variant of PWDD where no gradient search was used to optimize the scaling vector, only exceeded the majority class baseline on three datasets (*elephant*, *fox* and *maron*), demonstrating the importance of scaling features appropriately. PWDD *Scaling First* also performed quite poorly, indicating that undesirable scaling vectors were chosen. Moreover, PWDD *Scaling First* exhibited larger training times than even maxDD on the image datasets, indicating that optimizing the scaling vector only can result in hard optimization problems when the corresponding candidate target point is not appropriate.

Table 6. Percentage Accuracy for QuickDD Variants vs EM-DD

Dataset	EM-DD	QuickDD No Scaling	QuickDD Scaling Only	QuickDD Iterative	EM-DD Scaling Only	QuickDD Scaling Once
musk1	85.4±11.6	49.0±4.9 •	86.1±11.4	86.7±11.1	84.1±12.2	86.4±10.4
musk2	85.6±9.8	62.0±7.3	86.1±11.0	87.4±11.3	83.7±10.5	87.2±11.4
muta-atoms	72.2±8.4	64.5±4.8 •	70.9±8.2	68.5±8.2	75.6±9.5	68.5±8.2
muta-bonds	73.0±10.1	73.6±7.0	74.0±9.4	76.7±8.3	71.8±9.5	76.7±8.3
muta-chains	73.5±11.1	75.5±7.2	80.4±8.5	78.5±7.9	73.4±8.9	78.4±7.9
elephant	75.9±10.3	72.7±9.6	80.7±9.1 30	81.1±8.8	76.2±9.9	81.8±8.9
fox	60.3±8.5	53.7±11.3	60.3±11.2 57	64.0±10.3	61.0±10.4	64.0±10.3
tiger	71.9±10.3	60.0±8.9	74.3±10.1	75.1±10.1	72.5±10.5	75.5±9.6
maron	93.4±11.0	61.4±14.3 •	96.2±8.4	96.2±8.4	89.4±13.5	96.8±8.4

∘, •: significant increase or decrease vs EM-DD; number in small font: completed runs.

Table 7. Training Times in CPU seconds (s), minutes (m) or hours (h) for QuickDD Variants vs EM-DD

Dataset	EM-DD	QuickDD No Scaling	QuickDD Scaling Only	QuickDD Iterative	EM-DD Scaling Only	QuickDD Scaling Once
musk1	2.7m±35.2s	0.4s±0.0s •	8.2m±39.6s ∘	7.7s±2.9s •	36.2s±7.7s •	1.7s±0.6s •
musk2	29.6m±27.7m	36.2s±4.9s •	22.7h±8.9h ∘	6.0m±3.5m •	8.1m±7.1m •	4.2m±3.0m •
muta-a	0.9s±0.4s	0.6s±0.0s •	58.0s±3.1s ∘	3.3s±1.9s ∘	0.3s±0.1s •	0.7s±0.0s •
muta-b	6.4s±2.2s	4.5s±0.1s •	14.3m±39.2s ∘	49.9s±25.6s ∘	3.3s±0.8s •	5.0s±0.2s •
muta-c	16.9s±3.7s	10.4s±0.2s •	55.4m±3.2m ∘	1.3m±44.4s ∘	12.0s±2.5s •	11.8s±0.4s •
elephant	20.1m±3.9m	3.1s±0.1s •	80.7h±54.8h 30	10.6m±13.1m	1.5h±45.8m ∘	4.8m±6.7m •
fox	14.9m±3.5m	2.5s±0.0s •	29.3h±26.3h 57	2.7m±3.3m •	16.5m±7.1m	1.2m±1.6m •
tiger	9.5m±2.4m	1.9s±0.0s •	8.7h±4.4h ∘	1.5m±1.4m •	8.2m±4.8m	39.7s±35.6s •
maron	2.5s±0.1s	0.5s±0.0s •	55.7s±1.7s ∘	4.4s±1.6s ∘	0.9s±0.0s •	1.2s±0.0s •

∘, •: statistically significant increase or decrease vs EM-DD; number in small font: completed runs.

However, the *Iterative* and *Scaling Last* variants of PWDD were quite competitive with EM-DD overall, both in terms of classification accuracy and training time. Both achieved a significant win against EM-DD on the *elephant* data, while suffering a significant loss on *muta-atoms*, with no other significant differences. Both were significantly faster than EM-DD on all datasets except the three *mutagenesis* problems and *elephant*, exhibiting very fast training times on the two *musk* datasets. It is noteworthy that the single-iteration *Scaling Last* variant was very competitive with *Iterative* PWDD, where the maximum number of iterations was set to ten.

The results for QuickDD are summarized in Tables 6 and 7, and compared to EM-DD. We can see that except for the *No Scaling* variant — which performed poorly, as expected — and not withstanding the incomplete results for QuickDD *Scaling Only*, there were no significant differences for any of the QuickDD variants against EM-DD with respect to classification accuracy. Additionally, several QuickDD variants were superior in terms of training time.

The tables also show results for the EM-DD variant *Scaling Only*. It was significantly faster than the original EM-DD algorithm on six of the nine datasets, and only significantly slower on the *elephant* dataset, without any significant differences in classification accuracy.

QuickDD *Iterative* yielded an equal number of significant wins and losses for training time against EM-DD, but the wins were by a large margin on the slowest datasets, while the losses occurred only on datasets where the training times were already short.

Table 8. Percentage Accuracy for QuickDD Scaling Once and PWDD Scaling Last

Dataset	PWDD Scaling Last	QuickDD Scaling Once
musk1	85.5±11.5	86.4±10.4
musk2	84.9±10.2	87.2±11.4
muta-atoms	64.1±6.5	68.5±8.2
muta-bonds	73.2±8.5	76.7±8.3
muta-chains	80.3±7.8	78.4±7.9
elephant	82.6±8.8	81.8±8.9
fox	60.4±10.9	64.0±10.3
tiger	71.4±9.9	75.5±9.6
maron	94.4±9.9	96.8±8.4

No significant differences were observed.

Table 9. Training Times in CPU seconds (s) or minutes (m) for QuickDD Scaling Once and PWDD Scaling Last

Dataset	PWDD Scaling Last	QuickDD Scaling Once	
musk1	2.8s±0.6s	1.7s±0.6s	•
musk2	1.6m±45.8s	4.2m±3.0m	○
muta-atoms	0.8s±0.1s	0.7s±0.0s	
muta-bonds	9.4s±0.5s	5.0s±0.2s	•
muta-chains	21.3s±0.8s	11.8s±0.4s	•
elephant	19.3m±15.9m	4.8m±6.7m	•
fox	2.3m±3.7m	1.2m±1.6m	
tiger	2.2m±1.8m	39.7s±35.6s	•
maron	0.9s±0.0s	0.9s±0.0s	

○, • statistically significant increase.
or decrease vs PWDD Scaling Last.

Furthermore, QuickDD *Iterative* and *Scaling Once* both had a higher classification accuracy than EM-DD on eight of the nine datasets, though these differences were not individually statistically significant.

It is interesting to compare the runtime of QuickDD *Scaling Only* with that of maxDD from Table 3. The former was faster on the *musk* and *mutagenesis* datasets, but slower on the image datasets. This behaviour is similar to that of PWDD *Scaling First*, which we discussed above. In both cases, the dimensionality of the search space for the gradient optimization routine is halved relative to maxDD, but training time increases, implying the occurrence of harder optimization problems.

Similarly to the PWDD case, QuickDD *Iterative* performed just as well with one iteration (*Scaling Once*) as with a maximum of ten, with dramatic reductions in training time. Thus, repeated iterations appear unnecessary for both PWDD and QuickDD *Iterative*. This indicates that the initial scaling factor of 1.0 for all attributes may be sufficient for finding the location of a good hypothesis, perhaps aided by the dataset normalization step performed.

There were no significant differences in accuracy between the single-iteration versions of PWDD and QuickDD (Table 8), but QuickDD *Scaling Once* was faster that PWDD *Scaling Last* on eight of the nine datasets, (Table 9) with only one loss with respect to training time. Additionally, as Table 7 shows, QuickDD *Scaling Once* was significantly faster than EM-DD on seven of the nine datasets and still faster on the same seven datasets when the *Scaling Only* heuristic was applied in EM-DD. This is despite the fact that all points in all positive training bags were considered as candidate target points by the algorithm, while EM-DD only considered instances from three random bags as starting points. This shows that QuickDD *Scaling Once* is faster overall than all previous algorithms, while retaining the classification performance of the slower methods.

6 Boosting Diverse Density Learning

The above results show that simple heuristics can improve the runtime of diverse density learning. However, they do not increase accuracy in a significant manner. In this section, we discuss what modifications are required to successfully apply boosting to diverse density learning, and present experimental results demonstrating that significant increases in accuracy can be obtained in this manner. We use the *Real AdaBoost* algorithm described in [10]. In contrast to the original AdaBoost method [9], which is based on 0/1 predictions from the weak learner, this boosting method can exploit predictions that are class probability estimates.

As in the original AdaBoost, *Real AdaBoost* is a sequential process for learning an ensemble of weak classifiers. In each iteration, a weak classifier is learned based on a reweighted version of the training data. Initially, all examples receive the same weight. In subsequent iterations of the boosting process, the weight of an example, i.e. bag B in the context considered here, is updated based on the current hypothesis h using the following equation:

$$w := w \times e^{-0.5 \log \frac{Pr(\{+,-\}|B,h)}{1-Pr(\{+,-\}|B,h)}}$$

where $Pr(\{+,-\}|B,h)$ is the predicted class probability for the observed class of the bag (either $+$ or $-$). Thus, the square root of the predicted odds ratio for the observed class label determines the update. To reduce the likelihood of overfitting, the exponent can be moderated by multiplying it with a shrinkage value $s \in (0,1]$.

Boosting diverse density learning becomes computationally feasible by applying the QuickDD *Scaling Once* variant discussed above as the weak learner. *Real AdaBoost* requires the underlying learning algorithm to be able to deal with weighted examples, but it is straightforward to modify diverse density learning to do this by replacing the likelihood function with a weighted likelihood and adapting the gradient correspondingly. If w_i is the weight of a bag, then we now maximize:

$$\sum_i w_i^+ \log Pr(+|B_i^+, h) + \sum_i w_i^- \log Pr(-|B_i^-, h).$$

However, application of *Real AdaBoost* to the datasets considered above does not yield significant improvements in accuracy compared to applying stand-alone QuickDD *Scaling Once* itself directly to the data. We found that two further changes to the diverse density method are critical to render application of boosting successful:

1. **Symmetric learning.** Diverse density learning as discussed so far requires the user to decide prior to learning which class is to be treated as the positive class. The

Table 10. Percentage Accuracy for Boosted QuickDD *Scaling Once* and (optimized) MILES

Dataset	No boosting	10 boosting iterations	100 its. shrink. 0.5	Best MILES configuration
musk1	86.4±10.4	88.2±11.5	89.8±10.9	89.1
musk2	87.6±11.4	88.1±10.0	90.8±9.1	91.6
mutagenesis3-atoms	68.5±8.2	80.6±7.6 ○	84.7±7.2 ○	83.9
mutagenesis3-bonds	76.7±8.3	80.8±8.9	87.6±7.5 ○	86.3
mutagenesis3-chains	78.4±7.9	80.3±7.8	84.6±7.8	86.0
tiger	75.6±9.6	81.1±9.4	82.1±9.2	81.7
fox	64.0±10.3	62.2±9.1	64.4±8.7	64.9
elephant	81.7±9.0	84.5±8.2	86.9±7.9	84.1

○ statistically significant increase vs baseline
(no significance tests were performed wrt MILES-based results).

first modification is to eliminate this requirement: the basic algorithm is run twice, in each class treating one class as the positive class and the other class as the negative one. The final concept output is then the one of the two point-and-scaling concepts—one representing a negative target point and one representing a (traditional) positive one— that maximizes the (weighted) conditional loglikelihood. This means that different classes can be viewed as the positive class in different iterations of the boosting process.

2. **One-sided prediction.** Perhaps the most important change is to localize the influence of each diverse density classifier in the instance space. To this end, we change the diverse density model so that the probability predicted for the positive class can never drop below 0.5 (and, consequently, the probability for the negative class can never exceed 0.5). The new model is:

$$Pr(+|B_i, h) = 1 - 0.5 \times \prod_j (1 - Pr(+|B_{ij}, h))$$

This has the effect that the algorithm can abstain from making a prediction in the boosting process: a predicted probability of 0.5 means that the odds ratio becomes 1 in the above weight update. Thus, the influence of a weak classifier can be restricted to a small area around the concept that was found. The likelihood is optimized wrt this adjusted model.

Table 10 compares the accuracy of the boosted QuickDD *Scaling Once* algorithm with the above modifications to that of stand-alone QuickDD *Scaling Once*, which is the baseline in the left-most column, on the real-world datasets used in our study. Results for boosting with 10 iterations and no shrinkage, and boosting with 100 iterations and shrinkage 0.5 are included. To account for class imbalance, the boosting process was initialized with a model that predicts the class prior probabilities from the training data. No shrinkage was applied to the predictions of this initial model. The feature scaling was initialized to 100.0 for the mutagenesis datasets when boosting because poor results were obtained for value 1.0, most likely due to the more localized models being used. Note that runtime (not shown) is linear in the number of boosting iterations.

For reference the table also contains the best results obtained from different variants of the state-of-the-art MILES multi-instance learning method [4], taken from [8]. As discussed at the end of Section 3, MILES produces a similar model. The results for

MILES were generated under the same experimental conditions and are thus directly comparable. Note that these results are for the *best* configurations tried—in several cases the performance of the standard MILES approach could be improved by replacing the 1-norm support vector machine from [4] with another learning algorithm [8]—so they are likely to be optimistic. Despite this optimistic bias, we can see that boosted diverse density learning is highly competitive.

7 Conclusions

Our results show that PWDD *Iterative*, a previously proposed MI learning algorithm that has not received much attention in the literature, perhaps due to a lack of published empirical results, is very competitive with the more well-known EM-DD algorithm, both in terms of classification accuracy and training time. Moreover, we found that the repeated iteration of the algorithm is in fact unnecessary on the datasets we considered, as similar accuracy could be achieved with a single iteration of PWDD.

Our simplified QuickDD *Iterative* variant of PWDD, which provides convergence guarantees, improved results further. When restricted to a single iteration (QuickDD *Scaling Once*), the algorithm was very competitive with PWDD and EM-DD for classification accuracy, while enjoying faster training times. Our results show that instances from positive training bags are often a sufficient representation for the location of diverse density target points. This heuristic dramatically reduces the search space, enabling more efficient algorithms for learning diverse density concepts.

We also showed how boosting can be applied in conjunction with QuickDD *Scaling Once* to obtain state-of-the-art accuracy on the datasets investigated. Three changes to the algorithm were necessary to obtain improved accuracy using boosting: incorporation of bag weights, symmetric treatment of classes, and enabling one-sided prediction. With these changes, boosting diverse density learning appears to be a viable and practical alternative to other advanced methods for multi-instance learning.

References

1. Andrews, S., Tsochantaridis, I., Hofmann, T.: Support vector machines for multiple-instance learning. In: Neural Information Processing Systems, pp. 561–568. MIT Press, Cambridge (2003)
2. Andrews, S., Hofmann, T.: Multiple-instance learning via disjunctive programming boosting. In: Neural Information Processing Systems (2003)
3. Auer, P., Ortner, R.: A boosting approach to multiple instance learning. In: European Conference on Machine Learning, pp. 63–74. Springer, Heidelberg (2004)
4. Chen, Y., Bi, J., Wang, J.Z.: MILES: Multiple-instance learning via embedded instance selection. IEEE Pattern Analysis and Machine Intelligence 28(12), 1931–1947 (2006)
5. Chen, Y., Wang, J.Z.: Image categorization by learning and reasoning with regions. Journal of Machine Learning Research 5, 913–939 (2004)
6. Chevaleyre, Y., Zucker, J.D.: Solving multiple-instance and multiple-part learning problems with decision trees and rule sets. Application to the mutagenesis problem. In: Conference of the Canadian Society for Computational Studies of Intelligence, pp. 204–214. Springer, Heidelberg (2001)

7. Dietterich, T.G., Lathrop, R.H., Lozano-Perez, T.: Solving the multiple instance problem with axis-parallel rectangles. Artificial Intelligence 89(1–2), 31–71 (1997)
8. Foulds, J.R., Frank, E.: Revisiting multiple-instance learning via embedded instance selection. In: Proc. 21st Australasian Joint Conference on Artificial Intelligence, Auckland, New Zealand, pp. 300–310. Springer, Heidelberg (2008)
9. Freund, Y., Schapire, R.E.: Experiments with a new boosting algorithm. In: International Conference on Machine Learning, pp. 148–156. Morgan Kaufmann, San Francisco (1996)
10. Friedman, J., Hastie, T., Tibshirani, R.: Additive logistic regression: a statistical view of boosting. Annals of Statistics 28(2), 337–407 (2000)
11. Gärtner, T., Flach, P.A., Kowalczyk, A., Smola, A.: Multi-instance kernels. In: International Conference on Machine Learning, pp. 179–186. Morgan Kaufmann, San Francisco (2002)
12. Krogel, M.A., Wrobel, S.: Feature selection for propositionalization. In: International Conference on Discovery Science, pp. 430–434. Springer, Heidelberg (2002)
13. Maron, O.: Learning from ambiguity. Ph.D. thesis, Massachusetts Institute of Technology (1998)
14. Maron, O., Lozano-Pérez, T.: A framework for multiple-instance learning. In: Neural Information Processing Systems. MIT Press, Cambridge (1998)
15. Nadeau, C., Bengio, Y.: Inference for the Generalization Error. Machine Learning 52(3), 239–281 (2003)
16. Ray, S., Craven, M.: Supervised learning versus multiple instance learning: an empirical comparison. In: International Conference on Machine Learning, pp. 697–704. Omnipress (2005)
17. Reutemann, P.: Development of a Propositionalization Toolbox. Master's thesis, Albert Ludwigs University of Freiburg (2004)
18. Srinivasan, A., Muggleton, S., King, R., Sternberg, M.: Mutagenesis: ILP experiments in a non-determinate biological domain. In: Inductive Logic Programming, GMD-Studien, pp. 217–232 (1994)
19. Viola, P.A., Platt, J.C., Zhang, C.: Multiple instance boosting for object detection. In: Neural Information Processing Systems (2005)
20. Wang, J., Zucker, J.D.: Solving the multiple-instance problem: A lazy learning approach. In: International Conference on Machine Learning, pp. 1119–1125. Morgan Kaufmann, San Francisco (2000)
21. Weidmann, N., Frank, E., Pfahringer, B.: A two-level learning method for generalized multi-instance problems. In: European Conference on Machine Learning, pp. 468–479. Springer, Heidelberg (2003)
22. Witten, I.H., Frank, E.: Data Mining: Practical machine learning tools and techniques. Morgan Kaufmann, San Francisco (2005)
23. Xu, X.: Statistical Learning in Multiple Instance Problems. Master's thesis, University of Waikato (2003)
24. Xu, X., Frank, E.: Logistic regression and boosting for labeled bags of instances. In: Pacific-Asia Conference on Knowledge Discovery and Data Mining, pp. 272–281. Springer, Heidelberg (2004)
25. Zhang, Q., Goldman, S.: EM-DD: An improved multiple-instance learning technique. In: Neural Information Processing Systems, pp. 1073–1080. MIT Press, Cambridge (2002)
26. Zhang, Q., Yu, W., Goldman, S., Fritts, J.: Content-based image retrieval using multiple-instance learning. In: International Conference on Machine Learning, pp. 682–689. Morgan Kaufmann, San Francisco (2002)

Incremental Learning of Cellular Automata for Parallel Recognition of Formal Languages

Katsuhiko Nakamura and Keita Imada

School of Science and Engineering,
Tokyo Denki University, Hatoyama-machi, Saitama-ken,
350-0394 Japan
nakamura@rd.dendai.ac.jp

Abstract. Parallel language recognition by cellular automata (CAs) is currently an important subject in computation theory. This paper describes incremental learning of one-dimensional, bounded, one-way, cellular automata (OCAs) that recognize formal languages from positive and negative sample strings. The objectives of this work are to develop automatic synthesis of parallel systems and to contribute to the theory of real-time recognition by cellular automata. We implemented methods to learn the rules of OCAs in the Occam system, which is based on grammatical inference of context-free grammars (CFGs) implemented in Synapse. An important feature of Occam is incremental learning by a rule generation mechanism called bridging and the search for rule sets. The bridging looks for and fills gaps in incomplete space-time transition diagrams for positive samples. Another feature of our approach is that the system synthesizes minimal or semi-minimal rule sets of CAs. This paper reports experimental results on learning several OCAs for fundamental formal languages including sets of balanced parentheses and palindromes as well as the set $\{a^n b^n c^n \mid n \geq 1\}$.

1 Introduction

A cellular automaton (CA) is an array of regularly interconnected identical finite state machines called cells. The next state of a cell is determined by the states of its neighbor cells. Every cell synchronously updates its states. CAs have been used as theoretical models of parallel systems including biological systems and parallel computers. Among the research on CAs, parallel, real-time language recognition is especially important, because it is likely that languages are recognized in real time in the human mind, and CAs have been considered a standard parallel computation model.

In this paper, we discuss incremental learning of one-dimensional, bounded, one-way cellular automata (OCAs) that recognize formal languages. We implemented methods of learning OCAs in the Occam system. The objectives of this research are to develop automatic synthesis of parallel systems and to contribute to the theory of real-time recognition by cellular automata. This work is based on our method of incremental learning of context-free grammars (CFGs) and definite clause grammars (DCGs) implemented in Synapse [8,13,14]. In grammatical

B. Pfahringer, G. Holmes, and A. Hoffmann (Eds.): DS 2010, LNAI 6332, pp. 117–131, 2010.

inference, the learning system synthesizes production rules of a grammar from positive and negative sample strings; whereas, in learning CAs, the system synthesizes rules of local functions of CAs for recognizing a language from similar samples.

Parallel language recognition by one-dimensional CA was first investigated by Kasami et. al. [11] in the late 1960s. Real-time and linear-time language recognition by CA was introduced and discussed by Smith III [18] in 1972, and real-time recognition by OCAs by Dyer [3] in 1980. Since then, several studies including [6,21,19] have clarified the parallel recognition power of CA and OCA. Despite much investigation over more than three decades, many important problems in the limitations of recognition power still remain unsolved as described in Section 2.

We selected OCAs for learning the rule set of CAs for the following reasons.

- The state transition function of an OCA is represented by rules of the form $q\,r \to p$, which is simpler than the form $q\,r\,s \to p$ of two-way CA. This form is similar to Chomsky normal form of CFG used in Synapse and is convenient for the system to synthesize rules.
- OCAs can recognize a wide range of formal languages in real time.
- We can simulate real-time recognition in a CA by real-time recognition in an OCA as described in the next section.

The most important feature of our approach is incremental learning, in which the system can synthesize a rule set by adding rules to previously learned rules. This approach and the following heuristics are suggested by Balzer's work [1] on finding an 8-state solution for the firing squad synchronization problem.

- The system generates a rule of the form $q\,r \to p$ only when the pair q and r of states, appears in testing the transition for a positive samples.
- The system tries to generate the rule $q\,r \to p$ backward from p.
- The system first generates rules for short positive sample strings, and then produces rules for longer sample strings by adding rules.

Through incremental learning, the system can synthesize rule sets of any OCA recognizing a complex language, if we divide the language into appropriate subgroups. Incremental learning is implemented in Occam as well as in Synapse by rule generation called *bridging* and the search for rule sets. Bridging searches for any missing parts in an incomplete space-time transition diagram for a positive string, and synthesizes rule(s) that bridge the missing part.

Another important feature of our approach is that the learning system synthesizes minimal, or semi-minimal, rule sets. The small sets of rules are able to represent the mechanisms of CAs and the structures of the languages. Another benefit is the ease of checking the correctness of the synthesized rule sets.

Other than Balzer's work on the firing squad synchronization problem, several studies have been conducted on learning CAs including H. Juillé, Mitchell et al. [12] and F.C. Richard [16]. These works use a genetic algorithm to synthesize CAs and are different from our work in their objectives and methodology of learning.

This paper is organized as follows. Section 2 briefly describes definitions of CAs and OCAs and real-time language recognition. It also surveys previous researches on their parallel recognition power. Sections 3 and 4 describe the bridging rule generation procedure, search strategy and some extensions and heuristics in Occam system for synthesizing rule sets. Section 5 shows experimental results from learning CAs that recognize several formal languages by Occam and compares the results with those of learning CFGs by Synapse and those of SAT-based approach. Finally, Section 6 concludes the paper and describes future subjects of research.

2   Recognition of Languages by CA

2.1   CA, OCA and Their Language Recognition

A (two-way) *one-dimensional bounded cellular automaton* (*CA*) is a system $S = (K, \#, f, A)$, where:

- K is a finite set of *states*;
- $\# \in K$ is the *boundary state*;
- f, a *transition function*, is a mapping from $K \times K \times K$ into $K - \{\#\}$; and
- $A \subseteq K$ is a set of *accepting states*.

A *one-way cellular automaton* (*OCA*) is a CA $S = (K, \#, f, A)$ such that there exists a function $f' : K \times K \to K$ with $f(x, y, z) = f'(x, y)$ for all $x, y, z \in K$.

Informally, the next state of each cell i in a CA is determined by the states of its left and right neighbor cells and the state of the center cell i. In an OCA, the next state is determined by the states of its left neighbor cell and the center cell.

Any global state of the CA, called a *configuration*, is represented by a string $\#u\#$ of states with $u \in (K - \{\#\})^+$. The transition function f is extended to configurations by

$$f(\#a_1a_2 \cdots a_n\#) = \#b_1b_2 \cdots b_n\#$$

where $b_i = f(a_{i-1}, a_i, a_{i+1})$ for all $i \in \{1, 2, 3, \cdots, n\}$, and $a_0 = a_{n+1} = \#$. The function f^n of configurations is recursively defined by $f^0(c) = c$ and $f^n(c) = f(f^{n-1}(c))$, for all configurations c and $n \geq 0$.

A CA $S = (K, \#, f, A)$ *recognizes* a language $L \subset \Sigma^+$ in a *time* t (*by the rightmost cell*), if and only if $\Sigma \subseteq K$ and for all $w \in \Sigma^+, |w| \geq 2$, and

$$w \in L \Leftrightarrow f^t(\#w\#) \in \{\#\}K^+A\{\#\}.$$

The CA S recognizes L in *real time*, if and only if $t = |w| - 1$. The CA S recognizes L in *linear time*, if and only if there is a constant $c \geq 1$ with $t \leq c \cdot (|w| - 1)$. The languages recognized by a CA and an OCA are called a CA language (CAL) and an OCA language (OCAL), respectively.

The language recognized *by the leftmost cell*, called the *left* CAL or *left* OCAL, is defined similarly, and is equivalent to the reversal of the language recognized by the rightmost cell.

A two-way CA $S = (K, \#, f, A)$ recognizes a language $L \subseteq \Sigma^+$ in real time *by the center cell*, or in *minimal time*, if and only if $\Sigma \subseteq K$ and for all $w \in \Sigma^+, |w| \geq 2$, and

$$w \in L \Leftrightarrow f^{|w|/2}(\#w\#) \in \begin{cases} \{\#\}K^{|w|/2}A \, K^{|w|/2} \, \{\#\} & \cdots \text{ if } |w| \text{ is even.} \\ \{\#\}K^{|w|/2-1}AA \, K^{|w|/2-1} \, \{\#\} \cdots & \text{ if } |w| \text{ is odd.} \end{cases}$$

For real-time recognition by a CA, we can write $f^i(a_1a_2 \cdots a_n\#) = b_{i+1} \cdots b_n\#$ for all $i \in \{1, 2, \cdots, n-1\}$. For real-time recognition by an OCA, we also write $f^i(a_1a_2 \cdots a_n) = b_{i+1} \cdots b_n$.

Figs. 1 and 2 illustrate language recognition by CAs and OCAs. Among these several forms of recognition, the following facts [2,21] are especially important for the recognition by Occam.

1. For any CA S, we can construct an OCA S' such that for any language L, if S accepts L in real (minimal) time by the center cell, then S' recognizes L in real time and vice versa. This fact implies that the class of of languages recognized by OCA in real-time is closed under reversal.
2. For any CA S, we can construct an OCA S' such that for any language $L \subseteq \Sigma^+$, if S accepts a string $w \in L$ in real time, then S' accepts $w\,x^{|w|}$ with a symbol $x \notin \Sigma$ in real time. Note that we can transform the synthesized OCA into the CA.

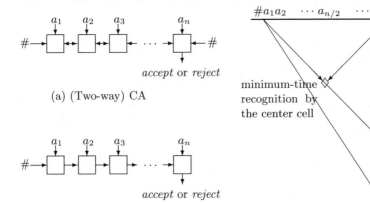

Fig. 1. Language recognition of CA and OCA by rightmost cells

Fig. 2. A space-time diagram illustrating minimum-time, real-time and linear-time recognition

2.2   Language Recognition Power of CA and OCA

Let $\mathcal{C}(L)$ denote the class of languages L. The results of previous research closely related to this work are summarized as follows.

1. $\mathcal{C}(\text{CAL})$ is equal to $\mathcal{C}(\text{deterministic context sensitive language})$ [11,17]. $\mathcal{C}(\text{OCAL})$ includes $\mathcal{C}(\text{context-free language})$ [11].
2. $\mathcal{C}(\text{real-time OCAL})$, as well as $\mathcal{C}(\text{real-time CAL})$, includes non-context-free languages such as $\{a^n b^n c^n \mid n \geq 1\}$ [3,17]. $\mathcal{C}(\text{real-time CAL})$ includes complex languages such as $\{a^n \mid n \text{ is a prime number}\}$ [5,17].
3. Both $\mathcal{C}(\text{real-time OCAL})$ and $\mathcal{C}(\text{real-time CAL})$ are closed under set operations [17].
4. $\mathcal{C}(\text{linear-time CAL})$ is closed under reversal [17].
5. $\mathcal{C}(\text{minimum-time CAL}) = \mathcal{C}(\text{real-time OCAL})$ and $\mathcal{C}(\text{real-time CAL}) = \mathcal{C}(\text{linear-time left OCAL})$ [2,21].

The relation between the real-time language recognition power of CA and OCA is especially important. The following languages are shown to be recognized by CA but not by OCA in real time.

- $\{1^{2^n} \mid n \geq 0\}$ [2].
- $\{uvu \mid u, v \in \{0,1\}^*, |u| \geq 1\}$ [19].
- the language $L_0 L_0$, where $L_0 = \{1^j 0^j \text{ or } 1^j 0y10^j \mid y \in \{0,1\}^*, j > 0\}$ [20].

As $L_0 L_0$ is a context-free language, $\mathcal{C}(\text{real-time OCAL})$ is not closed under concatenation, because L_0 is a real-time OCAL. Nevertheless, the limitation of the real-time language recognition power of OCAs has not been sufficiently clarified. Nakamura [15] showed a pumping lemma for recognition of cyclic strings by OCAs.

3   Learning Rule Sets of OCAs in Occam

This section describes the methods of learning the rules of OCAs implemented in Occam. The system is written in Prolog and is composed of rule generation and search for semi-minimal rule sets.

3.1   Representation of Cell States

In Occam, we represent cell states as follows.

1. The input symbols $s_1, s_2, \cdots, s_m$ are represented by the numbers $2, 3, \cdots, m$ respectively.
2. Each state of a cell is represented by a subset S of a set $\{1, 2, \cdots, M\}$ with $M \geq m$ (M is called the *max state*) and $1 \leq |S| \leq k$ for a constant k, called the *size* of the state. The state with $k \geq 2$ is called the *multiple state*, and the state with $k = 1$ is called the *singleton state*.
3. The state containing the number 1 is an accepting state.

A rule is of the form $qr \rightarrow p$, where q, r and p are numbers between 1 and M. The set of states of the OCA is a subset of the power set $2^{\{1,2,\cdots,M\}}$. For a set RS of rules, the transition function is represented by

$$f(Q, R) = \{p \,|\, (qr \rightarrow p) \in RS, q \in Q, r \in R\},$$

for any states Q and R.

One reason for using multiple states is that two or more accepting states are necessary for many CAs. We cannot construct some CAs (e.g., one that recognizes palindromes) with only one accepting state. Another reason for using multiple states is that we can extend the set of acceptable strings by simply adding rules. Using multiple states is related to the common method of using multiple layers in constructing a CA with some particular capability.

3.2   Rule Generation

The rule generation procedure first tries to test whether the current rule set derives an accepting state from an input string by generating a space-time transition diagram. If the test fails, then the bridging process generates rules for the OCA, which bridge any missing parts of the incomplete space-time transition diagram. The space-time transition diagram for an input string $a_1 a_2 \cdots a_n$ is represented by a triangular array of the states with indexes (i, j) where $0 \le j \le n-1$ and $j + 1 \le i \le n$.

Fig. 3 shows the rule generation procedure. This nondeterministic procedure receives a string $w = a_1 \cdots a_n$, a set S_N of negative samples and a set of rules in the global variable RS from the top-level search procedure, and returns a set of rules for an OCA, which derives the accepting state 1 from the string w but does not from any string in S_N. For a set RS of rules, each state of the array is determined by the operations in Step 2 of *GenerateRule*. The procedure *Bridge* looks for missing states in the triangular array T from the bottom $T[n, n-1]$ with the accepting state 1, and complete a space-time diagram in T by adding rules to RS that is consistent with the negative samples in S_N.

3.3   Searching for Rule Sets

Occam takes as input an ordered set S_P of positive samples, an ordered set S_N of negative samples and a set R_0 of optional initial rules for incremental learning. The system searches for any set RS of rules with $R_0 \subseteq RS$ that derives an accepting state from all of the strings in S_P but from no string in S_N.

The system scans every node in the search tree within a certain depth to find the minimal set of rules by using iterative deepening with the number of the rules. Fig. 4 shows the top-level global search procedure. The system controls the search by iteratively deepening the tree, the depth of which is the number K of rules. This control ensures that the procedure finds the minimal rule set, but the trade-off is that the system repeats the same search each time the limit is increased.

Procedure $GenerateRule(w, S_N)$ (*Comment: Generates rules and add them to the global variable RS that derives an accepting state from the string w but not from any negative sample in S_N.*)

Step 1 (*Initialize a triangular array T of states for $w = a_1 \cdots a_n$.*)
$T[i, 0] \leftarrow \{a'_i\}$ for all $1 \leq i \leq n$ (a'_i is the number for a_i).
$T[i, j] \leftarrow \emptyset$ for all $1 \leq j \leq n - 1, j + 1 \leq i \leq n$.

Step 2: (*Evolve the space-time diagram by the set of rule in RS.*)
For $j = 0$ to $n - 1$, for $i = j + 1$ to n,
$T[i, j] \leftarrow \{p \mid (q\,r \rightarrow p) \in RS, q \in T[i - 1, j - 1], r \in T[i, j - 1]\}$.
If $1 \in T[n, n - 1]$ then return.

Step 3: (*Bridging rule generation*)
Call $Bridge(1, T, n, n + 1, S_N)$. Return.

Procedure $Bridge(p, T, i, j, S_N)$ (*Complete a space-time diagram in the triangular array T with an element p in (i, j) by adding rules to RS that is consistent with negative sample in S_N. Note: The argument p is either a number or a variable.*)

Step 1: If $j = 0$, assign the number in $T[i, 0]$ to p; Return.
Else if $j \geq 1$, nondeterministically choose one of the following steps in order.
  1. If p is a variable, assign a number in $T[i, j]$ to p; Return.
  2. Proceed the next step.

Step 2: (*Find the numbers q and r.*)
Call $Bridge(X_q, T, i - 1, j - 1, S_N)$ to find q for the variable X_q.
Call $Bridge(X_r, T, i, j - 1, S_N)$ to find r for the variable X_r.

Step 3: (*Determine the number p.*)
Nondeterministically choose one of the following steps in order.
  1. If $(qr \rightarrow p) \in RS$, Return;
  2. If p is a variable, assign a number to p with $2 \leq p \leq max\_state$ (max_state is a predetermined parameter), and proceed the next step.

Step 4: (*Test a generated rule and add it to RS.*)
Add the rule $(qr \rightarrow p)$ to RS.
If S_N contains a string w such that RS derive the accepting state from w, then terminate (failure).

Step 5: (*Evolve the space-time diagram by RS.*)
For $j = 0$ to $n - 1$, for $i = j + 1$ to n,
$T[i, j] \leftarrow \{p \mid (q\,r \rightarrow p) \in RS, q \in T[i - 1, j - 1], r \in T[i, j - 1]\}$.
Return.

Fig. 3. Nondeterministic procedure for rule generation by bridging

---

Procedure $GlobalSearch(S_P, S_N, R_0)$ (*Comment: Finds a rule set that derives the accepting state from each positive sample in S_P but not from any negative sample in S_N. R_0 : a set of optional initial rules.*)

Step 1 (*Initialize variables.*)
$RS \leftarrow R_0$ (*The global variable RS holds the set of rules.*).
$K \leftarrow |R_0|$ (*The limit on the number of rules for iterative deepening.*).

Step 2: For each $w \in S_P$, call $GenerateRule(w, S_N)$.
If no set of rules is obtained within the limit $K \geq |RS|$, then add 1 to K and iterate this step.

Step 3: Output the rules in RS and terminate (Success).
To find multiple solutions, backtrack to the previous choice point.

Fig. 4. Top-level procedure for searching for rule sets in Occam

3.4   Example: Generation of Rules for Parentheses Language

Consider the learning of an OCA recognizing the balanced parenthesis language, that is, the set of strings composed of equal numbers of a's and b's such that every prefix does not have more b's than a's. For the first positive sample ab, the call $Bridge(1, T, 2, 1, S_N)$ synthesizes rule $2, 3 \rightarrow 1$, where a and b are represented by 2 and 3, respectively. For the second sample $aabb$, Step 2 of $GenerateRule$ generates the incomplete space-time diagram shown in Fig. 5 (a), where GR is the number of all generated rules. For this incomplete diagram, the operations of $Bridge$ synthesize four rules,

$$2, 2 \rightarrow 2; \quad 3, 3 \rightarrow 3; \quad 2, 1 \rightarrow 2 \text{ and } 1, 2 \rightarrow 3,$$

which satisfy no negative sample and complete the diagram shown in (b). Then, Occam tries to synthesize rules that satisfy all remaining positive samples but no negative samples. After generating 48 rules, the system fails in completing this job and backtracks to generate new rules for the sample $aabb$. The diagram in (c) is the result.

Fig. 6 shows the sequence of space-time diagrams generated by Occam that lead to the solution. Each marked state <s> in the diagrams denotes that a new rule for this state s is generated. After generating 66 rules, the system finds 13 rules, which satisfy all positive rules but no negative rules.

(a) GR = 1	(b) GR = 5	(c) GR = 48
[2] [2] [3] [3]	[2] [2] [3] [3]	[2] [2] [3] [3]
∅ [1] ∅	[2] [1] [2]	[2] [1] [3]
∅ ∅	[2] [3]	[2] [3]
[1]	[1]	[1]

Fig. 5. Sequence of space-time diagrams generated for string $aabb$

4   Extensions and Heuristics in Occam

To increase the power to learn complex OCAs, Occam adopts several heuristics and extensions.

4.1   Don't Care State in Rules

For reducing the number of rules, we incorporate *don't care states* into Occam. The special element 0 of states occurs only as one of the two elements in the left side of the rules and matches any number of any state. For example, the rule "$3, 0 \rightarrow 5$" applies to the pairs of numbers $(3, 1), (3, 2), (3, 3), \cdots$, and the rule "$0, 6 \rightarrow 7$" applies to $(1, 6), (2, 6), (3, 6), \cdots$.

(a) *aabb* (b) *abab* (c) *ababab* (d) *aaabbb*

```
[2] [2] [3] [3]     [2] [3] [2] [3]     [2] [3] [2] [3] [2] [3]     [2] [2] [2] [3] [3] [3]
   <2>[1]<3>           [1]<4>[1]           [1] [4] [1] [4] [1]         [2] [2] [1]<3>[3]
    <2><3>              <2><3>               [2] [3] [2] [3]             [2] [2]<3>[3]
     [1]                 [1]                  [1] [4] [1]                 [2] [1] [3]
                                               [2] [3]                     [2] [3]
                                                [1]                         [1]
```

(e) *aabbab* (f) *aabbaabb* (g) *aabaabbbab*

```
[2] [2] [3] [3] [2] [3]   [2] [2] [3] [3] [2] [2] [3] [3]   [2] [2] [3] [2] [2] [3] [3] [3] [2] [3]
   [2] [1] [3] [4] [1]       [2] [1] [3] [4] [2] [1] [3]       [2] [1] [4] [2] [1] [3] [3] [4] [1]
    [2] [3]<4>[3]             [2] [3] [4]<4>[2] [3]             [2] [2] [4] [2] [3] [3] [4] [3]
     [1] [4]<3>               [1] [4]<4>[4] [1]                 [2] [2] [4] [1] [3] [4] [3]
      [2] [3]                  [2] [4] [4] [3]                   [2] [2] [3] [3] [4] [3]
       [1]                      [2] [4] [3]                       [2] [1] [3] [4] [3]
                                 <2>[3]                            [2] [3] [4] [3]
                                  [1]                               [1] [4] [3]
                                                                     [2] [3]
                                                                      [1]
```

Fig. 6. Development of space-time diagrams in learning the parenthesis language (each symbol <s> denotes that a new rule for the state s is generated.)

At present, Occam cannot synthesize rules with the don't care state and only inputs initial rules of this type. We use the following initial rules in learning the sets of palindromes and strings with same number of *a*s and *b*s, shown in the next section.

$$2, 0 \rightarrow 4; \quad 3, 0 \rightarrow 5; \quad 4, 0 \rightarrow 4; \quad 5, 0 \rightarrow 5$$
$$0, 2 \rightarrow 6; \quad 0, 3 \rightarrow 7; \quad 0, 6 \rightarrow 6; \quad 0, 7 \rightarrow 7$$

These rules are used for propagating symbolic information from the input strings. The symbols *a* and *b* are represented by 2 and 3, respectively, and then by numbers 4 and 5, which move diagonally from left to right in the space-time diagram. The symbols *a* and *b* are represented also by numbers 6 and 7, respectively, which move vertically in the space-time diagram.

4.2   Hash Tables for Speeding-Up Search

Use of hash tables is a well-known technique in search programs, especially game-playing programs, to avoid identical partial search. We incorporate this technique into Occam as follow.

– A table contains hash codes of rule sets.
– Whenever the system generates a new rule, it checks whether the table contains the hash code of the set containing this rule. Unless the table has this code, the system adds the hash code to the table and at the same time adds the rule to the rule set.

We represent the hash code for the set of rules $R_1, R_2, \cdots R_k$ by $c(R_1) \oplus c(R_2) \oplus \cdots \oplus c(R_k)$, where $c(R)$ is the hash code of a rule R. By this method, we can obtain a hash code of a rule set by simply calculating the code of a rule R_k and the exclusive OR of this code and the code of the existing rule set $\{R_1, R_2, \cdots R_{k-1}\}$

By using this method, the overhead for hashing is small (usually a few percent). For finding OCAs with multiple rules set, the computation time is reduced to at most the order of one tenth in the case the limits on the numbers of rules and states are highly restricted.

4.3 Enumerating and Deriving Strings from Rule Sets

To examine synthesized OCAs, Occam has a function to enumerate strings from sets of rules. The system first generates the list $[X_1, X_2, \cdots, X_n]$ of variables in order of their length n, and then, it derives strings matching the lists from the rule set.

We can also use this function for deriving strings from rule sets to represent negative samples from their patterns. Some languages, for example, $\{a^n b^n c^n \mid n \geq 1\}$, require a large number of strings for negative samples. From the list $[\_,\_,\_,\_]$ of four anonymous variables in Prolog, we can represent the pattern of any string with length four such as $aaaa, aaab, aaba, \cdots, bbbb$, and by $[b,\_,\_,\_,\_,\_]$, the pattern of strings with the length six starting with b. For some languages, using the patterns of negative samples is very effective at reducing the number of negative samples and computation time.

5 Experimental Results

This section shows experimental results obtained by Occam Version 1.63 written in Prolog, using an Intel Core(TM) DUO processor with a 2.93 GHz clock and SWI-Prolog for Windows. We checked the correctness of several synthesized rule sets with a large number of samples and by making the rule sets enumerate strings. We used the number GR of all generated rules as an index of the size of the search tree, which does not depend on the number of samples and the processor environment as the computation time. In general, the number of samples does not strongly affect computation time provided that sufficient samples are given to the system. When the system succeeds in learning an OCA after using a number of positive samples for generating rules, it uses the remainder of the samples for checking the rule set.

5.1 Learning OCAs Results

Table. 1 shows the computation time and the sizes of rule sets for learning the rule sets of OCAs for fundamental languages including:

(a) the set $\{a^n b^n \mid n \geq 1\}$,
(b) the balanced parenthesis language,

Table 1. Synthesized rules for OCAs recognizing fundamental languages

language	space-time diagrams	the number of rules	the size of states	the number of states	GR	time in seconds
(a) $\{a^n b^n \mid n \geq 1\}$	Fig. 7 (1)	5	1	4	7	0.14
(b) parenthesis language	Fig. 6	13	1	4	66	0.53
(c) $\{a^n b^n c^n \mid n \geq 1\}$	Fig. 7 (2)	11	1	6	6663	90
(d) $\{a^n b^n \mid n \geq 1\}$	Fig. 8	21	6	12	8555	357
(e) set of palindromes	Fig. 9	12+8*	5	22	267	7.4
(f) $\{w \mid \#_a(w) = \#_b(w)\}$	–	12+8*	5	21	93	2.6

* The number of the initial rules.

(c) the set $\{a^n b^n c^n \mid n \geq 1\}$, a non-context-free language,
(d) the set $\{a^n b^n \mid n \geq 1\}$, which is the complement of language (a),
(e) the set of palindromes over $\{a, b\}$, and
(f) the set $\{w \mid \#_a(w) = \#_b(w)\}$, i.e., the set of strings with same number of as and bs.

The computation time and the numbers GR were obtained by setting the limits K in Fig. 4 on the numbers of rules to appropriate values. The search by the iterative deepening beginning with $K = 1$ needs several times more time and GR values.

Examples of state transitions for each OCA are shown in Figs. 6, 7, 8 and 9. The states for (a), (b) and (c) are singleton numbers with the size one, but those for (d) and (e) are multiple states with the size 6 and 5, respectively. We added the eight initial rules with the don't care states, shown in Section 4.1, to the OCAs for languages (e) and (f).

We gave Occam approximately 50 to 100 strings of negative and positive samples for each learning problem, except (c) that requires a small number (less than 13) of positive samples, abc, $aabbcc$, $aaabbbccc$, $\cdots$ but more than 200 negative samples. The positive samples are the first part of strings in the enumeration of the language in the order of length. The negative samples are those of the complement of the language.

Figs. 7 shows space-time diagrams of OCAs for the languages (a) and (c) in a compact format. The OCA for the language $\{a^n b^n \mid n \geq 1\}$ uses a signal, which is represented by a sequence of state changes with speed $1/2$ cells/time as shown in diagram (1). Diagrams (2) and (3) represent two different mechanisms of recognizing the set $\{a^n b^n c^n \mid n \geq 1\}$ by the OCAs obtained by Occam. Note that the OCA shown in (2) uses two kinds of signals with speeds $2/3$ and $1/3$; whereas, the other OCA (3) uses three kinds of signals with speeds 1, $1/2$ and $1/3$.

As the class of OCA languages are closed with Boolean operations, we can simply convert the OCA for the language (a) $\{a^n b^n \mid n \geq 1\}$ into an OCA for the language (d) by replacing the accepting state 1 by states other than 1. At this moment, however, Occam cannot find this simple rule set, and synthesized

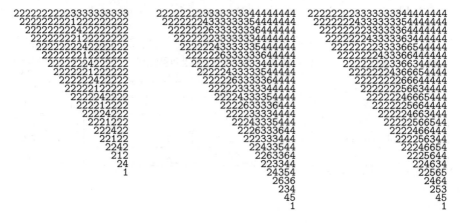

(1) *aaaaaaaaaaabbbbbbbbbb*   (2) *aaaaaaaaabbbbbbbbbcccccccc* (3) *aaaaaaaaabbbbbbbbbccccccccc*

Fig. 7. Space-time diagrams for recognizing strings of $\{a^n b^n \mid n \geq 1\}$ and $\{a^n b^n c^n \mid n \geq 1\}$ by synthesized rule sets

time								
0	[2]	[3]	[2]	[2]	[2]	[3]	[3]	[3]
1		[2]	[1,3,4,6]	[1,3]	[1,3]	[2]	[1,4]	[1,4]
2			[1,2]	[1,2,4,5]	[1,2,4,5]	[1,3,4,6]	[1]	[1,2,4]
3				[1,2,3,5]	[1,2,3,5]	[1,2,3,5]	[5]	[1,5]
4					[1,2,3,4,5,6]	[1,2,3,4,5,6]	[1]	[1]
5						[1,2,3,4,5,6]	[1,5]	[5]
6							[1,5]	[1]
7								[1,5]

Fig. 8. A space-time diagram for recognizing the string *abaaabbb* $\in \overline{\{a^n b^n \mid n \geq 1\}}$ (language (d)). (The states other than [2] and [5] contain 1 and are accepting states.).

21 rules for this language with 12 multiple states $[1], [2], [3], [5], [1,3], [1,4], [1.5],$ $[1,2,5], [1,2,3,5], [1,2,4,5], [1,3,4,6]$ and $[1,2,3,4,5,6]$.

5.2   Comparison with Learning CFGs

The results of learning rule sets of OCAs with Occam can be compared to those of learning CFGs with Synapse, which uses methods similar to Occam's for synthesizing rules. The results are summarized as follows.

- The CFGs have generally fewer rules in Chomsky normal form. For example, for the languages (a), (b), (d) and (e), Synapse synthesized both ambiguous and unambiguous CFGs, and the numbers of rules for the unambiguous CFGs are 3, 6, 11 and 10, respectively.
- Synapse synthesizes these grammars in shorter time, less GR and needs fewer samples. Language (d) is an exception: Synapse needs 185 seconds and

```
time
  0    [3]    [2]     [2]       [3]       [3]       [2]       [2]       [3]
  1         [5,6,9] [1,4,6,8] [4,7,10] [1,5,7,8]  [5,6,9] [1,4,6,8] [4,7,10]
  2                  [5,6]   [4,7,10]    [4,7]    [5,6,9]   [5,6]   [4,7,10]
  3                          [1,5,7]     [4,7]    [1,4,6]   [5,6]    [1,5,7]
  4                                    [5,7,10]    [4,6]   [4,6,9]    [5,7]
  5                                                [5,6]   [1,4,6]    [4,7]
  6                                                        [5,6]   [4,7,10]
  7                                                                 [1,5,7]
```

Fig. 9. A space-time diagram for recognizing a palindrome *baabbaab* in language (e). (The subsets containing 1 represent accepting states.).

generates more than 4×10^4 rules (GR) before synthesizing a CFG with 11 rules.

- Synapse can learn more complicated context-free languages such as the set of strings of *a*s and *b*s not of the form *ww* and the set of strings that have twice the number of *a*s than *b*s. It has not been shown that these two languages are real-time OCAL.

The reason for these differences is that Synapse synthesizes the grammar rules of the language; whereas, Occam synthesizes the parallel parser, or accepter, of the language, which recognizes all possible syntactic structures of the strings. Another difference is that Occam can synthesize non-context-free languages such as $a^n b^n c^n$, whereas Synapse is extended to learn the definite clause grammars (DCGs) of this language.

5.3   Comparison with Learning CAs by SAT

In addition to this work, we are currently working on another approach to learning CAs from samples based on the Boolean satisfiability problem (SAT). We used a similar method to synthesize minimal rule sets of CFGs from positive and negative samples [9].

In the SAT-based approach, the problem of synthesizing an OCA is represented by a Boolean formula using the following two predicates: $Rule(q, r, p)$ which means that there is a rule $q\,r \to p$ and $State(w, q)$ with the meaning that q is derived from the input string w. The Boolean formula describes positive and negative samples and how $State(w, q)$ is computed. By using a SAT solver, we obtained a solution consisting of a rule set as an assignment to variables in a table representing the local function of the OCA.

Currently, this SAT-based method synthesizes only rule sets recognizing languages (a) and (b) in the previous section from similar samples. Each of the rule sets is similar to the one that Occam has synthesized and the computation time is a little longer. An advantage of the SAT-based approach is that we can utilize the progress of SAT solvers. An advantage of Occam over the SAT-based approach is the use of incremental learning.

6   Conclusion

In this paper, we described methods for incremental learning of rule sets for one-way CAs (OCAs), recognizing formal languages and showed some experimental results. Although we have currently obtained a small number of rule sets for OCAs with the Occam system, the results are encouraging. The experimental results can be summarized as follows.

- Occam synthesized several minimal, or semi-minimal, rule sets of OCAs that recognize fundamental formal languages. The OCAs for the languages (d), (e) and (f) in the previous section would be the first nontrivial parallel systems synthesized by machine learning, which are not easy for human experts to construct.
- These results are comparable to those of learning CFGs by Synapse, which uses methods similar to Occam. Compared to learning CFGs, learning CAs generally requires more computation time and larger numbers of samples.
- There remain several simple languages that Occam has not synthesized because of computation time limitations.

Our work is related to the problem of how to construct, or program, parallel systems such as CAs. Although this problem has not been sufficiently investigated in general, constructing parallel systems seems to be more difficult than sequential systems with similar number of elements or rules. Machine learning is potentially an important approach to synthesizing parallel systems.

An approach to solve the computation time problem is incremental learning so that we divide the positive samples into subgroups and make the system learn the samples in the subgroups in order. We expect that it is possible to effectively learn complex OCAs by incremental learning with appropriate partitioning of the positive samples and ordering of the subgroups.

The most important future subject is to improve Occam so that the system can learn more complicated OCAs, recognize other languages including subsets of natural languages, and to extend the learning methods to OCAs with different types of recognition. Because languages recognized by OCAs in two times real time are equivalent to those by CA in real time as described in Section 2.1, this extension is not difficult. Other important future subjects include the following:

- applying our approaches to the learning of other CAs, especially, two-dimensional CAs for parallel recognition of two-dimensional patterns; and
- making good use of extended Occam to clarify parallel recognition power of OCAs and CAs.

Acknowledgments

The author would like to thank Yuki Kanke and Toshiaki Miyahara for their help in testing the Occam system. This work is partially supported by KAKENHI 21500148 and the Research Institute for Technology of Tokyo Denki University, Q09J-06.

References

1. Balzer, R.: An 8-State Minimal Time Solution to the Firing Squad Synchronization Problem. Information and Control 10, 22–42 (1967)
2. Choffrut, C., Culik II, K.: On real-time cellular automata and trellis automata. Acta Informatica 21, 393–407 (1984)
3. Dyer, C.: One-way bounded cellular automata. Inform. and Control 44, 54–69 (1980)
4. Hopcroft, J.E., Ullman, J.E.: Introduction to Automata Theory, Languages, and Computation. Addison-Wesley, Reading (1979)
5. Fisher, P.C.: Generation of primes by a one-dimensional real-time iterative array. Jour. of ACM 12, 388–394 (1965)
6. Ibarra, O.H., Palis, M.P., Kim, S.M.: Fast parallel language recognition by cellular automata. Theoret. Comput. Sci. 41, 231–246 (1985)
7. Ibarra, O.H., Jiang, T.: Relating the power of cellular arrays to their closure properties. Theoret. Comput. Sci. 57, 225–238 (1988)
8. Imada, K., Nakamura, K.: Towards Machine Learning of Grammars and Compilers of Programming Languages. In: Daelemans, W., Goethals, B., Morik, K. (eds.) ECML PKDD 2008, Part II. LNCS (LNAI), vol. 5212, pp. 98–112. Springer, Heidelberg (2008)
9. Imada, K., Nakamura, K.: Learning Context Free Grammars by SAT Solvers, Internat. Conf. on Machine Learning and Applications, IEEE DOI 10.1109/ICMLA, 28 267, 267-272 (2009)
10. Juillé, H., Pollack, J.B.: Coevolving the 'ideal' trainer: Application to the discovery of cellular automata rules. In: Koza, J.R., et al. (eds.) Proceedings of the Third Annual Conference on Genetic Programming 1998, pp. 519–527. Morgan Kaufmann, San Francisco (1998)
11. Kasami, T., Fujii, M.: Some results on capabilities of one-dimensional iterative logical networks. Electrical and Communication 51-C, 167–176 (1968)
12. Mitchell, M.: Crutchfield, J. P. and Hraber, P. T., Evolving cellular automata to perform computations: mechanism and impediments. Physica D 75, 361–391 (1994)
13. Nakamura, K., Matsumoto, M.: Incremental Learning of Context Free Grammars Based on Bottom-up Parsing and Search. Pattern Recognition 38, 1384–1392 (2005)
14. Nakamura, K.: Incremental Learning of Context Free Grammars by Bridging Rule Generation and Semi-Optimal Rule Sets. In: Sakakibara, Y., Kobayashi, S., Sato, K., Nishino, T., Tomita, E. (eds.) ICGI 2006. LNCS (LNAI), vol. 4201, pp. 72–83. Springer, Heidelberg (2006)
15. Nakamura, K.: Real-time recognition of cyclic strings by one-way and two-way cellular automata. IEICE Trans. of Information and Systems E88-D, 171–177 (2005)
16. Richards, F.C., Meyer, T.P., Packard, N.H.: Extracting cellular automaton rules from experimental data. Physica D 45, 189–202 (1990)
17. Smith III, A.R.: Real-time language recognition by one-dimensional cellular automata. Jour. Comput. and System Sci. 6, 233–253 (1972)
18. Smith III, A.R.: Cellular automata complexity trade-offs. Inform. and Control 18, 466–482 (1971)
19. Terrier, V.: On real time one-way cellular array. Theoret. Comput. Sci. 141, 331–335 (1995)
20. Terrier, V.: Languages not recognizable in real time by one-way cellular automata. Theoret. Comput. Sci. 156, 281–287 (1996)
21. Umeo, H., Morita, K., Sugata, K.: Deterministic one-way simulation of two-way real-time cellular automata and its related problems. Information Process. Lett. 14, 159–161 (1982)

Sparse Substring Pattern Set Discovery Using Linear Programming Boosting

Kazuaki Kashihara, Kohei Hatano, Hideo Bannai, and Masayuki Takeda

Department of Informatics, Kyushu University
{kazuaki.kashihara,hatano,bannai,takeda}@inf.kyushu-u.ac.jp

Abstract. In this paper, we consider finding a small set of substring patterns which classifies the given documents well. We formulate the problem as 1 norm soft margin optimization problem where each dimension corresponds to a substring pattern. Then we solve this problem by using LPBoost and an optimal substring discovery algorithm. Since the problem is a linear program, the resulting solution is likely to be sparse, which is useful for feature selection. We evaluate the proposed method for real data such as movie reviews.

1 Introduction

Text classification is an important problem in broad areas such as natural language processing, bioinformatics, information retrieval, recommendation tasks. Machine Learning has been applied to text classification tasks in various ways: SVMs and string kernels (n-gram kernels, subsequence kernels [15], mismatch kernels [14]) Boosting (e.g., Boostexter [21]).

In some applications regarding texts, not only classification accuracy but also what makes classification accurate is important. In other words, one might want to discover some knowledge from an accurate text classifier as well. For example, in classification task of biosequences, say, DNA or RNA, biologists want to know patterns in the data which make each sequence positive other than an accurate classifier. Simply put, one may want an accurate classifier associated with a set of patterns in the text. In particular, for the purpose of feature selection, it is desirable that such a set of patterns is small.

In this paper, we formulate the problem of finding a small set of patterns which induces an accurate classifier as 1-norm soft margin optimization over patterns. Roughly speaking, this problem is finding a linear combination of classifiers associated with patterns (or a hyperplane whose each component corresponds to a pattern) which maximizes the margin w.r.t. the given labeled texts as well as minimizing misclassification.

Our formulation has two advantages. The first advantage is accuracy of the resulting classifier. The large margin theory guarantees that linear classifier with large margin is likely to have low generalization error with high probability [20]. So, by choosing the class of patterns appropriately, solving the problem would

B. Pfahringer, G. Holmes, and A. Hoffmann (Eds.): DS 2010, LNAI 6332, pp. 132–143, 2010.

provide us an enough accurate classifier. The second advantage is that the resulting solution is often sparse since the 1-norm soft margin optimization is a linear program. In other words, many of patterns have zero weights in the obtained linear combination. This would help us to choose a small subset of patterns from the resulting classifier.

We solve the 1-norm soft margin optimization over patterns by combining LPBoost [4] and our pattern discovery algorithm. LPBoost is a boosting algorithm which provably solves the 1-norm soft margin optimization. Given a weak learning algorithm which outputs a "weak hypothesis", LPBoost iteratively calls the weak learning algorithm w.r.t. different distributions over training texts and obtains different weak hypotheses. Then it produces a final classifier as a linear combination of the weak hypotheses. In this work, we use our pattern discovery algorithm as the weak learning algorithm for LPBoost.

The pattern class we consider in this paper is that of all the possible substrings over some alphabet Σ. For substring patterns, we derive an efficient pattern discovery algorithm. A naive algorithm enumerates all the possible substrings appearing in the input texts and takes $O(N^2)$ time, where N is the length of total texts. On the other hand, ours runs in time $O(N)$. Our approach can be further extended by employing pattern discovery algorithms for other rich classes such as subsequence patterns [6] or VLDC patterns [9], which is our future work (See Shinohara's survey [22] for pattern discovery algorithms).

In our preliminary experiments, we apply our method for classification of movie reviews. In particular, for our data Movie-A, there are about 6×10^{13} possible substrings patterns. Our method outputs a classifier associated with a small set of substrings whose size is only about 800. Among such 800 patterns, we find interesting pattern candidates which explain positive and negative reviews.

Let us review some related researches. The bag of words model (BOW) has been popular in information retrieval and natural language processing. In this model, each text is regarded as a set of words appearing in the text, or equivalently, a weight vector where each component associates with a word and the value of each component is determined by the statistics of the word (say, frequency of the word in the text). The BOW model is often effective in classification of natural documents. However, we need to determine a possible set of words in advance, which is a nontrivial task. SVMs with string kernels (e.g., [24,23]) often provide us a state-of-the-art classification for texts. However, the solutions of kernelized SVMs do not have explicit forms of patterns.

Among related researches, the work of Okanohara and Tsujii [17] would be most related to ours. They consider a similar problem over substring patterns and they deal with logistic regression with 1-norm regularization. As we will show later, in our experiments, our method gains higher accuracy than they reported. Other related researches include the work of Saigo et al [19]. They consider 1-norm soft margin optimization over graph patterns and use LPBoost. Our framework is close to theirs, but we use different techniques for pattern discovery of substrings.

2   Preliminaries

2.1   1-Norm Soft Margin Optimization

Let $\mathcal{X}$ be the set of instances. We are given a set S of labeled instances $S = ((\boldsymbol{x}_1, y_1), \dots, (\boldsymbol{x}_m, y_m))$, where each instance $\boldsymbol{x}$ belongs to $\mathcal{X}$ and each label y_i is -1 or $+1$, and a set $\mathcal{H}$ of n hypotheses, i.e., a set of functions from $\mathcal{X}$ to $[-1, +1]$. The final classifier is a linear combination of hypotheses in $\mathcal{H}$, $\sum_{h \in \mathcal{H}} \alpha_h h + b$, where b is a constant called bias. Given an instance $\boldsymbol{x}$, the prediction is $\text{sign}(\sum_{h \in \mathcal{H}} \alpha_h h(\boldsymbol{x}) + b)$, where $\text{sign}(a)$ is $+1$ if $a > 0$ and -1, otherwise. Let $\mathcal{P}^k$ be the probability simplex, i.e., $\mathcal{P}^k = \{ \boldsymbol{p} \in [0, 1]^k, \sum_{i=1}^k p_i = 1 \}$. For a weighting $\boldsymbol{\alpha} \in \mathcal{P}^n$ over hypotheses in $\mathcal{H}$ and a bias b, its margin w.r.t. a labeled instance $(\boldsymbol{x}, y)$ is defined as $y(\sum_{h \in \mathcal{H}} \alpha_h h(\boldsymbol{x}) + b)$. If the margin of $\boldsymbol{\alpha}$ w.r.t. a labeled instance is positive, the prediction is correct, that is, $y = \text{sign}(\sum_{h \in \mathcal{H}} \alpha_h h(\boldsymbol{x}) + b)$.

The edge of a hypothesis $h \in \mathcal{H}$ for a distribution $\boldsymbol{d} \in \mathcal{P}^m$ over S is defined as

$$\mathbf{Edge}_{\boldsymbol{d}}(h) = \sum_{i=1}^m y_i d_i h(\boldsymbol{x}_i).$$

The edge of h can be viewed as accuracy w.r.t. the distribution $\boldsymbol{d}$. In fact, if the output of h is binary-valued ($+1$ or -1), $\mathbf{Edge}_{\boldsymbol{d}}(h) = 1 - 2\mathrm{Error}_{\boldsymbol{d}}(h)$, where $\mathrm{Error}_{\boldsymbol{d}}(h) \, is \, \sum_i d_i I(h(\boldsymbol{x}_i) = y_i)$, where $I(\cdot)$ is the indicator function such that $I(true) = 1$ and $I(false) = 0$.

The 1 norm soft margin optimization problem is formulated as follows (see, e.g., [4,25]):

$$\max_{\rho, \boldsymbol{\alpha}, \boldsymbol{\xi}, b} \rho - \frac{1}{\nu} \sum_{i=1}^m \xi_i \qquad (1)$$

$$\text{sub.to}$$

$$y_i \left(\sum_j \alpha_j h_j(\boldsymbol{x}_i) + b \right) \geq \rho - \xi_i \ (i = 1, \dots, m),$$

$$\boldsymbol{\alpha} \in \mathcal{P}^n, \ \boldsymbol{\xi} \geq \boldsymbol{0}.$$

That is, the problem is to find a weighting $\boldsymbol{\alpha}$ over hypotheses and a bias b which maximize the margin among given labeled instances as well as minimizing the sum of quantities (losses) by which the weighting misclassifies. Here, the parameter ν takes values in $\{1, \dots, m\}$ and it is fixed in advance. This parameter controls the tradeoff between maximization of the margin and minimization of losses.

By using Lagrangian duality (see, e.g., [3]), we can derive the dual problem as follows.

$$\min_{\gamma, d} \gamma \qquad (2)$$

$$\text{sub.to}$$

$$\mathbf{Edge}_d(h_j) = \sum_i d_i y_i h_j(\boldsymbol{x}_i) \le \gamma \ (j = 1, \dots, n),$$

$$d \le \frac{1}{\nu}\mathbf{1}, \ d \in \mathcal{P}^m,$$

$$d \cdot y = 0.$$

The dual problem is to find a distribution over instances for which the edges of hypotheses are minimized. In other words, the problem is to find the most difficult distribution for the hypotheses in $\mathcal{H}$.

It is well known that if the primal and dual problems are linear programs, they are equivalent to each other, i.e., if one solves one problem, one have also solved the other and vice versa. More precisely, let $(\rho^*, \boldsymbol{\alpha}^*, \boldsymbol{\xi}^*, b^*)$ be an optimizer of the primal problem (1) and let $(\gamma^*, \boldsymbol{d}^*)$ be an optimizer of the dual problem (2), respectively. Then, by the duality of the linear program, $\rho^* - \frac{1}{\nu} \sum_{i=1}^m \xi_i^* = \gamma^*$.

KKT conditions (see, e.g., [3]) implies that an optimal solution has the following property.

- If $y_i \left(\sum_j \alpha_j^* h_j(\boldsymbol{x}_i) + b^* \right) > \rho^*$, then $d_i^* = 0$.
- If $0 < d_i^* < 1/\nu$, then $y_i(\sum_j \alpha_j^* h_j(\boldsymbol{x}_i) + b^*) = \rho^*$.
- If $\xi_i^* > 0$, then $d_i^* = 1/\nu$.

That is, only such a labeled instance $(\boldsymbol{x}_i, y_i)$ that have margin no larger than ρ^* can have a positive weight $d_i^* > 0$. Further, note that the number of inseparable examples (for which $\xi_i^* > 0$) is at most ν. This property shows the sparsity of a dual solution. The primal solution has sparsity as well:

- If $\mathbf{Edge}_{d^*}(h_j) < \gamma^*$, $\alpha_j^* = 0$.

Similarly, only such a hypothesis h_j that $\mathbf{Edge}_{d^*}(h_j) = \gamma^*$ can have a positive coefficient $\alpha_j^* > 0$.

2.2 LPBoost

We review LPBoost [4] for solving the problem (2). Roughly speaking, LPBoost iteratively solves some restricted dual problems and gets a final solution.

The detail of LPBoost is given in Algorithm 1. Given the initial distribution d_1, LPBoost works in iterations. At each iteration t, LPBoost chooses a hypothesis h_t maximizing the edge w.r.t. d_t, and add a new constraint $\mathbf{Edge}_d(h_t) \le \gamma$. problem and solve the linear program and get d_{t+1} and γ_{t+1}.

In fact, given a precision parameter $\varepsilon > 0$, LPBoost outputs an ε-approximation.

Algorithm 1. LPBoost(S,ε)

---

1. Let $\boldsymbol{d}_1$ be the distribution over S such that $\boldsymbol{d}_1 \cdot \boldsymbol{y} = 0$ and $\boldsymbol{d}_1$ is uniform w.r.t. positive or negative instances only. Let $\gamma_1 = -1$.
2. For $t = 1, \ldots,$
   (a) Let $h_t = \arg\max_{h \in \mathcal{H}} \mathbf{Edge}_{\boldsymbol{d}_t}(h)$.
   (b) If $\mathbf{Edge}_{\boldsymbol{d}_t}(h_t) \leq \gamma_t + \varepsilon$, let $T = t - 1$ and break.
   (c) Otherwise, solve the soft margin optimization problem (2) w.r.t. the restricted hypothesis set $\{h_1, \ldots, h_t\}$. That is,

$$(\gamma_{t+1}, \boldsymbol{d}_{t+1}) = \arg\min_{\gamma, \boldsymbol{d} \in \mathcal{P}^m} \gamma$$

sub. to

$$\gamma_{\boldsymbol{d}}(h_j) \leq \gamma \quad (j = 1, \ldots, t)$$

$$\boldsymbol{d} \leq \frac{1}{\nu}\mathbf{1}, \boldsymbol{d} \cdot \boldsymbol{y} = 0.$$

3. Output $f(\boldsymbol{x}) = \sum_{t=1}^{T} \alpha_t h_t(\boldsymbol{x})$, where each α_t $(t = 1, \ldots, T)$ is a Lagrange dual of the soft margin optimization problem (2).

Theorem 1 (Demiriz et al. [4]). *LPBoost outputs a solution whose objective is an ε-approximation of an optimum.*

2.3   Strings

Let Σ be a finite *alphabet* of size σ. An element of Σ^* is called a *string*. Strings x, y and z are said to be a *prefix*, *substring*, and *suffix* of the string $u = xyz$. The length of any string u is denoted by $|u|$. Let ε denote the empty string, that is, $|\varepsilon| = 0$. Let $\Sigma^+ = \Sigma^* - \{\varepsilon\}$. The i-th character of a string u is denoted by $u[i]$ for $1 \leq i \leq |u|$, and the substring of u that begins at position i and ends at position j is denoted by $u[i:j]$ for $1 \leq i \leq j \leq |u|$. For a set of strings S, let $\|S\| = \sum_{s \in S} |s|$.

2.4   Our Problem

We consider the 1 norm soft margin optimization problem for string data sets, where each hypothesis corresponds to a string pattern. That is, we are given a set of labeled documents (strings), and each substring $p \in \Sigma^*$ corresponds to a hypothesis $h_p \in \mathcal{H}$, and $h_p(\boldsymbol{x})$ for $x \in \Sigma^*$ is defined as follows:

$$h_p(\boldsymbol{x}) = \begin{cases} 1 & p \text{ is a substring of } \boldsymbol{x} \\ -1 & p \text{ is not a substring of } \boldsymbol{x} \end{cases}.$$

Thus, our "weak" learner will solve the following problem. Given a set of labeled strings $S = ((x_1, y_m), \ldots, (x_m, y_m)) \subset \Sigma^* \times \{-1, +1\}$, and a distribution $\boldsymbol{d} \in \mathcal{P}^m$

over S, find a string $p \in \Sigma^*$ such that

$$p = \arg \max_{q \in \Sigma^*} \mathbf{Edge}_d(q) = \sum_{i=1}^{m} y_i d_i h_q(x_i) \qquad \tag{3}$$

To solve this problem optimally and efficiently, we make use of the suffix array data structure [16] as well as other related data structures described in the next section.

3   Algorithms

3.1   Data Structures

Below, we describe the data structures used in our algorithm.

The suffix tree of a string T is a compacted trie of all suffixes of T. For any node v in the suffix tree, let $path(v)$ denote the string corresponding to the path from the root to node v. We assume that the string ends with a unique character '\$' not appearing elsewhere in the string, thus ensuring that the tree contains $|T|$ leaves, each corresponding to a suffix of T. The suffix tree and generalized suffix tree are very useful data structures for algorithms that consider the substrings of a given string or set of strings. Each node v in the suffix tree corresponds to a substring of the input strings, and the leaves represent occurrences of the substring $path(v)$ in the string.

The generalized suffix tree for a set of strings $(T_1, \ldots, T_m)$, can be defined as the suffix tree for the string $T = T_1 \$_1 \cdots T_m \$_m$, where each $\$_i$ $(1 \le i \le m)$ is a unique character not appearing elsewhere in the strings, and each edge is terminated with the first appearance of any $\$_i$. We assume that the leaves of the generalized suffix tree are labeled with the document index i.

It is well known that suffix trees can be constructed in linear time [26]. In practice, it is more efficient to use a data structure called suffix arrays which require less memory. The suffix array of string T is a permutation of all suffixes of T so that the suffixes are lexicographically sorted. More precisely, the suffix array of T is an array $SA[1, \ldots, |T|]$ containing a permutation of $\{1, \ldots |T|\}$, such that $T[SA[i]:|T|] \preceq T[SA[i+1]:|T|]$, for all $1 \le i < s$, where $\preceq$ denotes the lexicographic ordering on strings.

It is well known that the suffix array for a given string can be built in time linear of its length [10,12,13].

Another important array often used together with the suffix array is the height array. Let $LCP[i] = lcp(T[SA[i]:|T|], T[SA[i+1]:|T|])$ be the *height array* $LCP[1, |T|]$ of T, where $lcp(T_{SA[i]}, T_{SA[i+1]})$ is the length of the longest common prefix between $T[SA[i]:|T|]]$ and $T[SA[i+1]:|T|]]$. The height array can also be constructed in linear time [11]. Also, by using the suffix array and height arrays we can simulate a bottom-up post-order traversal on the suffix tree [11]. Most other algorithms on suffix trees can be efficiently implemented using the suffix and height arrays [1].

Figure 1 shows an example of a suffix array and suffix tree for the string BANANA.

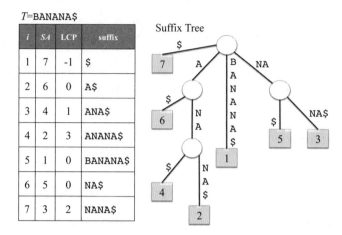

Fig. 1. Suffix array (left) and suffix tree (right) for string $T = $ BANANA$. The column SA shows the suffix array, the column LCP shows the height array, the column 'suffix' shows the suffixes starting at position i.

3.2   Finding the Optimal Pattern

We briefly describe how we can find the substring $p \in \Sigma^*$ to maximize Equation (3) in linear time.

First, we note that it is sufficient to consider strings which correspond to nodes in the generalized suffix tree of the input strings. This is because for any string corresponding to a path that ends in the middle of an edge of the suffix tree, the string which corresponds to the path extended to the next node will occur in the same set of documents and, hence, its edge score would be the same. Figure 2 shows an example.

Also, notice that for any substring $p \in \Sigma^*$, we have

$$
\begin{aligned}
\mathbf{Edge}_d(p) &= \sum_{i=1}^{m} y_i d_i h_p(x_i) \\
&= \sum_{\{i:h_p(x_i)=1\}} y_i d_i - \sum_{\{i:h_p(x_i)=-1\}} y_i d_i \\
&= \sum_{\{i:h_p(x_i)=1\}} y_i d_i - \left(\sum_{i=1}^{m} y_i d_i - \sum_{\{i:h_p(x_i)=1\}} y_i d_i \right) \\
&= 2 \cdot \sum_{\{i:h_p(x_i)=1\}} y_i d_i - \sum_{i=1}^{m} y_i d_i.
\end{aligned}
$$

Since $\sum_{i=1}^{m} y_i d_i$ can be easily computed, we need only to compute $\sum_{\{i:h_p(x_i)=1\}} y_i d_i$ for each p to compute its edge score.

This value can be computed for each string $path(v)$ corresponding to a node v in the generalized suffix tree, basically using the linear time algorithm for solving a generalized version of the color set size problem [7,2]. When each document is assigned arbitrary numeric weights, the algorithm computes for each node v of the generalized suffix tree, the sum of weights of the documents that contain $path(v)$ as a substring. For our problem, we need only to assign the weight $y_i d_i$ to each document.

The main algorithm and optimal pattern discovery algorithms are summarized in Algorithm 2 and Algorithm 3. It is easy to see that FindOptimalSubstring-Pattern(...) runs in linear time: The algorithm of [2] runs in linear time. Also, since the number of nodes in a generalized suffix tree is linear in the total length of the strings, line 3 can also be computed in linear time.

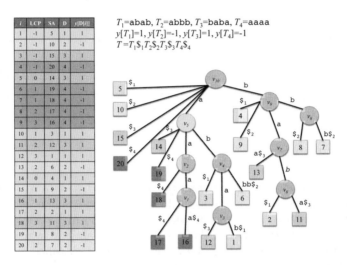

i	LCP	SA	D	y|D[i]|
1	-1	5	1	1
2	-1	10	2	-1
3	-1	15	3	1
4	-1	20	4	-1
5	0	14	3	1
6	1	19	4	-1
7	1	18	4	-1
8	2	17	4	-1
9	3	16	4	-1
10	1	3	1	1
11	2	12	3	1
12	3	1	1	1
13	2	6	2	-1
14	0	4	1	1
15	1	9	2	-1
16	1	13	3	1
17	2	2	1	1
18	3	11	3	1
19	1	8	2	-1
20	2	7	2	-1

T_1=abab, T_2=abbb, T_3=baba, T_4=aaaa
$y[T_1]$=1, $y[T_2]$=-1, $y[T_3]$=1, $y[T_4]$=-1
$T = T_1\$_1 T_2\$_2 T_3\$_3 T_4\$_4$

Fig. 2. Finding the substring that gives the maximum edge on four documents T_1, T_2, T_3, and T_4, with labels $\boldsymbol{y} = (1, -1, 1, -1)$ and weights $\boldsymbol{d} = (0.3, 0.1, 0.2, 0.4)$. The generalized suffix tree is depicted on the right, and corresponding suffix arrays and height arrays are depicted on the left. D holds the document index assigned to each leaf. For example, $\mathbf{Edge}_d(path(v_1)) = 0.3 * 1 * (-1) + 0.1 * (-1) * (-1) + 0.2 * 1 * (-1) + 0.4 * (-1) * 1 = -0.8$. The optimal patterns are $path(v_3)$ = 'aba' and $path(v_6)$ = 'bab' giving an edge of 1.

Algorithm 2. Compute 1 norm soft margin optimal problem for string

1: **Input:** Data $S = ((T_1, y_1), \ldots, (T_m, y_m))$, parameter ε.
2: Construct suffix array SA and LCP array for string $T = T_1\$_1 \cdots T_m\$_m$.
3: Run Algorithm 1 (LPBoost(S,ε)) using FindOptimalSubstringPattern(SA, LCP, $\boldsymbol{y}$, $\boldsymbol{d}$) for line 2(a).

Algorithm 3. FindOptimalSubstringPattern(SA, LCP, $\boldsymbol{y}$, $\boldsymbol{d}$)

1: $wtot := \sum y_i d_i$;
2: Calculate $w_v = \sum_{\{i:h_{path(v)}(x_i)=1\}} y_i d_i$ for each node v of the generalized suffix tree, using SA, LCP and algorithm of [2];
3: $vmax := \arg\max_v \mathbf{Edge}_{\boldsymbol{d}}(path(v)) = \arg\max_v (2w_v - wtot)$;
4: **return** $path(vmax)$;

Table 1. Detail of the data sets

Corpus	# of docs	total length
MOVIE-A	2000	7786004
MOVIE-B	7440	213970

Table 2. Percentage of correct classifications in classification task

Corpus	LPSSD	SVM+Ngram	normalized SVM+Ngram	OT [17]
MOVIE-A	91.25%	85.75%	89.25%	86.5%
MOVIE-B	78.50%	73.80%	74.80%	75.1%

4   Experiments

We conducted sentiment classification experiments for two data sets, MOVIE-A and MOVIE-B. MOVIE-A is a dataset by Bo Pang and Lillian Lee [18][1]. The data consists of reviews of various movies, with 1000 positive reviews, and 1000 negative reviews. MOVIE-B is a dataset by Ifrim et al. [8] [2] which consisting of reviews taken from the IMDB database, for movies classified as 'Crime' or 'Drama'. There are 3720 reviews for each genre. Table 1 shows simple statistics of the data.

We examined the performance of our approach using 10 cross validations. More precisely, at each trial, we split each of the positive and negative data randomly so that 4/5 is training, and 1/5 is test. Then we train our method for the training data and measure the accuracy of the obtained classifier over the test data. We average the accuracy over 10 trials. The parameter ν for our method is set as $\nu/m = 0.1$, which, roughly speaking, means that we estimate the level of noise in the data as 10%.

Table 2 shows the results of our method, as well as several other methods. "SVM + Ngram" denotes the support vector machine using an ngram kernel, and "normalized SVM +Ngram" denotes a version which uses normalization (normalized SVM + Ngram). The scores shown for these methods are for ngrams

---

[1] http://www.cs.cornell.edu/People/pabo/movie-review-data/, polarity dataset v2.0.

[2] http://www.mpi-inf.de/~ifrim/data/kdd08-datasets.zip, KDD08-datasets/IMDB.

of length $n = 7$, which gave the best score. The score for "OT [17]" is the score of a 10-fold cross validation taken directly from their paper.

Table 3 shows substrings with MOVIE-A's top 10 largest weights in the final weighting α. It also shows number of documents in which the pattern occurs. Our method found some interesting patterns, such as *est movie, or best, s very e,* and *s perfect*. Table 4 shows MOVIE-A's some of the context of the occurrence of these patterns.

Table 3. MOVIE-A's Top 10 substrings with largest weight (α)

pattern	α	#occ in positive	#occ in negative
iase	0.01102	13	1
ronicle	0.00778	21	4
s very e	0.00776	15	2
ents r	0.00659	9	1
or best	0.00642	21	5
e of your s	0.00633	8	0
finest	0.00615	44	5
ennes	0.00575	28	8
un m	0.00567	13	1
s insid	0.00564	14	5

Table 4. MOVIE-A's Context of some substrings in top 100 largest weightings (α)

Pattern	est movie	s perfect	or best	s very e
Context	b*est movie*	is *perfect*	act*or best*	is *very* effective
	funn*est movie*	thi*s perfectly*	awor*d for best*	is *very* entertain
	great*est movie*	seem*s perfectly*	nominate*d for best*	is *very* enjoyable

Pattern	fun	o entertain	much like t	s a fine
Context	*fun*	s*o entertaining*	*much like the*	is *a fine*
	*fun*ny	t*o entertain*	*much like* their	deliver*s a fine*
	*fun*niest	t*o entertaining*	*much like* tis	doe*s a fine*
	*fun*nest	t*o entertain*ment	*much like* titanic	contribut*s a fine*

5   Conclusion and Future Work

We considered 1-norm soft margin optimization over substring patterns. We solve this problem by using a combination of LPBoost and an optimal substring pattern discovery algorithm. In our preliminary experiments on data sets concerning movie reviews, our method actually found some interesting pattern candidates. Also, the experimental results showed that our method achieves higher accuracy than other previous methods.

There is much room for improvements and future work. First, our method might become more scalable by employing faster solvers for 1-norm soft margin optimization, e.g., Sparse LPBoost [5]. Second, extending the pattern class to more richer ones such as VLDC patterns [9] would be interesting. Finally, applying our method to DNA or RNA data would be promising.

References

1. Abouelhoda, M.I., Kurtz, S., Ohlebusch, E.: Replacing suffix trees with enhanced suffix arrays. Journal of Discrete Algorithms 2(1), 53–86 (2004)
2. Bannai, H., Hyyrö, H., Shinohara, A., Takeda, M., Nakai, K., Miyano, S.: An $O(N^2)$ algorithm for discovering optimal Boolean pattern pairs. IEEE/ACM Transactions on Computational Biology and Bioinformatics 1(4), 159–170 (2004)
3. Boyd, S., Vandenberghe, L.: Convex Optimization. Cambridge University Press, Cambridge (2004)
4. Demiriz, A., Bennett, K.P., Shawe-Taylor, J.: Linear programming boosting via column generation. Mach. Learn. 46(1-3), 225–254 (2002)
5. Hatano, K., Takimoto, E.: Linear programming boosting by column and row generation. In: Gama, J., Costa, V.S., Jorge, A.M., Brazdil, P.B. (eds.) DS 2009. LNCS, vol. 5808, pp. 401–408. Springer, Heidelberg (2009)
6. Hirao, M., Hoshino, H., Shinohara, A., Takeda, M., Arikawa, S.: A practical algorithm to find the best subsequence patterns. Theoretical Computer Science 292(2), 465–479 (2003)
7. Hui, L.: Color set size problem with applications to string matching. In: Apostolico, A., Galil, Z., Manber, U., Crochemore, M. (eds.) CPM 1992. LNCS, vol. 644, pp. 230–243. Springer, Heidelberg (1992)
8. Ifrim, G., Bakir, G.H., Weikum, G.: Fast logistic regression for text categorization with variable-length n-grams. In: KDD, pp. 354–362 (2008)
9. Inenaga, S., Bannai, H., Shinohara, A., Takeda, M., Arikawa, S.: Discovering best variable-length-don't-care patterns. In: Lange, S., Satoh, K., Smith, C.H. (eds.) DS 2002. LNCS (LNAI), vol. 2534, pp. 86–97. Springer, Heidelberg (2002)
10. Kärkkäinen, J., Sanders, P.: Simple linear work suffix array construction. In: Baeten, J.C.M., Lenstra, J.K., Parrow, J., Woeginger, G.J. (eds.) ICALP 2003. LNCS, vol. 2719, pp. 943–955. Springer, Heidelberg (2003)
11. Kasai, T., Lee, G., Arimura, H., Arikawa, S., Park, K.: Linear-time longest-common-prefix computation in suffix arrays and its applications. In: Amir, A., Landau, G.M. (eds.) CPM 2001. LNCS, vol. 2089, pp. 181–192. Springer, Heidelberg (2001)
12. Kim, D.K., Sim, J.S., Park, H., Park, K.: Linear-time construction of suffix arrays. In: Baeza-Yates, R., Chávez, E., Crochemore, M. (eds.) CPM 2003. LNCS, vol. 2676, pp. 186–199. Springer, Heidelberg (2003)
13. Ko, P., Aluru, S.: Space efficient linear time construction of suffix arrays. In: Baeza-Yates, R., Chávez, E., Crochemore, M. (eds.) CPM 2003. LNCS, vol. 2676, pp. 200–210. Springer, Heidelberg (2003)
14. Leslie, C.S., Eskin, E., Weston, J., Noble, W.S.: Mismatch string kernels for svm protein classification. In: Advances in Neural Information Processing Systems 15 (NIPS 2002), pp. 1417–1424 (2002)
15. Lodhi, H., Saunders, C., Shawe-Taylor, J., Cristianini, N., Watkins, C.J.C.H.: Text classification using string kernels. Journal of Machine Learning Research 2, 419–444 (2002)

16. Manber, U., Myers, G.: Suffix arrays: a new method for on-line string searches. SIAM J. Computing 22(5), 935–948 (1993)
17. Okanohara, D., Tsujii, J.: Text categorization with all substring features. In: Proc. 9th SIAM International Conference on Data Mining (SDM), pp. 838–846 (2009)
18. Pang, B., Lee, L.: A sentimental education: Sentiment analysis using subjectivity summarization based on minimum cuts. In: Proceedings of the ACL (2004)
19. Saigo, H., Nowozin, S., Kadowaki, T., Kudo, T., Tsuda, K.: gboost: a mathematical programming approach to graph classification and regression. Machine Learning 75(1), 69–89 (2009)
20. Schapire, R.E., Freund, Y., Bartlett, P., Lee, W.S.: Boosting the margin: a new explanation for the effectiveness of voting methods. The Annals of Statistics 26(5), 1651–1686 (1998)
21. Schapire, R.E., Singer, Y.: Boostexter: A boosting-based system for text categorization. Machine Learning 39, 135–168 (2000)
22. Shinohara, A.: String pattern discovery. In: Ben-David, S., Case, J., Maruoka, A. (eds.) ALT 2004. LNCS (LNAI), vol. 3244, pp. 1–13. Springer, Heidelberg (2004)
23. Teo, C.H., Vishwanathan, S.V.N.: Fast and space efficient string kernels using suffix arrays. In: ICML, pp. 929–936 (2006)
24. Vishwanathan, S.V.N., Smola, A.J.: Fast kernels for string and tree matching. In: NIPS, pp. 569–576 (2002)
25. Warmuth, M.K., Glocer, K.A., Vishwanathan, S.V.: Entropy regularized lpboost. In: Freund, Y., Györfi, L., Turán, G., Zeugmann, T. (eds.) ALT 2008. LNCS (LNAI), vol. 5254, pp. 256–271. Springer, Heidelberg (2008)
26. Weiner, P.: Linear pattern-matching algorithms. In: Proc. of 14th IEEE Ann. Symp. on Switching and Automata Theory, pp. 1–11 (1973)

Discovery of Super-Mediators of Information Diffusion in Social Networks

Kazumi Saito[1], Masahiro Kimura[2], Kouzou Ohara[3], and Hiroshi Motoda[4]

[1] School of Administration and Informatics, University of Shizuoka
52-1 Yada, Suruga-ku, Shizuoka 422-8526, Japan
k-saito@u-shizuoka-ken.ac.jp
[2] Department of Electronics and Informatics, Ryukoku University
Otsu 520-2194, Japan
kimura@rins.ryukoku.ac.jp
[3] Department of Integrated Information Technology, Aoyama Gakuin University
Kanagawa 229-8558, Japan
ohara@it.aoyama.ac.jp
[4] Institute of Scientific and Industrial Research, Osaka University
8-1 Mihogaoka, Ibaraki, Osaka 567-0047, Japan
motoda@ar.sanken.osaka-u.ac.jp

Abstract. We address the problem of discovering a different kind of influential nodes, which we call "super-mediator", i.e. those nodes which play an important role to pass the information to other nodes, and propose a method for discovering super-mediators from information diffusion samples without using a network structure. We divide the diffusion sequences in two groups (lower and upper), each assuming some probability distribution, find the best split by maximizing the likelihood, and rank the nodes in the upper sequences by the F-measure. We apply this measure to the information diffusion samples generated by two real networks, identify and rank the super-mediator nodes. We show that the high ranked super-mediators are also the high ranked influential nodes when the diffusion probability is large, i.e. the influential nodes also play a role of super-mediator for the other source nodes, and interestingly enough that when the high ranked super-mediators are different from the top ranked influential nodes, which is the case when the diffusion probability is small, those super-mediators become the high ranked influential nodes when the diffusion probability becomes larger. This finding will be useful to predict the influential nodes for the unexperienced spread of new information, e.g. spread of new acute contagion.

1 Introduction

There have been tremendous interests in the phenomenon of influence that members of social network can exert on other members and how the information propagates through the network. Social networks (both real and virtual) are now recognized as an important medium for the spread of information. A variety of information that includes news, innovation, hot topics, ideas, opinions and even malicious rumors, propagates in the form of so-called "word-of-mouth" communications. Accordingly, a considerable amount of studies has been made for the last decade [1–20].

B. Pfahringer, G. Holmes, and A. Hoffmann (Eds.): DS 2010, LNAI 6332, pp. 144–158, 2010.
© Springer-Verlag Berlin Heidelberg 2010

Among them, widely used information diffusion models are the *independent cascade (IC)* [1, 8, 13] and the *linear threshold (LT)* [4, 5] and their variants [6, 14–18]. These two models focus on different information diffusion aspects. The IC model is sender-centered and each active node *independently* influences its inactive neighbors with given diffusion probabilities. The LT model is receiver-centered and a node is influenced by its active neighbors if their total weight exceeds the threshold for the node. Which model is more appropriate depends on the situation and selecting the appropriate one is not easy [18].

The major interests in the above studies are finding influential nodes, i.e. finding nodes that play an important role of spreading information as much as possible. This problem is called *influence maximization problem* [8, 10]. The node influence can only be defined as the expected number of active nodes (nodes that have become influenced due to information diffusion) because the diffusion phenomenon is stochastic, and estimating the node influence efficiently is still an open problem. Under this situation, solving an optimal solution, i.e. finding a subset of nodes of size K that maximizes the expected influence degree with K as a parameter, faces with combinatorial explosion problem and, thus, much of the efforts has been directed to finding algorithms to efficiently estimate the expected influence and solve this optimization problem. For the latter, a natural solution is to use a greedy algorithm at the expense of optimality. Fortunately, the expected influence degree is submodular, i.e. its marginal gain diminishes as the size K becomes larger, and the greedy solution has a lower bound which is 63% of the true optimal solution [8]. Various techniques to reduce the computational cost have been attempted including bond percolation [10] and pruning [14] for the former, and lazy evaluation [21], burnout [15] and heuristics [22] for the latter.

Expected influence degree is approximated by the empirical mean of the influence degree of many independent information diffusion simulations, and by default it has been assumed that the degree distribution is Gaussian. However, we noticed that this assumption is not necessarily true, which motivated to initiate this work. In this paper, we address the problem of discovering a different kind of influential nodes, which we call "super-mediator", i.e. those nodes which play an important role in passing the information to other nodes, try to characterize such nodes, and propose a method for discovering super-mediator nodes from information diffusion sequences (samples) without using a network structure. We divide the diffusion samples in two groups (lower and upper), each assuming some probability distribution, find the best split by maximizing the likelihood, and rank the nodes in the upper sequences by the F-measure (more in subsection 3.2).

We tested our assumption of existence of super-mediators using two real networks[1] and investigated the utility of the F-measure. As before, we assume that information diffusion follows either the independent cascade (IC) model or the linear threshold (LT) model. We first analyze the distribution of influence degree averaged over all the initial nodes[2] based on the above diffusion models, and empirically show that it becomes a

---

[1] Note that we use these networks only to generate the diffusion sample data, and thus are not using the network structure for the analyses.

[2] Each node generates one distribution, which is approximated by running diffusion simulation many times and counting the number of active nodes at the end of simulation.

power-law like distribution for the LT model, but it becomes a mixture of two distributions (power-law like distribution and lognormal like distributions) for the IC model. Based on this observation, we evaluated our super-mediator discovery method by focusing on the IC model. It is reasonable to think that the super mediators themselves are the influential nodes, and we show empirically that the high ranked super-mediators are indeed the high ranked influential nodes, i.e. the influential nodes also play a role of super-mediator for the other source nodes, but this is true only when the diffusion probability is large. What we found more interesting is that when the high ranked super-mediators are different from the top ranked influential nodes, which is the case when the diffusion probability is small, those super-mediators become the high ranked influential nodes when the diffusion probability becomes larger. We think that this finding is useful to predict the influential nodes for the unexperienced spread of new information from the known experience, e.g. spread of new acute contagion from the spread of known moderate contagion for which there are abundant data.

The paper is organized as follows. We start with the brief explanation of the two information diffusion models (IC and LT) and the definition of influence degree in section 2, and then describe the discovery method based on the likelihood maximization and F-measure in section 3. Experimental results are detailed in section 4 together with some discussion. We end this paper by summarizing the conclusion in section 5.

2   Information Diffusion Models

We mathematically model the spread of information through a directed network $G = (V, E)$ without self-links, where V and E ($\subset V \times V$) stand for the sets of all the nodes and links, respectively. For each node v in the network G, we denote $F(v)$ as a set of child nodes of v, i.e. $F(v) = \{w; (v, w) \in E\}$. Similarly, we denote $B(v)$ as a set of parent nodes of v, i.e. $B(v) = \{u; (u, v) \in E\}$. We call nodes *active* if they have been influenced with the information. In the following models, we assume that nodes can switch their states only from inactive to active, but not the other way around, and that, given an initial active node set H, only the nodes in H are active at an initial time.

2.1   Independent Cascade Model

We recall the definition of the IC model according to [8]. In the IC model, we specify a real value $p_{u,v}$ with $0 < p_{u,v} < 1$ for each link (u, v) in advance. Here $p_{u,v}$ is referred to as the *diffusion probability* through link (u, v). The diffusion process unfolds in discrete time-steps $t \geq 0$, and proceeds from a given initial active set H in the following way. When a node u becomes active at time-step t, it is given a single chance to activate each currently inactive child node v, and succeeds with probability $p_{u,v}$. If u succeeds, then v will become active at time-step $t + 1$. If multiple parent nodes of v become active at time-step t, then their activation attempts are sequenced in an arbitrary order, but all performed at time-step t. Whether or not u succeeds, it cannot make any further attempts to activate v in subsequent rounds. The process terminates if no more activations are possible.

2.2   Linear Threshold Model

In the LT model, for every node $v \in V$, we specify a *weight* ($\omega_{u,v} > 0$) from its parent node u in advance such that $\sum_{u \in B(v)} \omega_{u,v} \leq 1$. The diffusion process from a given initial active set H proceeds according to the following randomized rule. First, for any node $v \in V$, a *threshold* θ_v is chosen uniformly at random from the interval $[0, 1]$. At time-step t, an inactive node v is influenced by each of its active parent nodes, u, according to weight $\omega_{u,v}$. If the total weight from active parent nodes of v is no less than θ_v, that is, $\sum_{u \in B_t(v)} \omega_{u,v} \geq \theta_v$, then v will become active at time-step $t + 1$. Here, $B_t(v)$ stands for the set of all the parent nodes of v that are active at time-step t. The process terminates if no more activations are possible.

2.3   Influence Degree

For both models on G, we consider information diffusion from an initially activated node v, i.e. $H = \{v\}$. Let $\varphi(v; G)$ denote the number of active nodes at the end of the random process for either the IC or the LT model on G. Note that $\varphi(v; G)$ is a random variable. We refer to $\varphi(v; G)$ as the *influence degree* of node v on G. Let $\mathcal{E}(v; G)$ denote the expected number of $\varphi(v; G)$. We call $\mathcal{E}(v; G)$ the *expected influence degree* of node v on G. In theory we can simply estimate $\mathcal{E}$ by the simulations based on either the IC or the LT model in the following way. First, a sufficiently large positive integer M is specified. Then, the diffusion process of either the IC or the LT model is simulated from the initially activated node v, and the number of active nodes at the end of the random process, $\varphi(v; G)$, is calculated. Last, $\mathcal{E}(v; G)$ for the model is estimated as the empirical mean of influence degrees $\varphi(v; G)$ that are obtained from M such simulations.

From now on, we use $\varphi(v)$ and $\mathcal{E}(v)$ instead of $\varphi(v; G)$ and $\mathcal{E}(v; G)$, respectively if G is obvious from the context.

3   Discovery Method

3.1   Super-Mediator

As mentioned in section 1, we address the problem of discovering a different kind of influential nodes, which we call "super-mediator". These are the nodes which appear frequently in long diffusion sequences with many active nodes and less frequently in short diffusion sequences, i.e. those nodes which play an important role to pass the information to other nodes. Figure 1 (a) shows an example of information diffusion samples. In this figure, by independently performing simulations $5,000$ times based on the IC model, we plotted $5,000$ curves for influence degree of a selected information source node with respect to time steps[3]. From this figure, we can observe that 1) due to its stochastic nature, each diffusion sample varies in a quite wide range for each simulation; and 2) some curves clearly exhibit sigmoidal behavior in part, in each of which the influence degree suddenly becomes relatively high during a certain time interval.

---

[3] The network used to generate these data is the blog network (see subsection 4.1).

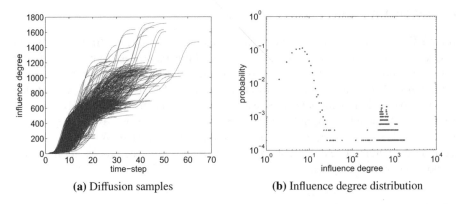

(a) Diffusion samples   (b) Influence degree distribution

Fig. 1. Information diffusion from some node in the blog network for the IC model ($p = 0.1$)

In Figure 1 (b), we plotted the distribution of the final influence degree for the above $5,000$ simulations. From this figure, we can observe that there exist a number of bell-shaped curves (which can be approximated by quadratic equations) in a logarithmic scale for each axis, which suggests that the influence degree distribution consists of several lognormal like distributions. Together with the observation from Figure 1 (a), we conjecture that super-mediators appear as a limited number of active nodes in some lognormal components with relatively high influence degree. Therefore, in order to discover these super-mediator nodes from information diffusion samples, we attempt to divide the diffusion samples in two groups (lower and upper), each assuming some probability distribution, find the best split by maximizing the likelihood, and rank the nodes in the upper samples by the F-measure.

3.2   Clustering of Diffusion Samples

Let $S(v) = \{1, 2, \cdots, M(v)\}$ denote a set of indices with respect to information diffusion samples for an information source node v, i.e. $\{d_1(v), d_2(v), \cdots, d_{M(v)}(v)\}$. Here note that $d_m(v)$ stands for a set of active nodes in the m-th diffusion sample. As described earlier, in order to discover super-mediator nodes, we consider dividing $S(v)$ into two groups, $S_1(v)$ and $S_2(v)$, which are the upper group of samples with relatively high influence degree and the lower group, respectively. Namely, $S_1(v) \cup S_2(v) = S(v)$ and $\min_{m \in S_1(v)} |d_m(v)| > \max_{m \in S_2(v)} |d_m(v)|$. Although we can straightforwardly extend our approach in case of k-groups division, we focus ourselves on the simplest case ($k = 2$) because of ease of both evaluation of basic performance and the following derivation. By assuming the independence of each sample drawn from either the upper or the lower group, we can consider the following likelihood function.

$$\mathcal{L}(S(v); S_1(v), \Theta) = \prod_{k \in \{1,2\}} \prod_{m \in S_k(v)} p(m; \theta_k), \tag{1}$$

where $p(m; \theta_k)$ denotes some probability distribution with the parameter set θ_k for the m-th diffusion sample, and $\Theta = \{\theta_1, \theta_2\}$. If it is assumed that the influence degree distribution consists of lognormal components, we can express $p(m; \theta_k)$ by

$$p(m; \theta_k) = \frac{1}{\sqrt{2\pi\sigma_k^2}|d_m(v)|} \exp\left(-\frac{(\log|d_m(v)| - \mu_k)^2}{2\sigma_k^2}\right), \tag{2}$$

where $\theta_k = \{\mu_k, \sigma_k^2\}$. Then, based on the maximum likelihood estimation, we can identify the optimal upper group $\hat{S}_1(v)$ by the following equation.

$$\hat{S}_1(v) = \arg\max_{S_1(v)}\left\{\mathcal{L}(S; S_1(v), \hat{\Theta})\right\}, \tag{3}$$

where $\hat{\Theta}$ denotes the set of maximum likelihood estimators.

Below we describe our method for efficiently obtaining $\hat{S}_1(v)$ by focusing on the
case that $p(m; \theta_k)$ is the lognormal distribution defined in Equation (2), although the
applicability of the method is not limited to this case. For a candidate upper group
$S_1(v)$, by noting the following equations of the maximum likelihood estimation,

$$\hat{\mu}_k = \frac{1}{|S_k(v)|} \sum_{m \in S_k(v)} \log|d_m(v)|, \quad \hat{\sigma}_k^2 = \frac{1}{|S_k(v)|} \sum_{m \in S_k(v)} (\log|d_m(v)| - \hat{\mu}_k)^2, \tag{4}$$

we can transform Equation (3) as follows.

$$\hat{S}_1(v) = \arg\max_{S_1(v)}\left\{2\log\mathcal{L}(S(v); S_1(v), \hat{\Theta})\right\} = \arg\max_{S_1(v)}\left\{-\sum_{k \in \{1,2\}} |S_k| \log\left(\hat{\sigma}_k^2\right)\right\}. \tag{5}$$

Therefore, when a candidate upper group $S_1(v)$ is successively changed by shifting its
boundary between $S_1(v)$ and $S_2(v)$, we can efficiently obtain $\hat{S}_1(v)$ by simply updating
the sufficient statistics for calculating the maximum likelihood estimators. Here, we
define the following operation to obtain the set of elements with the maximum influence
degree,

$$\eta(S(v)) = \left\{m; |d_m(v)| = \max_{m \in S(v)} \{|d_m(v)|\}\right\}, \tag{6}$$

because there might exist more than one diffusion sample with the same influence degree. Then, we can summarize our algorithm as follows.

1. Initialize $S_1(v) \leftarrow \eta(S(v))$, $S_2(v) \leftarrow S(v) \setminus \eta(S(v))$, and $\hat{L} \leftarrow -\infty$.
2. Iterate the following procedure:
2-1. Set $S_1(v) \leftarrow S_1(v) \cup \eta(S_2(v))$, and $S_2(v) \leftarrow S_2(v) \setminus \eta(S_2(v))$.
2-2. If $S_2(v) = \eta(S_2(v))$, then terminate the iteration.
2-3. Calculate $L = -\sum_{k \in \{1,2\}} |S_k(v)| \log(\hat{\sigma}_k^2)$.
2-4. If $\hat{L} < L$ then set $\hat{L} \leftarrow L$ and $\hat{S}_1(v) \leftarrow S_1(v)$
3. Output $\hat{S}_1(v)$, and terminate the algorithm.

We describe the computational complexity of the above algorithm. Clearly, the number of iterations performed in step 2 is at most $(M(v) - 2)$. On the other hand, when
applying the operator $\eta(\cdot)$ in steps 1 and 2.1 (or 2.2), by classifying each diffusion

sample according to its influence degree in advance, we can perform these operations with computational complexity of $O(1)$. Here note that since the influence degree is a positive integer less than or equal to $|V|$, we can perform the classification with computational complexity of $O(M(v))$. As for step 2.3, by adding (or removing) statistics calculated from $\eta(S_2(v))$, we can update the maximum likelihood estimators $\hat{\Theta}$ defined in Equation (4) with computational complexity of $O(1)$. Therefore, the total computational complexity of our clustering algorithm is $O(M(v))$. Note that the above discussion can be applicable to a more general case for which the sufficient statistics of $p(m; \theta_k)$ is available to its parameter estimation.

A standard approach to the above clustering problem might be applying the EM algorithm by assuming a mixture of lognormal components. However, this approach is likely to confront the following drawbacks: 1) due to the local optimal problem, a number of parameter estimation trials are generally required by changing the initial parameter values, and we cannot guarantee the global optimality for the final result; 2) since many iterations are required for each parameter estimation trial, we need a substantially large computational load for obtaining the solution, which results in a prohibitively large processing time especially for a large data set; and 3) in case that a data set contains malicious outlier samples, we need a special care to avoid some unexpected problems such as degradation of $\hat{\sigma}_k^2$ to 0. Actually, our preliminary experiments based on this approach suffered from these drawbacks. In contrast, our proposed method always produces the optimal result with computational complexity of $O(M(v))$.

3.3 Super-Mediator Discovery

Next, we describe our method for discovering super-mediator nodes. Let $D = \{d_m(v); v \in V, m = 1, \cdots, M(v)\}$ denote a set of observed diffusion samples. By using the above clustering method, we can estimate the upper group $\hat{S}_1$ for each node $v \in V$. For $\hat{S}_1(v)$, we employ, as a natural super-mediator score for a node $w \in V$, the following F-measure $F(w; v)$, a widely used measure in information retrieval, which is the harmonic average of recall and precision of a node w for the node v. Here the recall means the number of samples that include the node w in the upper group divided by the total number of samples in the upper group, and the precision means the number of samples that include a node w in the upper group divided by the total number of the node w in the samples.

$$F(w; v) = \frac{2|\{m; m \in \hat{S}_1(v), w \in d_m(v)\}|}{|\hat{S}_1(v)| + |\{m; m \in S(v), w \in d_m(v)\}|}. \tag{7}$$

Note that instead of the F-measure, we can employ the other measures such as the Jaccard coefficients, but for our objective that discovers characteristic nodes appearing in $\hat{S}_1(v)$, we believe that the F-measure is most basic and natural. Then, we can consider the following expected F-measure for D.

$$\mathcal{F}(w) = \sum_{v \in V} F(w; v) r(v), \tag{8}$$

where $r(v)$ stands for the probability that the node v becomes an information source node, which can be empirically estimated by $r(v) = M(v) / \sum_{v \in V} M(v)$. Therefore, we can discover candidates for the super-mediator nodes by ranking the nodes according to the above expected F-measure.

In order to confirm the validity of the F-measure and characterize its usefulness, we compare the ranking by the F-measure with the rankings by two other measures, and investigate how these rankings are different from or correlated to each other considering several situations. The first one is the expected influence degree defined in Section 2.3. From observed diffusion samples D, we can estimate it as follows.

$$\mathcal{E}(w) = \frac{1}{M(w)} \sum_{m=1}^{M(w)} |d_m(w)|. \tag{9}$$

The second one is the following measure:

$$\mathcal{N}(w) = \sum_{v \in V} |\{m; w \in d_m(v)\}| r(v). \tag{10}$$

This measure ranks high those nodes that are easily influenced by many other nodes.

4 Experimental Evaluation

4.1 Data Sets

We employed two datasets of large real networks, which are both bidirectionally connected networks. The first one is a trackback network of Japanese blogs used in [13] and has $12,047$ nodes and $79,920$ directed links (the blog network). The other one is a network of people derived from the "list of people" within Japanese Wikipedia, also used in [13], and has $9,481$ nodes and $245,044$ directed links (the Wikipedia network).

Here, according to [17], we assumed the simplest case where the parameter values are uniform across all links and nodes, i.e. $p_{u,v} = p$ for the IC model. As for the LT model, we assumed $\omega_{u,v} = q|B(v)|^{-1}$, and adopted q ($0 \leq q \leq 1$) as the unique parameter for a network instead of $\omega_{u,v}$ as in [18]. According to [8], we set p to a value smaller than $1/\bar{d}$, where $\bar{d}$ is the mean out-degree of a network. Thus, the value of p was set to 0.1 for the blog network and 0.02 for the Wikipedia network. These are the base values, but in addition to them, we used two other values, one two times larger and the other two times smaller for our analyses, i.e. 0.02 and 0.05 for the blog network, and 0.04 and 0.01 for the Wikipedia network. We set the base value for q to be 0.9 for the both networks to achieve reasonably long diffusion results. Same as p, we also adopted two other values, one two times larger and the other two times smaller. Since the double of 0.9 exceeds the upper-bound of q, i.e. 1.0, we used 1.0 for the larger value, and we used 0.45 for the smaller one.

For each combination of these values, information diffusion samples were generated for the corresponding model on each network using each node in the network as the initial active node. In our experiments, we set $M = 10,000$, which means $10,000$ information diffusion samples were generated for each initial active node. Then, we analyzed them to discover super-mediators. To efficiently generate those information diffusion samples and estimate the expected influence degree $\mathcal{E}$ of an initial active node,

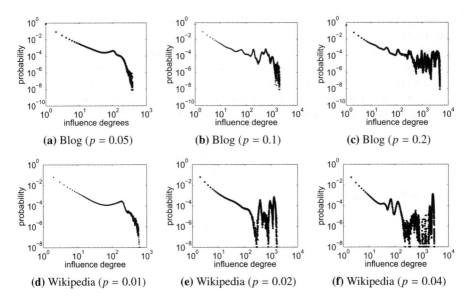

Fig. 2. The average influence degree distribution of the IC model

we adopted the method based on the bond percolation proposed in [14]. Note that we only use these two networks to generate the diffusion sample data which we assume we observed. Once the data are obtained, we no more use the network structure.

4.2 Influence Degree Distribution

First, we show the influence degree distribution for all nodes. Figure 2 is the results of the IC model and Fig. 3 is the results of the LT model. $M(= 10,000)$ simulations were performed for each initial node $v \in V$ and this is repeated for all the nodes in the network. Since the number of the nodes $|V|$ is about 10,000 for both the blog and the Wikipedia networks, these results are computed from about one hundred million diffusion samples and exhibits global characteristics of the distribution. We see that the distribution of the IC model consists of lognormal like distributions for a wide range of diffusion probability p with clearer indication for a larger p. Here it is known that if the variance of the lognormal distribution is large, it can be reasonably approximated by a power-law distribution [23]. On the contrary, we note that the distribution of the LT model is different and is a monotonically decreasing power-law like distribution. This observation is almost true of the distribution for an individual node v except that the distribution has one peak for the LT model. One example is already shown in Fig 1 (b) for the IC model. Figures 4 and 5 show some other results for the both models. In each of these figures the most influential node for the parameter used was chosen as the initial activated source node v. From this observation, the discovery model we derived in subsections 3.2 and 3.3 can be straightforwardly applied to the IC model by assuming that the probability distribution consists of lognormal components and the succeeding experiments were performed for the IC model. However, this does not necessarily

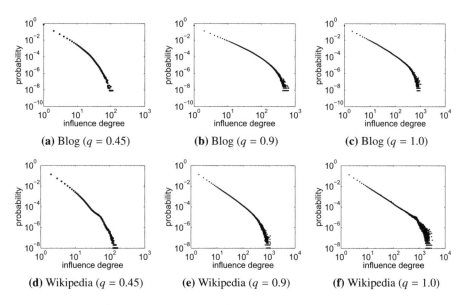

Fig. 3. The average influence degree distribution of the LT model

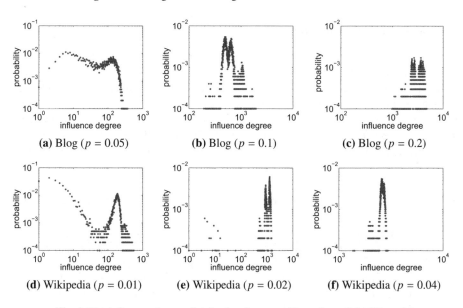

Fig. 4. The influence degree distribution for a specific node v of the IC model

mean that the notion of super-mediator is only applicable to the IC model. Finding a reasonable and efficient way to discover super-mediator nodes for the LT model is our on-going research topic. Further, the assumption of dividing the groups into only two need be justified. This is also left to our future work.

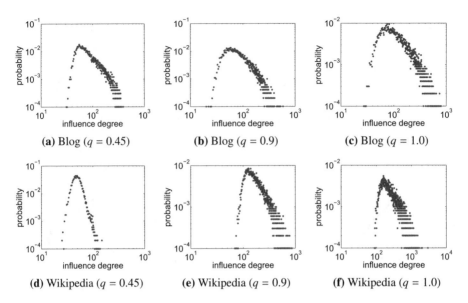

Fig. 5. The influence degree distribution for a specific node v of the LT model

4.3 Super-Mediator Ranking

Tables 1, 2 and 3 summarize the ranking results. Ranking is evaluated for two different values of diffusion probability ($p = 0.1$ and $p = 0.05$ for the blog data, and $p = 0.02$ and 0.01 for the Wikipedia data) and for the three measures mentioned in subsection 3.3. Rank by all the measures is based on the value rounded off to three decimal places. So the same rank appears more than once. The first two (Tables 1 and 2) rank the nodes by $\mathcal{F}$ for $p = 0.1$ and 0.05 (blog data) and $p = 0.02$ and 0.01 (Wikipedia data), respectively, and compare each ranking with those by $\mathcal{E}$ and $\mathcal{N}$. From these results we observe that

Table 1. Comparison of the ranking by $\mathcal{F}$ with rankings by $\mathcal{E}$ and $\mathcal{N}$ for a large diffusion probability

(a) Blog network ($p = 0.1$)				(b) Wikipedia network ($p = 0.02$)			
Ranking by $\mathcal{F}$		Ranking by $\mathcal{E}, \mathcal{N}$		Ranking by $\mathcal{F}$		Ranking by $\mathcal{E}, \mathcal{N}$	
Ranking	Node ID	$\mathcal{E}$	$\mathcal{N}$	Ranking	Node ID	$\mathcal{E}$	$\mathcal{N}$
1	146	2	2	1	790	1	1
1	155	1	1	1	8340	2	2
3	140	3	3	3	323	3	3
3	150	4	4	3	279	4	4
5	238	5	5	5	326	5	5
5	278	6	6	6	772	6	6
5	240	7	7	6	325	7	7
5	618	10	8	8	1407	8	8
9	136	8	9	9	4924	9	9
9	103	9	10	10	3149	11	10

Table 2. Comparison of the ranking by $\mathcal{F}$ with rankings by $\mathcal{E}$ and $\mathcal{N}$ for a small diffusion probability

(a) Blog network ($p = 0.05$)				(b) Wikipedia network ($p = 0.01$)			
Ranking by $\mathcal{F}$		Ranking by $\mathcal{E}, \mathcal{N}$		Ranking by $\mathcal{F}$		Ranking by $\mathcal{E}, \mathcal{N}$	
Ranking	Node ID	$\mathcal{E}$	$\mathcal{N}$	Ranking	Node ID	$\mathcal{E}$	$\mathcal{N}$
1	155	26	28	1	790	167	168
2	146	29	29	2	279	199	198
3	140	41	44	2	4019	1	1
4	150	63	66	4	3729	2	2
5	238	92	93	4	7919	3	3
6	618	79	81	4	1720	7	4
6	240	113	112	4	4465	5	6
8	103	84	86	4	1712	6	7
8	490	95	96	9	4380	4	5
8	173	88	89	9	3670	9	8

Table 3. Comparison of the ranking by $\mathcal{E}$ for a high diffusion probability with rankings by $\mathcal{E}, \mathcal{F}$, and $\mathcal{N}$ for a low diffusion probability

(a) Blog network					(b) Wikipedia network				
Ranking by $\mathcal{E}$ for $p = 0.1$		Ranking by $\mathcal{E}, \mathcal{F}, \mathcal{N}$ for $p = 0.05$			Ranking by $\mathcal{E}$ for $p = 0.02$		Ranking by $\mathcal{E}, \mathcal{F}, \mathcal{N}$ for $p = 0.01$		
Ranking	Node ID	$\mathcal{E}$	$\mathcal{F}$	$\mathcal{N}$	Ranking	Node ID	$\mathcal{E}$	$\mathcal{F}$	$\mathcal{N}$
1	155	26	1	28	1	790	167	1	168
2	146	29	2	29	2	8340	200	9	201
3	140	41	3	44	3	323	196	14	200
4	150	63	4	66	4	279	199	2	198
5	238	92	5	93	5	326	212	24	206
6	278	161	18	154	6	325	231	51	236
7	240	113	6	112	7	772	242	41	235
8	136	83	8	85	8	1407	257	80	264
9	103	84	8	86	9	4924	305	111	298
10	618	79	6	81	10	2441	279	103	287

when the diffusion probability is large all the three measures ranks the nodes in a similar way. This means that the influential nodes also play a role of super-mediator for the other source nodes. When the diffusion probability is small, the Wikipedia data still shows the similar tendency but the blog data does not. We further note that $\mathcal{E}$ and $\mathcal{N}$ rank the nodes in a similar way regardless of the value of diffusion probability. This is understandable because the both networks are bidirectional. In summary, when the diffusion probability is large, all the three measures are similar and the influential nodes also play a role of super-mediator for the other source nodes.

The third one (Table 3) ranks the nodes by $\mathcal{E}$ for $p = 0.01$ (blog data) and $p = 0.02$ (Wikipedia data) and compares them with the rankings by the three measures for $p = 0.05$ (blog data) and $p = 0.01$ (Wikipedia data). The results say that the influential nodes are different between the two different diffusion probabilities, but what is strikingly interesting to note is that the nodes that are identified to be influential (up to 10th) at a large diffusion probability are almost the same as the nodes that rank high by $\mathcal{F}$ at a small diffusion probability for the blog data. This correspondence is not that clear for the Wikipedia data but the correlation of the rankings by $\mathcal{E}$ (at a large diffusion probability) and $\mathcal{F}$ (at a small diffusion probability) is much larger than the corresponding correlation by the other two measures ($\mathcal{E}$ and $\mathcal{N}$). This implies that the super-mediators at a small diffusion probability become influential at a large diffusion probability. Since the F-measure can be evaluated by the observed information sample data alone and there is no need to know the network structure, this fact can be used to predict which nodes become influential when the diffusion probability switches from a small value for which we have enough data to a large value for which we do not have any data yet.

4.4   Characterization of Super-Mediator and Discussions

If we observe that some measure evaluated for a particular value of diffusion probability gives an indication of the influential nodes when the value of diffusion probability is changed, it would be a useful measure for finding influential nodes for a new situation. It is particularly useful when we have abundant observed set of information diffusion samples with normal diffusion probability and we want to discover high ranked influential nodes in a case where the diffusion probability is larger. For example, this problem setting corresponds to predicting the influential nodes for the unexperienced rapid spread of new information, e.g. spread of new acute contagion, because it is natural to think that we have abundant data for the spread of normal moderate contagion.

The measure based on $\mathcal{E}$ ranks high those nodes that are also influential where the diffusion probability is different from the current value if nodes are not sensitive to the diffusion probability, i.e. a measure useful to estimate influential nodes from the known results when the diffusion probability changes under such a condition. The measure based on $\mathcal{N}$ ranks high those nodes that are easily influenced by many other nodes. It is a measure useful to estimate influential nodes from the known results if they are the nodes easily influenced by other nodes. In our experiments, the influential nodes by $\mathcal{E}$ for the much larger diffusion probability, i.e. $p = 0.2$ (blog data) and $p = 0.04$ (Wikipedia data) were almost the same as the high ranked ones by any one of the three measures $\mathcal{E}$, $\mathcal{N}$ and $\mathcal{F}$ for $p = 0.1$ (blog data) and $p = 0.02$ (Wikipedia data), although we have to omit the details due to the space limitation.

In the previous subsection we showed that the super-mediators at a small diffusion probability become influential at a large diffusion probability. In a situation where there are relatively large number of active nodes, the probability that more than one parent try to activate their same child increases, which mirrors the situation where the diffusion probability is effectively large. It is the super-mediators that play the central role in these active node group under such a situation. This would explain why the super-mediators at a small diffusion probability become influential nodes at a large diffusion probability.

5   Conclusion

We found that the influence degree for the IC model exhibits a distribution which is a mixture of two distributions (power-law like distribution and lognormal like distribution). This implied that there are nodes that may play different roles in information diffusion process. We made a hypothesis that there should be nodes that play an important role to pass the information to other nodes, and called these nodes "super-mediators". These nodes are different from what is usually called "influential nodes" (nodes that spread information as much as possible). We devised an algorithm based on maximum likelihood and linear search which can efficiently identify the super-mediator node group from the observed diffusion sample data, and proposed a measure based on recall and precision to rank the super-mediators. We tested our hypothesis by applying it to the information diffusion sample data generated by two real networks. We found that the high ranked super-mediators are also the high ranked influential nodes when the diffusion probability is large, i.e. the influential nodes also play a role of super-mediator for the other source nodes, but not necessarily so when the diffusion probability is small, and further, to our surprise, that when the high ranked super-mediators are different from the top ranked influential nodes, which is the case when the diffusion probability is small, those super-mediators become the high ranked influential nodes when the diffusion probability becomes larger. This finding will be useful to predict the influential nodes for the unexperienced spread of new information from the known experience, e.g. prediction of influential nodes for the spread of new acute contagion for which we have no available data yet from the abundant data we already have for the spread of moderate contagion.

Acknowledgments

This work was partly supported by Asian Office of Aerospace Research and Development, Air Force Office of Scientific Research under Grant No. AOARD-10-4053, and JSPS Grant-in-Aid for Scientific Research (C) (No. 20500147).

References

1. Goldenberg, J., Libai, B., Muller, E.: Talk of the network: A complex systems look at the underlying process of word-of-mouth. Marketing Letters 12, 211–223 (2001)
2. Newman, M.E.J., Forrest, S., Balthrop, J.: Email networks and the spread of computer viruses. Physical Review E 66, 35101 (2002)
3. Newman, M.E.J.: The structure and function of complex networks. SIAM Review 45, 167–256 (2003)
4. Watts, D.J.: A simple model of global cascades on random networks. Proceedings of National Academy of Science, USA 99, 5766–5771 (2002)
5. Watts, D.J., Dodds, P.S.: Influence, networks, and public opinion formation. Journal of Consumer Research 34, 441–458 (2007)
6. Gruhl, D., Guha, R., Liben-Nowell, D., Tomkins, A.: Information diffusion through blogspace. SIGKDD Explorations 6, 43–52 (2004)
7. Domingos, P.: Mining social networks for viral marketing. IEEE Intelligent Systems 20, 80–82 (2005)
8. Kempe, D., Kleinberg, J., Tardos, E.: Maximizing the spread of influence through a social network. In: Proceedings of the 9th ACM SIGKDD International Conference on Knowledge Discovery and Data Mining (KDD 2003), pp. 137–146 (2003)

9. Leskovec, J., Adamic, L.A., Huberman, B.A.: The dynamics of viral marketing. In: Proceedings of the 7th ACM Conference on Electronic Commerce (EC 2006), pp. 228–237 (2006)

10. Kimura, M., Saito, K., Nakano, R.: Extracting influential nodes for information diffusion on a social network. In: Proceedings of the 22nd AAAI Conference on Artificial Intelligence (AAAI 2007), pp. 1371–1376 (2007)

11. Kimura, M., Saito, K., Nakano, R., Motoda, H.: Extracting influential nodes on a social network for information diffusion. In: Data Mining and Knowledge Discovery, vol. 20, pp. 70–97. Springer, Heidelberg (2010)

12. Kimura, M., Saito, K., Motoda, H.: Minimizing the spread of contamination by blocking links in a network. In: Proceedings of the 23rd AAAI Conference on Artificial Intelligence (AAAI 2008), pp. 1175–1180 (2008)

13. Kimura, M., Saito, K., Motoda, H.: Blocking links to minimize contamination spread in a social network. ACM Trans. Knowl. Discov. Data 3(2), Article 9, 9:1–9:23 (2009)

14. Kimura, M., Saito, K., Motoda, H.: Efficient estimation of influence functions fot SIS model on social networks. In: Proceedings of the 21st International Joint Conference on Artificial Intelligence, IJCAI 2009 (2009)

15. Saito, K., Kimura, M., Motoda, H.: Discovering influential nodes for sis models in social networks. In: Gama, J., Costa, V.S., Jorge, A.M., Brazdil, P.B. (eds.) DS 2009. LNCS (LNAI), vol. 5808, pp. 302–316. Springer, Heidelberg (2009)

16. Kimura, M., Saito, K., Nakano, R., Motoda, H.: Finding influential nodes in a social network from information diffusion data. In: Proceedings of the International Workshop on Social Computing and Behavioral Modeling (SBP 2009), pp. 138–145 (2009)

17. Saito, K., Kimura, M., Ohara, K., Motoda, H.: Learning continuous-time information diffusion model for social behavioral data analysis. In: Zhou, Z.-H., Washio, T. (eds.) ACML 2009. LNCS, vol. 5828, pp. 322–337. Springer, Heidelberg (2009)

18. Saito, K., Kimura, M., Ohara, K., Motoda, H.: Behavioral analyses of information diffusion models by observed data of social network. In: Chai, S.-K., Salerno, J.J., Mabry, P.L. (eds.) Advances in Social Computing. LNCS, vol. 6007, pp. 149–158. Springer, Heidelberg (2010)

19. Goyal, A., Bonchi, F., Lakshhmanan, L.V.S.: Learning influence probabilities in social networks. In: Proceedings of the third ACM international conference on Web Search and Data Mining, pp. 241–250 (2010)

20. Bakshy, E., Karrer, B., Adamic, L.A.: Social influence and the diffusion of user-created content. In: Proceedings of the tenth ACM conference on Electronic Commerce, pp. 325–334 (2009)

21. Leskovec, J., Krause, A., Guestrin, C., Faloutsos, C., VanBriesen, J., Glance, N.: Cost-effective outbreak detection in networks. In: Proceedings of the 13th ACM SIGKDD International Conference on Knowledge Discovery and Data Mining (KDD 2007), pp. 420–429 (2007)

22. Chen, W., Wang, Y., Yang, S.: Efficient influence maximization in social networks. In: Proceedings of the 15th ACM SIGKDD International Conference on Knowledge Discovery and Data Mining (KDD 2009), pp. 199–208 (2009)

23. Mitzenmacher, M.: A brief history of generative models for power law and lognormal distributions. Internet Mathematics 1, 226–251 (2004)

Integer Linear Programming Models for Constrained Clustering

Marianne Mueller and Stefan Kramer

Technische Universität München, Institut für Informatik, 85748 Garching, Germany

Abstract. We address the problem of building a clustering as a subset of a (possibly large) set of candidate clusters under user-defined constraints. In contrast to most approaches to constrained clustering, we do not constrain the way observations can be grouped into clusters, but the way candidate clusters can be combined into suitable clusterings. The constraints may concern the type of clustering (e.g., complete clusterings, overlapping or encompassing clusters) and the composition of clusterings (e.g., certain clusters excluding others). In the paper, we show that these constraints can be translated into integer linear programs, which can be solved by standard optimization packages. Our experiments with benchmark and real-world data investigates the quality of the clusterings and the running times depending on a variety of parameters.

1 Introduction

Constraint-based mining approaches aim for the incorporation of domain knowledge and user preferences into the process of knowledge discovery [6]. This is mostly supported by inductive query languages [5, 12, 16, 13, 4]. One of the most prominent and important instances of constraint-based mining is constrained clustering. Since its introduction (incorporating pairwise constraints into k-Means [20]), constrained clustering has been extended to various types of constraints and clustering methods [3]. Most of the approaches constrain the way observations can be grouped into clusters, i.e., they focus on *building clusters* under constraints. In this paper, we consider a different problem, namely that of *building clusterings* from a (possibly large) set of candidate clusters under constraints. In other words, we address the following problem: Given a set of candidate clusters, find a subset of clusters that satisfies user-defined constraints and optimizes a score function reflecting the quality of a clustering. Clearly, both approaches are not mutually exclusive, but just represent different aspects of finding a good clustering under constraints. In fact, the problem of constructing suitable clusterings requires a suitable set of cluster candidates, which can be the result of, e.g., constrained clustering under pairwise constraints [20] or itemset classified clustering [19].

The process of building suitable clusterings can be constrained by the user in various ways: The constraints may concern the completeness of a clustering, the disjointness of clusters, or they may concern the number of times examples are covered by clusters. Moreover, some clusters may preclude others in a clustering,

B. Pfahringer, G. Holmes, and A. Hoffmann (Eds.): DS 2010, LNAI 6332, pp. 159–173, 2010.

or clusters may require others. Constraints of the latter type can be formulated as logical formulae. The quality of a clustering to be optimized can then be defined as the mean quality of the clusters, their median quality, or their minimum quality. The paper presents a set of possible constraints along those lines and shows how they can be mapped onto integer linear program models. In this way, users can obtain tailor-made clusterings without being concerned with the technical details, much in the spirit of constraint-based mining and inductive query languages in general [6, 5]. Doing so, it is also possible to take advantage of a huge body of literature on the subject and advanced optimization tools.

2   Constrained Clustering

In this section, we present the possible constraints that can be applied to a clustering. First, we have to introduce some notation: Let $X = \{e_1, ..., e_m\}$ be a set of examples, and B a set of base clusters (i.e., cluster candidates potentially to be included in a clustering). Furthermore, the number of examples is denoted by $m = |X|$, and the number of base clusters by $n = |B|$. Then a set of clusters $C = \{C_1, ..., C_k\} \subseteq B$ denotes a clustering. Moreover, we are given an objective function $f : 2^B \to \mathbb{R}$ which scores a given clustering according to its quality. Finally, we are given constraints $\Phi(C)$ that restrict the admissible subsets of B, either with respect to the sets of instances (e.g., whether they are overlapping or encompassing one another) or with respect to known interrelationships between the clusters (e.g., cluster C_1 precludes cluster C_2 in a clustering). The overall goal is then to find a clustering $C \subseteq B$ satisfying the constraints $\Phi(C)$ and optimizing the objective function f.

Next, we discuss the different types of constraints on clusterings. First, we present *set-level constraints*, i.e., constraints in the form of logical formulae that control the clusters that can go into a clustering depending on other clusters. Second, we present *clustering constraints*, that is, constraints that determine the form of a clustering, for instance, whether the clusters are allowed to overlap. Third, three different *optimization constraints* [13] will be introduced, which determine the objective function to be optimized.

2.1   Set-Level Constraints

In the following, let a *literal* Lit_i be either a constraint $C_j \in C$ or $C_j \notin C$. For convenience, we will call $C_j \in C$ an *unnegated literal* and $C_j \notin C$ a *negated literal*. Then we can have three types of constraints:

- Conjunctive constraints $setConstraint(C, Lit_1 \wedge Lit_2 \wedge ... \wedge Lit_l)$: This type of constraint ensures that certain clusters are included or excluded from a clustering. Clusters referred to by unnegated literals have to be included, clusters referred to by negated literals have to be excluded.
- Disjunctive constraints $setConstraint(C, Lit_1 \vee Lit_2 \vee ... \vee Lit_l)$: This type of constraint ensures that at least one of the conditions holds for a clustering. If all literals are unnegated, for instance, it is possible to state that at least one of the listed clusters has to participate in a clustering.

- Clausal constraints $setConstraint(C, Lit_1 \wedge Lit_2 \wedge ... \wedge Lit_{l-1} \rightarrow Lit_l)$: This type of constraint states that if all conditions from the left-hand side are satisfied, then also the condition on the right-hand side has to be satisfied. For instance, it is possible to say that the inclusion of one cluster has to imply the inclusion of another (e.g., $C_i \in C \rightarrow C_j \in C$) or that one clusters makes the inclusion of another impossible (e.g., $C_i \in C \rightarrow C_j \notin C$).

These constraints can be translated easily into linear constraints (see Section 3). Arbitrary Boolean formulae could, in principle, be supported as well. However, they would require the definition of new variables, thus complicating the definition of the optimization problem. Note that *instance-level constraints* in the style of Wagstaff *et al.* [20], *must-link* and *cannot-link*, can easily be taken into account as well: However, here the constraints would effectively reduce the set of clusters that are put into the set of candidate clusters B. If we enforced those constraints on some set of candidate clusters B, then we would discard those individual clusters not satisfying the constraints, and restrict B to some subset $B' \subseteq B$ in the process.

Also note that the set of base clusters B can be the result of other data mining operations. For instance, consider the case of *itemset classified clustering* [19], where the potential clusters in one feature space (view) are restricted to those that can be described by frequent itemsets in another feature space (view). This is the setting that will be explored in Section 4 on experimental results.

2.2 Clustering Constraints

This section describes constraints that determine the basic characteristics of clusterings:

- $completeness(C, minCompl, maxCompl)$: This constraint determines the degree of completeness of a clustering. More formally, it ensures that for clustering C, it holds that $minCompl \leq \frac{|\cup_{C_j \in C} C_j|}{|X|} \leq maxCompl$.
- $overlap(C, minOverlap, maxOverlap)$: This constraint determines the allowed degree of overlap between clusters of a clustering. Let $coverage(e_i, C)$ be a function determining the number of clusters in a clustering C containing example e_i: $coverage(e_i, C) = |\{C_j | C_j \in C \wedge e_i \in C_j\}|$. Furthermore, let $numberOverlaps(C) = \sum_{e_i \in X,\ coverage(e_i,C)>1}(coverage(e_i, C) - 1)$ be a function counting the number of times instances are covered more than once (with multiple overlaps of one instance counted multiple times). Then the constraint $overlap(C, minOverlap, maxOverlap)$ is satisfied if $minOverlap \leq \frac{numberOverlaps(C)}{|X|} \leq maxOverlap$.[1]
- $encompassing(C, Flag)$: This constraint determines whether the clusters are allowed to encompass each other, i.e., whether the clusters are allowed to

---

[1] This is related to the notion of disjointness of clusters, but minimum overlap and maximum overlap seem more intuitive than minimum disjointness and maximum disjointness.

form a hierarchy. If $Flag = no$, then it holds that there is no pair $C_i, C_j \in C$ such that $C_i \subset C_j$.

- $numberClusters(C, minK, maxK)$: This constraint restricts the number of clusters that can be part of a solution: $minK \leq |C| \leq maxK$.
- $exampleCoverage(C, minCoverage, maxCoverage)$: This constraint limits the number of times an example can be covered by clusters. Formally, it holds that for each $e_i \in X : minCoverage \leq coverage(e_i, C) \leq maxCoverage$.

2.3   Optimization Constraints

This section introduces the optimization constraints [13] used in our approach. The quality of a clustering is defined as an aggregate over the qualities of its clusters.

- $maxMeanQuality(C)$: This constraint implies that the mean quality of the clusters contained in a clustering is optimized.
- $maxMinQuality(C)$: This constraint implies that the minimum quality of the clusters contained in a clustering is optimized.
- $maxMedianQuality(C)$: This constraint implies that the median quality of the clusters contained in a clustering is optimized.

Those optimization constraints are just the most basic ones that are conceivable. In fact, it is easy to combine the quality of a clustering with any of the above clustering constraints, for instance, not excluding overlapping clusters, but penalizing too much overlap. The same is possible for completeness (penalizing lack of completeness). Note that if cluster quality is defined as within-cluster distance, it is necessary to invert this quantity (e.g., by changing the sign) for our purposes. When translated into an integer linear model, the optimization constraints determine the objective function used.

2.4   Combining Constraints

Given the constraints introduced above, it is now possible to combine them in queries for clusterings. More precisely, a query can now be formed by a pair $(\Phi, f())$, where Φ is a logical conjunction of set-level and clustering constraints, and $f()$ is one of the three optimization constraints from the previous section.

As an example, consider the following query: $q = (numberClusters(C, 2, 5) \wedge overlap(C, 0.1, 0.2) \wedge completeness(C, 0.9, 1.0) \wedge setConstraint(C, C_1 \in C \rightarrow C_2 \in C) \wedge setConstraint(C, C_1 \in C \rightarrow C_3 \notin C), maxMeanQuality(C))$. It aims to find a clustering containing between 2 and 5 clusters, with an allowed overlap between 0.1 and 0.2, a desired completeness between 0.9 and 1.0, and such that the inclusion of cluster C_1 implies the inclusion of C_2 and the exclusion of C_3. The possible clusterings are optimized with respect to their mean cluster quality.

3   Method

In this section, we describe how to translate the introduced constraints into linear constraints. In Section 3.1 and 3.2 we focus on the clustering constraints

Table 1. Problem with 7 examples and 5 base clusters. Matrix A and vector w for the given promblem.

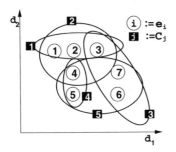

A:	C_1	C_2	C_3	C_4	C_5
e_1	1	1	0	0	0
e_2	1	1	0	0	0
e_3	1	1	1	0	0
e_4	0	1	0	1	1
e_5	0	0	0	1	1
e_6	0	0	1	0	1
e_7	0	0	1	0	1
w:	1.33	1.3	1.55	1	1.75

$(i) := e_i$
$5 := C_j$

restricted to the optimization constraint $maxMeanQuality(C)$, in Section 3.3 we handle alternative optimization constraints and in Section 3.4 we present the translation of set-level constraints. The constraints are used to form an integer linear program, which can then be solved by any package for ILP optimization.

First, we define an $m \times n$-matrix A (m being the number of examples and n being the number of base clusters as introduced above) with:

$$a_{ij} = \begin{cases} 1 \text{ if cluster } C_j \text{ contains example } e_i, \\ 0 \text{ otherwise,} \end{cases} \qquad (1)$$

and $w \in \mathbb{R}^n$, where w_j is the *within-cluster distance*[2], which is defined as the mean of the pairwise distances between the examples covered by the cluster C_j.

Table 1 shows an example for $m = 7$ examples and $n = 5$ candidate clusters. The matrix A and the vector w are displayed in Table 1. Here, w_j is the mean of the pairwise Euclidian distances between the examples covered by cluster C_j.

3.1 Modeling Clustering Constraints: Disjoint Clustering

Let our objective be to determine the disjoint clustering C with the minimal mean within-cluster distance, i.e., C has to satisfy *(completeness(C, minCompl, 1) $\wedge$ overlap(C, 0, 0), maxMeanQuality(C))*. This is also known as the Weighted Set Packing Problem [14], which is NP-complete. This task can be defined as the optimization problem shown in Table 3.

The goal is to minimize the mean of the within-cluster distance (w) over all k selected clusters, which is equivalent to maximizing the mean of the inner cluster similarities ($w_{max} - w$) of all k selected clusters, where $w_{max} := \max_j w_j$.

---

[2] Note that minimizing the within-cluster distance is equivalent to maximizing the between-cluster distance for a fixed k. In our setting, k varies between $minK$ and $maxK$. As our clustering approach is strongly constrained by the given set of base clusters, there is no or only little bias towards a k near $maxK$. For instance, smaller base clusters may not provide the required completeness, thus, a solution with larger base clusters and a smaller k may be preferred.

Table 2. Optimal clusterings for the example introduced in Table 1

minCompl	maxOverlap	selected sets	mean(w)
1	0	$\{C_1, C_5\}$	1.54
6/7	0	$\{C_1, C_5\}$	1.54
5/7	0	$\{C_1, C_4\}$	1.17
1	1/7	$\{C_1, C_3, C_4\}$	1.29
1	2/7	$\{C_2, C_3, C_4\}$	1.28
5/7	2/7	$\{C_2, C_4\}$	1.15

Table 3. Optimization task for determining the optimal disjoint clustering

maximize	$\frac{1}{k}(w_{max} - w)^T x$
subject to	(i) $Ax \leq 1$
	(ii) $Ax \geq y$ (v) $x \in \{0,1\}^n$
	(iii) $1^T x = k$ (vi) $y \in \{0,1\}^m$
	(iv) $1^T y \geq m \cdot minCompl$

Since we would like to formulate a linear program, it is not possible to optimize over the variable k that appears in the denominator of the objective function. However, we can keep the problem linear by treating k as a constant and resolve the optimization problem with varying values for k.

We introduce a vector x expressing which clusters are selected (i),(v):

$$x_j = \begin{cases} 1 \text{ if cluster } C_j \text{ is selected,} \\ 0 \text{ otherwise.} \end{cases} \qquad (2)$$

The vector y contains information about which examples are covered by the selected clusters: If $y_i = 1$ then e_i is covered by a selected cluster (ii).[3]

The clustering is further subject to the constraints that each example must not be covered by more than one clustering (i) and that at least $m \cdot minCompl$ examples have to be covered (iv). For the example in Table 1, we obtain the solutions presented in the first three rows of Table 2.

3.2   Modeling Clustering Constraints: Clustering with Overlaps

We can relax the constraint of disjointness by allowing some of the selected clusters to overlap. This means that some examples can be covered by more than one cluster: C has to satisfy (completeness(C, minCompl, 1) $\wedge$ overlap(C, 0,maxOverlap), maxMeanQuality(C)). This can be realized by the optimization task in Table 4.

---

[3] Note: if we set $minCompl < 1$, it is possible that $y_i = 0$ for some example e_i, even though e_i is covered by the selected clusters. However, this does not affect the solution.

Table 4. Optimization task for determining the optimal clustering with up to *maxOverlap* multiply covered examples

maximize	$\frac{1}{k}(w_{max} - w)^T x$	
subject to	(i) $Ax = y$	(vi) $\mathbf{1}^T v \leq n \cdot maxOverlap$
	(ii) $\mathbf{1}^T x = k$	(vii) $x \in \{0,1\}^n$
	(iii) $z \geq \mathbf{1} - y$	(viii) $y \in \mathbb{N}_0^m$
	(iv) $\mathbf{1}^T z \leq (m - m \cdot minCompl)$	(ix) $v \in \mathbb{N}_0^m$
	(v) $v \geq y - \mathbf{1}$	(x) $z \in \{0,1\}^m$

The goal is still to maximize the mean of the inner cluster similarity of the k selected sets. Again, we have the constraint that at least $m \cdot minCompl$ examples have to be covered (iv). In this setting we allow that some examples can be covered by more than one set. We restrict the number of allowed overlaps to *maxOverlap* (vi). This yields an integer-valued vector y, where y_i = number of sets that cover example e_i ($y(i) = coverage(e_i, C)$). Furthermore, we need the vector v, where v_i = number of overlaps of the example e_i.[4] To model the constraint that demands at least $minCompl$ examples to be covered, we introduce a vector z such that if $z_i = 0$ then example e_i is covered.

For the example in Table 1, we obtain the solutions presented in the lower part of Table 2 where $maxOverlap > 0$.

3.3   Modeling Optimization Constraints

So far we have shown integer linear models that determine the optimal clustering C with respect to $maxMeanQuality(C)$. In this section we will show how to model the optimization constraints $maxMinQuality(C)$ and $maxMedianQuality(C)$. The optimization task in Table 4 is modified in the following way:

$maxMinQuality(C)$: Instead of maximizing the mean $\frac{1}{k}(w_{max} - w)^T x$, the aim is now to maximize the objective function d_{min}, that is the lowest inner cluster similarity. For this purpose, we introduce the additional constraint $d_{min} \leq w_{max} - w_j x_j$ for each j, making sure that d_{min} takes the intended value. With this objective, we can remove the constraint $\mathbf{1}^T x = k$.

$maxMedianQuality(C)$: Here, the objective is to maximize d_{med}, which means to maximize the median quality. Again, we can remove the constraint $\mathbf{1}^T x = k$. We need to introduce the following additional constraints:

(xi) $x_l \in \{0,1\}^n$	(xiv) $x_l + x_r \leq \mathbf{1}$	(xvii) $(w_{max} - w_j)x_{l_j} \leq d_{med}, \; \forall j$
(xii) $x_r \in \{0,1\}^n$	(xv) $\mathbf{1}^T x_r - \mathbf{1}^T x_l \leq 1$	(xviii) $d_{med} \leq w_{max} - w_j x_{r_j}, \; \forall j$
(xiii) $x = x_l + x_r$	(xvi) $\mathbf{1}^T x_l - \mathbf{1}^T x_r \leq 0$	

---

[4] If the optimal C contains less overlaps than $n \cdot maxOverlap$, v_i may take higher values than the actual # overlaps of e_i. (v) However, this does not affect the solution.

The intuitive explanation for the introduced vectors x_l and x_r is as follows. To determine the median inner cluster similarity we partition the selected clusters into two sets whose cardinalities differ by at most 1 (constraints (xv) and (xvi)). The clusters that have a lower quality than the median-cluster quality are those having an x_l value of 1 (constraint (xvii)), and the clusters that have a higher quality than the median-cluster quality are those having an x_r value of 1 (constraint (xviii)).

Note that for the case of an even number of sets, this model maximizes the upper median, i.e., the set at the $(\frac{k}{2}+1)$th position. If the goal is to maximize the lower median, i.e., the set at the $\frac{k}{2}$th position, we need to modify the constraints (xv) and (xvi) to: (xv) $\mathbf{1}^T x_r - \mathbf{1}^T x_l \geq 1$ and (xvi) $\mathbf{1}^T x_r - \mathbf{1}^T x_l \leq 2$.

3.4 Modeling Set-Level Constraints

Finally, we explain how set-level constraints can be dealt with, and based on this, how the $encompassing(C, Flag)$ constraint can be solved. For convenience, we define a transformation operator τ on literals, which gives $\tau(C_j \in C) = x_j$ in case of unnegated and $\tau(C_j \notin C) = (1 - x_j)$ in case of negated literals.

- Conjunctive constraints $setConstraint(C, Lit_1 \wedge Lit_2 \wedge ... \wedge Lit_l)$. This states that certain clusters have to be included or cannot be included in a clustering. These constraints can directly be transformed into equality constraints of the form $x_j = 1$ for $C_j \in C$ and $x_j = 0$ for $C_j \notin C$.
- Disjunctive constraints $setConstraint(C, Lit_1 \vee Lit_2 \vee ... \vee Lit_l)$. This gives rise to an additional linear constraint of the following form: $\sum_{i=1}^{l} \tau(Lit_i) \geq 1$.
- Clausal constraints $setConstraint(C, Lit_i \wedge Lit_2 \wedge ... \wedge Lit_{l-1} \rightarrow Lit_l)$. This gives rise to an additional linear constraint $\tau(Lit_l) - (\sum_{i=1}^{l-1} \tau(Lit_i)) \geq 2 - l$. For instance, $setConstraint(C, C_1 \in C \wedge C_2 \notin C \rightarrow C_3 \in C)$ gives rise to the constraint $x_3 - x_1 - (1 - x_2) \geq -1$, i.e., $x_3 - x_1 + x_2 \geq 0$, which is only violated for $x_1 = 1, x_2 = 0, x_3 = 0$.

Using these set-level constraints, it is now possible to solve the $encompassing(C, Flag)$ constraint: If $Flag = no$, then for each $C_i \in B$ and $C_j \in B$ with $C_i \subset C_j$, the following set constraint has to be set: $setConstraint(C, C_i \notin C \vee C_j \notin C)$. In other words, it will be translated into a linear constraint $(1 - x_i) + (1 - x_j) \geq 1$, i.e., $x_i + x_j \leq 1$.

4   Experiments and Results

We implemented the two linear models of Table 3 and Table 4 and tested their performance on three datasets. For optimization, we use the Xpress-Optimizer [8], which combines common methods, such as the simplex method, cutting plane methods, and branch and bound algorithms [18].

For the first batch of experiments we use a dataset on dementia patients provided by the psychiatry and nuclear medicine departments of Klinikum rechts der Isar of Technische Universität München. It consists of two types of data: structured data (demographic information, clinical data, including neuropsychological test results) and image data (PET scans showing the patient's cerebral metabolism).[5] We include 257 data records.

Our experimental setting is similar to the usual approach in medicine: select a subset of patients fulfilling specific predefined criteria and compare the images associated with those patients. Automating this process results in determining frequent itemsets based on the structured non-image data. This can easily be achieved with the Apriori algorithm [1]. First we select a subset of patients covered by one frequent itemset. Then we evaluate the similarity of their PET scans by the mean of the pairwise weighted Euclidean distance [15]. In this way, we obtain a mean distance w_j for each itemset C_j. To tackle outliers, i.e., PET scans that are very distant from all other PET scans, we remove those data records, before generating itemsets. Otherwise each outlier affects the mean distance of each itemset it is covered by. For the given dataset, three outliers are removed. For a relative $minSupport$ of 0.1, we obtain $5,447$ itemsets.

We measure how the mean within-cluster distance[6] performs for different parameter settings. For each setting, we run the optimization task for all $k \in \{2, \ldots, \lfloor (maxOverlap + 1) \cdot n \cdot minSupport \rfloor \}$[7] and decide for the best solution of these runs. For example, for our data with $m = 254$, a $minSupport$ of 0.1 and $maxOverlap = 0.15$, we run it for all $k \in \{2, \ldots, 10\}$. We test on six different values for $maxOverlap$ $(0, 0.025, 0.05, 0.1, 0.15, 0.2)$ with $minCompl$ varying from 1.0 to 0.4. For each parameter setting, we are interested in the clustering with the lowest $mean(w)$, comparing the solutions of different k values (Figure 1). For most parameter settings, the best clustering consists of 3 or 4 clusters. Only when we allow an overlap > 0.1, we obtain optimal clusterings with $k = 5$. Figure 2 shows the $mean(w)$ of the best clustering for each parameter setting. The left diagram shows results for disjoint clusterings $(maxOverlap=0)$ and the right one results for clustering with overlaps. Overall, our experiments show that the quality of the resulting clustering can be increased by relaxing the completeness constraint, allowing overlapping clusters, and allowing smaller base clusters.

4.1   Scalability

In a second batch of experiments, we focus on the scalability of our approach. This is known to be a big challenge for integer linear programs. We experiment on two publicly available datasets on thyroid disease[8] ($m = 2,659$ examples, 16 categorical and 6 numerical attributes[9]) and on forest cover type[10] ($m =$

---

[5] For a detailed description of the data see [15].

[6] Note: our base clusters have a minimal size of $n \cdot minSupport$. Thus, we avoid that the clustering can consist of singleton clusters with a within-cluster distance of 0.

[7] Due to the minimal cluster size, there is no solution possible for a larger k.

[8] http://archive.ics.uci.edu/ml/machine-learning-databases/ thyroid-disease/

[9] We discretized the numerical attribute age into five (equally sized) values. Also, we removed examples with no numerical attribute values.

[10] http://archive.ics.uci.edu/ml/datasets/Covertype

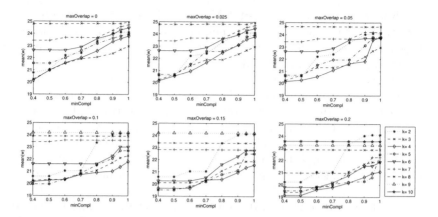

Fig. 1. Mean within-cluster distance (y-axis) for varying $maxOverlap$, $minCompl$ (x-axis), and k parameters and $minSupport = 0.1$

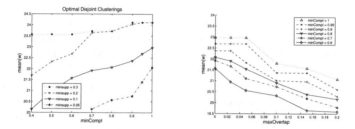

Fig. 2. Mean within-cluster distance (y-axis) for varying $minCompl$ (left diagram) of the optimal disjoint clusterings with varying $minSupport$ and for varying $maxOverlap$ (right diagram) with a $minSupport$ of 0.1

10,000 examples, 44 binary and 10 numerical attributes). The pairwise Euclidian distances are computed on the normalized numerical attributes. On the thyroid (cover type) dataset, we determine $n=12,134$ ($n=709$) frequent itemsets with a $minSupport$ of 0.1 (0.05) on the categorical attributes. We did experiments on the entire thyroid dataset and two subsets: one with $m=500$ examples and $n=12,134$ base clusters, and the other one with $m=1,000$ examples and again $n=12,134$ base clusters.

As in the previous experiments, we tested different values for k, $minCompl$, and $maxOverlap$. For all tested parameter settings on the dementia data set, time varies from 0.4 seconds (s) to almost 15 hours per problem instance, with 10 minutes on average and 31.7 s for the median.[11] On the thyroid dataset (see Table 5), the majority (86.8%) of problems is solved in less than 500 s and 36% in less than 60 s (minimum is 7.4 s). We interrupt runs that take longer than six hours and mark them with an ×.

---

[11] On a 512 MB RAM (900 MHz) machine.

Table 5. Running times (in seconds) of the thyroid dataset for varying parameters. ∗ indicates that no solution exists for these parameters. × indicates that the experiment was stopped after six hours.

Parameters			number of examples			Parameters			number of examples		
k	$maxO$	$minC$	500	1000	2659	k	$maxO$	$minC$	500	1000	2659
2	0	1	8.8	15.1	70.3	2	0.05	1	25.1	40.4	84.6
4	0	1	*8.1	*13.5	*49.8	4	0.05	1	*21.4	*38.8	*101.5
5	0	1	10.5	14.7	58.2	5	0.05	1	26	42	92.9
2	0	0.9	31.7	58.3	135	2	0.1	1	21.3	40.4	107.1
4	0	0.9	*105.4	*169.2	*420.0	4	0.1	1	*28.8	*63.3	*186.7
5	0	0.9	20.8	159	264.9	5	0.1	1	42.6	69.3	206.7
2	0	0.8	38.6	168.9	176.4	2	0.05	0.9	157.5	296.5	346.8
4	0	0.8	356.6	116.3	547.5	4	0.05	0.9	×	×	×
5	0	0.8	23.7	53.9	281.7	5	0.05	0.9	119.1	14382.3	×

Table 6. Running times for cover type ($mO = maxOverlap$ $mC = minCoverage$)

k	mO mC	mO mC	mO mC	mO mC	mO mC	mO mC	mO mC
	0 1	0 0.9	0 0.8	0.05 1	0.1 1	0.05 0.9	0.1 0.9
2	0.8	3.9	4.0	2.1	2.1	5368.5	641.4
4	0.8	3.7	4.0	2.1	2.1	×	1168.6
5	0.8	3.6	3.7	2.1	2.2	6441.8	×

The following observations from more than 500 runs on the benchmark (thyroid and cover type) and the real-world dataset (dementia) shed some light on the behavior of the optimizer depending on some key parameters:

- The running times appear to scale roughly linearly in the number of base clusters n (see Figure 3) and the number of examples m (see rows in Table 5).
- Relaxing the $maxOverlap$ constraint from zero (disjoint clusters) to slightly larger values typically leads to an (often sharp) increase in the running times. However, they may decrease again for larger values (compare Figure 4).
- A similar observation can be made for $minCompl$: Relaxing the completeness requirement from one to slightly smaller values typically leads to an (often sharp) increase in the running times. However, further reducing $minCompl$ often does not change the running times too much (compare Figures 3 and 4). Setting $minCompl = 1$ is easier to solve because the problem solver has to search only in the solution space where $y_i = 1$ for all i (compare (iv) in Table 3).
- On the thyroid and the cover type dataset, allowing overlaps in combination with a lower $minCompl$ increases the runtime dramatically (compare results for $minCompl = 0.9$ and $maxOverlap = 0.05$ in Table 5 and 6). For those settings no optimal solution could be obtained in reasonable time.
- The behavior in terms of k (between $minK$ and $maxK$) is highly non-monotonic: A problem instance may be extremely hard for a certain k,

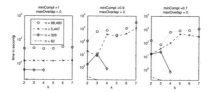

Fig. 3. Runtime (in seconds) of disjoint clustering on dementia data for different k and different numbers of base clusters n

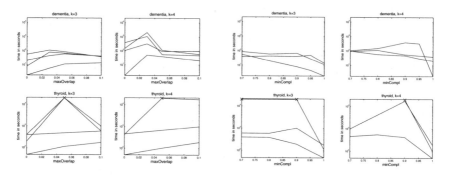

Fig. 4. Effect of varying $maxOverlap$ values on the runtime for $k = 3$ (first column) and $k = 4$ (second column) and $minCompl \in \{0.7, 0.8, 0.9, 1.0\}$. Effect of varying $minCompl$ values on the runtime for $k = 3$ (third column) and $k = 4$ (fourth column) and $maxOverlap \in \{0, 0.05, 0.1\}$.

whereas it may become easy again for $k + 1$. This may be explained by the "puzzle" that has to be solved: It may be impossible to reach a certain required completeness for a smaller number of larger "tiles". However, given a larger number of smaller "tiles", it may become possible again. The precise behavior is clearly dependent on the available base clusters.

Experiments with set-level constraints (detailed results not shown) indicate that they either make a problem insolvable (which can be determined very fast) or do not impact running times, because they constitute only a very small fraction of the constraints. Simple set-level constraints like $C_j \in C$, however, simplify the set of base levels a priori and thus speed up the overall process.

In summary, our experiments on three datasets showed that the running times depend on the structure of the problems (k, $minCompl$, $maxOverlap$) stronger than on their size (dimensions of the A matrix, number of constraints). Although the majority of tested problem instances was computable within a reasonable time, we found some instances that were more difficult to compute than others. For the latter cases, it may be an option to set a runtime limit and output a near-optimal solution.

5   Related Work

Constrained clustering has been extensively investigated over the past few years. Basu *et al.* [3] give a detailed overview of the state of the art in this area. Many contributions focus on incorporating background knowledge in the form of instance-level constraints (e.g., [20]). More recent work investigates set-level and other types of constraints (e.g, [9]). Davidson *et al.* [9, 10] study the computational complexity of finding a feasible solution for clustering with constraints and show that finding a feasible solution is NP-complete for a combination of instance-level and cluster-level constraints. Clustering has been approached with linear programs before [11, 17]. However, these approaches start from a number of instances they want to assign to clusters, whereas our approach starts from a set of possible base clusters. Demiriz *et al.* [11] use linear programs to solve k-means with the constraint that each cluster has to contain a minimum number of points. This approach is extensible to pairwise constraints. In their experiments they show that it is feasible to solve constrained clustering by linear programming even for large datasets.

Itemset classified clustering has been introduced by Sese *et al.* [19]. They start from a dataset with feature attributes and objective attributes. In our case, the PETs-voxels correspond to the objective attributes and the psychological data corresponds to the feature attributes. Sese *et al.* focus on 2-clusterings and maximize the interclass variance between the two groups. Our approach handles a more general setting and can also find k-clusterings with $k > 2$.

Although it may appear related at first glance, the approach is different from clustering approaches for association rules (e.g., by An *et al.* [2]) in its goal of constrained clustering (not summarization of pattern mining results). However, set covering and set packing approaches may also be useful for summarizing itemsets and association rules.

Chaudhuri *et al.* [7] mention the weighted set packing problem in the context of finding a partition consisting of valid groups maximizing a benefit function. In contrast to our approach, they solve the problem in a greedy fashion and thus do not aim for a global optimum.

6   Discussion and Conclusion

We presented an approach to constrained clustering based on integer linear programming. The main assumption is that a (possibly large) set of candidate clusters is given in advance, and that the task is then to construct a clustering by selecting a suitable subset. The construction of a suitable clustering from candidate clusters can be constrained in various ways. *Clustering constraints* allow specifying the degree of completeness of a clustering, the allowed overlap of clusters, and whether encompassing clusters are acceptable (hierarchical clusterings). In contrast, *set-level constraints* let the user explicitly state logical formulae that must hold for clusterings to be valid, for instance, that two clusters are mutually exclusive or that one cluster requires another in a clustering. Set-level constraints restrict the combinations of admissible clusters without reference to the instances. The overall quality of a clustering can be optimized in various ways: by optimizing the minimum, the mean and the median of the individual clusters' qualities in a clustering.

Our provided framework is very general and flexible and hence can be adapted to the user's needs. The user may start with the default values of $minCompl = 1.0$ and $maxOverlap = 0$. In case she wants to increase the quality of the resulting clustering, she can relax the completeness constraint to a lower value of $minCompl$ and/or allow for some multiply covered examples by increasing the $maxOverlap$ parameter. Additionally, set-level constraints may be used to exclude or include certain clusters.

Given such a set of constraints, it is then possible to map it onto a program for integer linear programming. In this sense, the presented work stands in the tradition of other approaches to constraint-based mining and inductive query languages, where the technical complexity of the task is hidden from the users and they can still freely combine mining primitives according to their interests and preferences [6, 5, 12, 16, 13, 4].

Although integer linear programming is known to be an NP hard problem, there are fast solvers available today, making use of a wide range of different solution strategies and heuristics. Generally speaking, the base set of candidate clusters still has to be relatively small (compared to the power set of instances) to keep the optimization feasible. Vice versa, the set of constraints should not be excessively large. Contrary to the intuition, however, that the running times should depend heavily on the number of instances and the number of available clusters, we found in our experiments that the scalability in these two parameters was not as critical as expected. Other parameters, like the degree of allowed overlap, showed a much greater impact on the running times.

From a more general point of view, it is clear that problem instances with excessive running times exist, and we also encountered such instances in our experiments. In future work, we plan to address this issue by near-optimal solutions, which are an option offered by many optimization packages like Xpress-Optimizer. One possible approach could be based on a user-defined time limit: If the optimal solution can be found within the time frame, it is returned and flagged as optimal. If the time limit is exceeded and a solution was found, the best solution could be returned and flagged as near-optimal. If no solution was found within the given time, the user is informed about this outcome. In this way, the system remains transparent about the quality of its solutions.

References

[1] Agrawal, R., Srikant, R.: Fast Algorithms for Mining Association Rules. In: Proceedings of the 20th VLDB Conference, pp. 487–499 (1994)

[2] An, A., Khan, S., Huang, X.: Objective and subjective algorithms for grouping association rules. In: Third International Conference on Data Mining, pp. 477–480 (2003)

[3] Basu, S., Davidson, I., Wagstaff, K.: Constrained Clustering: Algorithms, Applications and Theory. Chapman & Hall/CRC Press, Boca Raton (2008)

[4] Bonchi, F., Giannotti, F., Pedreschi, D.: A Relational Query Primitive for Constraint-Based Pattern Mining. In: Constraint-Based Mining and Inductive Databases, pp. 14–37 (2004)

[5] Boulicaut, J.F., Masson, C.: Data mining query languages. In: Maimon, O., Rokach, L. (eds.) The Data Mining and Knowledge Discovery Handbook, pp. 715–727 (2005)

[6] Boulicaut, J.F., Jeudy, B.: Constraint-based data mining. In: Maimon, O., Rokach, L. (eds.) The Data Mining and Knowledge Discovery Handbook, pp. 399–416 (2005)

[7] Chaudhuri, S., Sarma, A.D., Ganti, V., Kaushik, R.: Leveraging Aggregate Constraints for Deduplication. In: Proceedings of the International Conference on Management of Data (SIGMOD), pp. 437–448 (2007)

[8] Dash Optimization: XPRESS-MP, http://www.dash.co.uk

[9] Davidson, I., Ravi, S.: Clustering with Constraints: Feasibility Issues and the k-Means Algorithm. In: Proceedings of the Fifth SIAM International Conference on Data Mining (SDM 2005), pp. 138–149 (2005)

[10] Davidson, I., Ravi, S.: The complexity of non-hierarchical clustering with instance and cluster level constraints. Data Mining and Knowledge Discovery 14(1), 25–61 (2007)

[11] Demiriz, A., Bennett, K., Bradley, P.S.: Using assignment constraints to avoid empty clusters in k-means clustering. In: Basu, S., Davidson, I., Wagstaff, K. (eds.) Constrained Clustering: Algorithms, Applications and Theory (2008)

[12] De Raedt, L.: A Perspective on Inductive Databases. SIGKDD Explorations 4(2), 66–77 (2002)

[13] Dzeroski, S., Todorovski, L., Ljubic, P.: Inductive Queries on Polynomial Equations. In: Boulicaut, J.F., De Raedt, L., Mannila, H. (eds.) Constraint-Based Mining and Inductive Databases, pp. 127–154. Springer, Heidelberg (2004)

[14] Garey, M.R., Johnson, D.S.: Computers and Intractability. Freeman, New York (1979)

[15] Hapfelmeier, A., Schmidt, J., Mueller, M., Perneczky, R., Kurz, A., Drzezga, A., Kramer, S.: Interpreting PET Scans by Structured Patient Data: A Data Mining Case Study in Dementia Research. In: Eighth IEEE International Conference on Data Mining, pp. 213–222 (2008)

[16] Nijssen, S., De Raedt, S.: IQL: A Proposal for an Inductive Query Language. In: Džeroski, S., Struyf, J. (eds.) KDID 2006. LNCS, vol. 4747, pp. 189–207. Springer, Heidelberg (2007)

[17] Saglam, B., Sibel, F., Sayin, S., Turkay, M.: A mixed-integer programming approach to the clustering problem with an application in customer segmentation. European Journal of Operational Research 173(3), 866–879 (2006)

[18] Schrijver, A.: Theory of Linear and Integer Programming. John Wiley&Sons, West Sussex (1998)

[19] Sese, J., Morishita, S.: Itemset Classified Clustering. In: Proceedings of the 8th European Conference on Principles and Practice of Knowledge Discovery in Databases, pp. 398–409 (2004)

[20] Wagstaff, K., Cardie, C., Rogers, S., Schrödl, S.: Constrained K-means Clustering with Background Knowledge. In: Proceedings of the Eighteenth International Conference on Machine Learning, pp. 577–584 (2001)

Efficient Visualization of Document Streams

Miha Grčar[1], Vid Podpečan[1], Matjaž Juršič[1], and Nada Lavrač[1,2]

[1] Jožef Stefan Institute, Ljubljana, Slovenia
[2] University of Nova Gorica, Nova Gorica, Slovenia

Abstract. In machine learning and data mining, multidimensional scaling (MDS) and MDS-like methods are extensively used for dimensionality reduction and for gaining insights into overwhelming amounts of data through visualization. With the growth of the Web and activities of Web users, the amount of data not only grows exponentially but is also becoming available in the form of streams, where new data instances constantly flow into the system, requiring the algorithm to update the model in near-real time. This paper presents an algorithm for document stream visualization through a MDS-like distance-preserving projection onto a 2D canvas. The visualization algorithm is essentially a pipeline employing several methods from machine learning. Experimental verification shows that each stage of the pipeline is able to process a batch of documents in constant time. It is shown that in the experimental setting with a limited buffer capacity and a constant document batch size, it is possible to process roughly 2.5 documents per second which corresponds to approximately 25% of the entire blogosphere rate and should be sufficient for most real-life applications.

1 Introduction

Visualization is an extremely useful tool for gaining overviews and insights into overwhelming amounts of data. Handling vast streams is a relatively new challenge emerging mainly from the self-publishing activities of Web users (e.g. blogging[1], twitting, and participating in discussion forums). Furthermore, news streams (e.g. Dow Jones, BusinessWire, Bloomberg, Reuters) are growing in number and rate, which makes it impossible for the users to systematically follow the topics of their interest.

This paper discusses an adaptation of a document space visualization algorithm for document stream visualization. A document space is a high-dimensional bag-of-words space in which documents are represented as feature vectors. To visualize a document space, feature vectors need to be projected onto a 2-dimensional canvas so that the distances between the planar points reflect the cosine similarities between the corresponding feature vectors.

---

[1] Technorati <http://technorati.com/> tracks approximately 100 million blogs, roughly 15 million of them are active. Around 1 million blog posts are published each day (i.e. around 10 each second).

B. Pfahringer, G. Holmes, and A. Hoffmann (Eds.): DS 2010, LNAI 6332, pp. 174–188, 2010.
© Springer-Verlag Berlin Heidelberg 2010

When visualizing static document spaces, the dataset can be fairly large (e.g. a couple of millions of documents) thus it is important that it can be processed in a time that is still acceptable by the application (e.g. a couple of hours). On the other hand, when dealing with streams, new documents constantly flow into the system, requiring the algorithm to update the visualization in near-real time. In this case, we want to ensure that the throughput of the visualization algorithm suffices for the stream's document rate.

The contribution of our work is most notably a new algorithm for document stream visualization. In addition, we implicitly show that a set of relatively simple "tricks" can suffice for transforming an algorithm for static data processing to an algorithm for large-scale stream processing. Specifically, we rely on the "warm start" of the iterative optimization methods and on parallelization through pipelining. The former means that we use the solution computed at time $t - 1$ as the initial guess at time t, which results in faster convergence (see Sections 4.2, 4.3, and 4.5), while the latter refers to breaking up the algorithm into independent consecutive stages that can be executed in parallel (see Section 3).

The paper is organized as follows. In Section 2, we first discuss related work. Section 3 presents the algorithm for document space visualization that we adapt for document stream visualization in Section 4. We present the experimental results in Section 5. Specifically, we measure the throughput of different stages of the stream visualization pipeline. Section 6 concludes the paper with a summary and provides ideas for future work.

2   Related Work

The proposed stream visualization algorithm belongs to the family of temporal pooling algorithms [1]. These techniques maintain a buffer (a pool) of data instances: new instances constantly flow into the buffer, while outdated instances flow out of the buffer. The content of the buffer is visualized to the user and the visualization is at all times synchronized with the dynamic content of the buffer. In [1], the authors discuss TextPool, a system for document stream visualization based on temporal pooling. They extract salient terms from the buffer and construct the term co-occurrence graph. They employ a force-directed graph layout algorithm to visualize the graph to the user. Our work differs mostly in the fact that we initially[2] layout documents rather than words and that we employ MDS-like projection rather than force-based graph layout. Furthermore, while [1] focuses on the perception of motion, we provide technical details and performance evaluation. Even so, our "use" of motion fits relatively well with the guidelines provided in [1].

In contrast to temporal pooling, some researchers consider visualizing a document collection aligned with a timeline and thus emphasizing the dynamics of trends. ThemeRiver [2] visualizes thematic variations over time for a given time period. The visualization resembles a river of colored "currents" representing

---

[2] Note that we too can visualize salient words on top of the computed document layout.

different topics. A current narrows or widens to indicate a decrease or increase in the strength of the corresponding topic. The topics of interest are predefined by the analyst as a set of keywords. The strength of a topic is computed as the number of documents containing the corresponding keyword. Shaparenko et al. build on top of this idea to analyze the dataset of NIPS[3] publications [3]. They identify topics automatically by employing the k-means clustering algorithm. Our approach differs from "topic flows" mainly in the fact that we are not concerned with visualizing topic trends through time but rather topics (represented by dense clouds of documents) and their interrelatedness (represented by proximity of document clouds) in a real-time online fashion. We therefore aim to support real-time surveillance rather than temporal analysis.

Krstajić et al. [4] just recently presented a system for large-scale online visualization of news collected by the European Media Monitor (EMM). EMM collects and preprocesses news from several news sources, most notably it extracts named entities which are then used in the visualization phase. Each named entity corresponds to one topic in a ThemeRiver-like visualization. They complement their topic flow visualization with a named-entity co-occurrence graph visualization. The system processes roughly 100,000 news per day in an online fashion. Our work differs from [4] mostly in the fact that we do not rely on the data being preprocessed with a named-entity extractor. Furthermore, as already pointed out, we employ a MDS-like projection of documents onto a planar canvas rather than visualizing the term or entity co-occurrence graph.

Document Atlas [5] employs document space visualization to provide an overview of a static document corpus. It is based on Latent Semantic Indexing (LSI) [6] for dimensionality reduction and Multi-Dimensional Scaling (MDS) [7] for projection onto a canvas. In this paper, we mostly build on top of the work on static document corpora visualization presented by Paulovich et al. [8]. In the following section, we thus discuss their work in more detail.

3   Document Corpora Visualization Pipeline

The static document corpus visualization algorithm presented in [8] utilizes several methods to compute a layout. In this section, we make explicit that these methods can be percieved as a pipeline, which makes an important reinterpretation when designing algorithms for large-scale processing of streams. Throughout the rest of this section, we present our own implementation of each of the pipeline stages. The visualization pipeline is illustrated in Figure 1. In contrast to the work presented in [8], we provide details on the document preprocessing, argue for a different way of selecting representative instances, and concretize the algorithms for projection of representative instances, neighborhoods computation, and least-squares interpolation, respectively.

---

[3] Conference on Neural Information Processing Systems (NIPS) is a machine learning and computational neuroscience conference held every December in Vancouver, Canada.

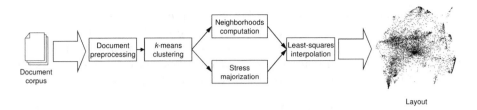

Fig. 1. Document space visualization pipeline

3.1 Document Preprocessing

To preprocess documents (i.e. convert them into a bag-of-words representation), we followed a typical text mining approach [9]. The documents were tokenized, stop words were removed, and the tokens (i.e. words) were stemmed. Bigrams were considered in addition to unigrams. If a term appeared in the corpus less than 5 times, it was removed from the vocabulary. In the end, TF-IDF vectors were computed and normalized. From each vector, the lowest weighted terms of which cumulative weight accounted for 20% of the overall cumulative weight were removed (i.e. their weights were reset to 0).

3.2 *k*-Means Clustering

To segment the document space, we implemented the *k*-means clustering algorithm [10]. The purpose of the clustering step is to obtain "representative" instances. In [8], it is suggested to take the medoids of the clusters as the representative instances. However, we decided to take the centroids rather than the medoids. In the least-squares interpolation process (the final stage of the visualization pipeline), each non-control point is required to be directly or indirectly linked to a control point. If the control points are represented by the centroids, each non-control point is guaranteed to have at least one non-orthogonal neighbor which is a control point. This prevents the situations in which a point or a clique of points is not linked to a control point and thus cannot be positioned. We believe that this change to the original algorithm results in visually more pleasing layouts (we do not provide experimental evidence to support or reject this claim as this is beyond the scope of this work).

k-means clustering is an iterative process. In each iteration, the quality of the current partition is computed as the average cosine similarity between a document instance and the centroid to which the instance was assigned. If the increase in quality, from one iteration to another, is below a predefined threshold (ε^{CL}), the clustering process is stopped.

3.3 Stress Majorization

In the final stage of the pipeline, the least-squares solver interpolates between coordinates of the projected representative instances in order to determine planar locations of the other instances. Since the number of representative instances is relatively low, it is possible to employ computationally expensive methods to project them onto a planar canvas. We therefore resorted to the stress majorization method which monotonically decreases the stress function in each iteration [11]. The stress function (energy of the model) is given in Eq. 1 and reflects the quality of the layout: the lower the stress the better the layout.

$$stress = \sum_{i<j} d_{i,j}^{-2} \left(\|\mathbf{p}_i - \mathbf{p}_j\| - d_{i,j} \right)^2 \tag{1}$$

In the equation, $\mathbf{p}_i$ is the location (both coordinates) of point i, $\|\mathbf{p}_i - \mathbf{p}_j\|$ is the Euclidean distance between points i and j, and $d_{i,j}$ is the optimal distance between points i and j. In our case, $d_{i,j}$ equals to $1 - \cosSim(\mathbf{v}_i, \mathbf{v}_j)$, that is the cosine distance between data instances $\mathbf{v}_i$ and $\mathbf{v}_j$. The iterative majorization process for Eq. 1 results in $\mathcal{O}(kn^3)$ time complexity where k is the number of iterations and n is the number of points. Therefore, we employed the localized variant as described in [11], where in each iteration, the positions of all points except one are fixed. The localized variant of stress majorization is described in Eq. 2.

$$\mathbf{p}_i = \frac{\sum_{i<j} d_{i,j}^{-2} \left(\mathbf{p}_j + \frac{d_{i,j}(\mathbf{p}_i - \mathbf{p}_j)}{\|\mathbf{p}_i - \mathbf{p}_j\|} \right)}{\sum_{i<j} d_{i,j}^{-2}} \tag{2}$$

Similar to k-means clustering, stress majorization is an iterative process. If the reduction in stress, from one iteration to another, is below a predefined threshold (ε^{SM}), the layout computation process is stopped.

3.4   Neighborhoods Computation

For the interpolation step, it is also necessary to determine k nearest neighbors of each data instance. The basic idea of the algorithm is simple: for each data instance, (a) compute the similarities to all other instances and (b) select k nearest instances from the list.

Part (b) of the naive algorithm can be efficiently implemented by choosing one of the best performing selection algorithms (e.g. the Median of Medians algorithm) which are guaranteed to have $\mathcal{O}(n)$ worst case time complexity (here, n denotes the number of instances). Provided that we need to execute this selection for each data instance, we get $\mathcal{O}(n^2)$ combined time complexity.

Efficient implementation of part (a) is more intriguing and is possible due to the fact that we use the cosine similarity measure to determine similarities between data instances. Computing cosine similarity between two instances is equivalent to computing the dot product of the two corresponding vectors (provided that the vectors are normalized). When multiplying two arbitrary vectors, the standard implementation of the dot product has a time complexity proportional to the average length of a sparse vector in a document collection, $\mathcal{O}(\text{avgLen}(\mathbf{v}_i))$. Since we need to compute n dot products for each instance, the time complexity sums up to $\mathcal{O}(n^2\text{avgLen}(\mathbf{v}_i))$ for all the instance pairs.

It is possible to reduce the time complexity of (a) and therefore of the whole k nearest neighbors computation to $\mathcal{O}(n^2 \text{avgMatch}(\mathbf{v}_i, \mathbf{v}_j))$ by employing Algorithm 1, where avgMatch$(\mathbf{v}_i, \mathbf{v}_j)$ stands for the number of elements (words) that two vectors (documents) share on average in a document collection. In our practical experiments, we noticed that avgMatch$(\mathbf{v}_i, \mathbf{v}_j)$ is usually an order of magnitude smaller than avgLen$(\mathbf{v}_i)$ which represents a substantial speed boost in practice.

Algorithm 1. Fast algorithm for computing neighborhoods

1. Build an inverted index invIdx so that $\text{invIdx}(e_k) = \{\mathbf{v}_j : \mathbf{v}_j \in \mathbf{C}, e_k \in \mathbf{v}_j\}$, where e_i represents a particular vector element (i.e. a word), $\mathbf{v}_i$ a particular vector (i.e. a document), and $\mathbf{C}$ the document corpus.
2. For each instance $\mathbf{v}_i$ do
   (a) create a dense vector $\mathbf{d}_i$ of length $|\mathbf{C}|$ and fill it with zeros,
   (b) for each $e_k \in \mathbf{v}_i$ do
       for each vector $\mathbf{v}_j \in \text{invIdx}(e_k)$ do $\mathbf{d}_i[j] = \mathbf{d}_i[j] + \text{weight}(e_k, \mathbf{v}_i) \cdot \text{weight}(e_k, \mathbf{v}_j)$,
   (c) select k nearest neighbors of $\mathbf{v}_i$ with respect to $\mathbf{d}_i$.

Note that when computing neighborhoods in the document corpus visualization pipeline, we need to add the representative instances (i.e. the centroids resulting from the k-means process) to the dataset. The centroids are thus treated in the exact same way as the documents. If n^D is the number of documents in the corpus and n^C is the user-defined number of centroids, then $n = n^D + n^C$ when computing neighborhoods.

3.5   Least-Squares Interpolation

The final stage of the pipeline employs a least-squares solver to compute the layout of the non-control points by interpolating between the coordinates of the control points. To construct a system of linear equations required for the interpolation process, we need the coordinates of the control points (obtained by the stress majorization algorithm) and the k nearest neighbors of the document instances and centroids (the neighborhoods are computed by the k-NN algorithm). The basic idea is that each point can then be described as the center of its neighbors as given in Eq. 3 [12]. In Eq. 3, (x_i, y_i) denotes a planar point and R_i denotes the set of its k nearest neighbors (note that a point is not its own nearest neighbor).

$$x_i = \tfrac{1}{|R_i|} \sum_{(x_j, y_j) \in R_i} x_j$$
$$y_i = \tfrac{1}{|R_i|} \sum_{(x_j, y_j) \in R_i} y_j \tag{3}$$

Eq. 3 can be expressed as a system of sparse linear equations (denoted with $\mathbf{AX} = \mathbf{B}$ in Figure 2). Matrix $\mathbf{A}$ is an $n \times n$ matrix, $n = n^D + n^C$, in which each row represents Eq. 3 for a particular point. $\mathbf{X}$ is a vector of length n and contains pairs of coordinates (x_i, y_i) which represent the solution of the system, i.e. the coordinates of all the points. **B** is a vector of length n and contains values $(0, 0)$.

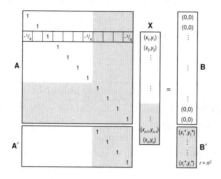

Fig. 2. System of sparse linear equations for projection of vectors onto a plane. The shaded sections denote the rows and columns that correspond to the centroids (i.e. representative instances).

In addition to equations $\mathbf{AX} = \mathbf{B}$ which describe the neighborhoods, we add equations which incorporate the coordinates of the control points (denoted with $\mathbf{A'X} = \mathbf{B'}$ in Figure 2). Such combined system is given in Eq. 4. Matrix $\mathbf{A'}$ is a $r \times n$ matrix, $r = n^C$, in which each row represents a control point. $\mathbf{B'}$ is a vector of length r and contains the actual coordinates of the control points, i.e. (x_i^*, y_i^*).

$$\begin{bmatrix} \mathbf{A}_{n \times n} \\ \mathbf{A'}_{r \times n} \end{bmatrix} \mathbf{X}_{n \times 2} = \begin{bmatrix} \mathbf{B}_{n \times 2} \\ \mathbf{B'}_{r \times 2} \end{bmatrix} \tag{4}$$

In our document stream visualization framework, the LSQR solver, developed by Paige and Saunders [13], was used to solve Eq. 4. The solution is a set of planar points corresponding to the high-dimensional feature vectors.

4   Visualization of Document Streams

This section discusses the adaptations of the document corpus visualization pipeline for document stream visualization. All stages of the pipeline are modified in a way which allow fast sequential updates, thereby allowing us to efficiently process document streams. The online document stream visualization pipeline is illustrated in Figure 3. The document stream flows into the buffer of limited capacity and thus outdated documents are gradually removed from the buffer following the FIFO (first in, first out) principle (the buffer can thus be perceived as a queue). The model required for the visualization and the visualization itself are at all times synchronized with the content of the buffer.

Before going into details of how separate stages of the pipeline are implemented, let us establish common notions required for the configuration of the pipeline:

- Let n^C denote the number of clusters computed in the k-means clustering process. This corresponds exactly to the number of representative instances (i.e. centroids) that are positioned using the stress majorization procedure.
- Let u_i be the number of documents that enter the buffer and t_i the number of documents that are removed from the buffer at time step i. For the sake of simplicity, we will assume that $u \doteq u_i \doteq t_i$ at each i.
- Let n^N be the number of closest neighbors that are assigned to each instance. The neighborhoods are used to construct the system of linear equations. Furthermore, let n^E be the extended neighborhood size, satisfying $n^E \geq n^N$. We will use extended neighborhoods to speed up the process of updating neighborhoods (see Section 4.4).
- Let n_i^Q be the number of instances in the buffer at time step i. For the sake of simplicity, we will assume that $n^Q \doteq n_i^Q$ at each i.
- Finally, let ε^{CL}, ε^{SM}, and ε^{LS} be the stopping criteria for the clustering algorithm, stress majorization method, and least-squares interpolation method, respectively.

In the next subsections, we provide online variants of the document preprocessor, k-means algorithm, stress majorization optimization method, k-nearest neighbors algorithm, and least-squares interpolation method.

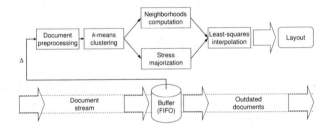

Fig. 3. Document stream visualization pipeline

4.1 Online Document Preprocessing

Online document preprocessing can be seen as a queue of term frequency (TF) bag-of-word vectors. When a number of vectors are removed from the queue, the vocabulary is updated accordingly: global document frequency (DF) values are decreased appropriately. If a global DF value reaches zero, the corresponding word is removed from the vocabulary. When a batch of new TF vectors is enqueued, on the other hand, global DF values are increased accordingly and new words (i.e. those not yet contained in the vocabulary) are added to the vocabulary.

At any time, any TF vector in the queue can be converted into its normalized TF-IDF representation by taking the global DF values into account. In this process, the original TF vector is not altered and remains at its original position in the queue. Note that a single TF vector can have many different TF-IDF representations, depending on the state of the vocabulary at the time a TF-IDF vector is computed (when the queue changes, the global DF values normally change; this results in different TF-IDF values in the affected vectors). In the presented online visualization process, each TF-IDF vector is computed immediately after the corresponding TF vector is enqueued.

4.2   Online k-Means Clustering

The online k-means clustering algorithm takes into account the centroids and the assignments of instances to the centroids from the preceding step. After the centroids are updated due to the removal of the outdated instances and assignment of the newly arrived instances (this is a relatively fast operation), the online k-means clustering algorithm proceeds with the usual k-means loop.

Assuming that the perturbation of the buffer is small and the set of data instances is much larger ($u \ll n^Q$), the centroids are proven to be stable [14] which means that the k-means algorithm will converge rapidly on the perturbed set of data instances. Specifically, if the perturbation is limited by $\mathcal{O}(\sqrt{n^Q})$ where n^Q is the number of data instances in the buffer, the online variant of the k-means algorithm is expected to converge rapidly.

4.3   Online Stress Majorization

Taking into account the stability and rapid convergence of the online variant of the k-means clustering algorithm, it is easy to see that stress majorization of the set of representative data instances (centroids) is also fast. Since the perturbation of particles (centroids) in our stress majorization optimization problem is small, the overall increase in stress is small as well, which guarantees that only very small number of recomputations of particles' positions (according to Eq. 4) is needed.

4.4   Online Neighborhoods Computation

In the first step, the neighborhoods are computed in the standard way by using the algorithm discussed in Section 3.4. For each instance i, the n^E most similar neighbors are retained in a sorted list L_i. In addition, each instance i holds the references to the instance with the highest queue index, max_i, and the instance with the lowest queue index, min_i, in the extended list of neighbors. Finally, let $minSim_i$ denote the minimum similarity between i and any $j \in L_i$. In each subsequent online step, these sorted lists are updated so that they contain at least n^N closest neighbors; max_i, min_i, and $minSim_i$ are updated with respect to the changes in the lists.

The online k-NN procedure starts by removing the outdated instances from the queue. The outdated instances are, on the one hand, the outdated bags-of-words and, on the other, the centroids that have been changed in the online k-means step. Next, the removed instances need to be removed from the lists of neighbors as well. The algorithm thus goes through all the remaining instances and for each instance first checks if min_i or max_i was removed from the queue. If and only if this is true, L_i is thoroughly examined and all the instances that were removed from the queue are also removed from the list. max_i, min_i, and $minSim_i$ are updated with respect to the changes in L_i. If the size of the list falls below n^N, a special flag $fullUpdate_i$ is set, indicating that the list should be fully updated.

Next, the newly arrived instances need to be enqueued and the lists need to be updated accordingly. Apart from the new bags-of-words obtained from the stream, the updated centroids are also enqueued. The online k-NN procedure then goes through all the enqueued instances and updates the corresponding lists of nearest neighbors. For each instance i, the algorithm first checks if i is a newly enqueued instance or if the flag $fullUpdate_i$ is set. If and only if one of these two conditions is true, L_i is computed in a usual way as explained in Section 3.4. The list is sorted and the n^E nearest neighbors are retained. On the other hand, if L_i is not required to be fully updated, only the similarities between i and each of the newly enqueued instances are computed (again, by using the algorithm discussed in Section 3.4) and put into a list, L_i'. Then, if and only if at least one of the computed similarities is greater than $minSim_i$, the lists L_i and L_i' are merged and the resulting list is sorted and trimmed so that it contains at most n^E neighbors. In either case, after the list changes, max_i, min_i, and $minSim_i$ are updated to reflect the changes in the list.

The discussed procedure results in updated neighborhoods. Each neighborhood contains between n^N and n^E nearest neighbors. The first n^N most similar neighbors of each instance are passed on to the next stage of the pipeline, where the system of linear equations is constructed.

4.5   Online Coordinates Interpolation

Modifying the coordinates interpolation step to work with streams is a relatively trivial task. We construct the system of linear equations in exactly the same way as in the original visualization algorithm.

In addition, we take the coordinates from the previous step into account when solving the system in the least-squares sense. In this work, we employ the LSQR algorithm [13] which is based on a conjugate gradient iterative method that starts with an initial guess for the solution and iteratively modifies the solution vector towards the optimal solution. In our online visualization process, the coordinates of points at time step $i + 1$ are similar to those at time step i. This results from the fact that most of the data instances and similarities between them are unchanged and thus the instances tend to move only marginally from their previous positions. Since the coordinates correspond to the solution of the least-squares solver, the coordinates from the preceding step can be used as a good initial guess for the solution. The only set of instances to which we are unable to assign coordinates from the preceding step corresponds to the batch of documents that entered the system at step i. We simply initialize that part of the solution vector to zeros.

4.6 Boundary Cases

Special care needs to be taken at the start when the buffer is filling up. While the number of instances in the buffer is smaller than or equal to the number of desired control points, $n_i^Q \leq n^C$, the k-means clustering step is skipped and the instances are projected onto a plane with the online variant of the stress majorization algorithm. There is no need for the interpolation step via the least-squares solver thus the last stage of the pipeline is skipped as well. Immediately after n_i^Q exceeds n^C, the instances processed at time step $i - 1$ are perceived as initial centroids for the k-means procedure and the online variants of k-means and stress majorization are executed. The neighborhoods computation and least-squares interpolation are performed in a standard (i.e. offline) way, setting up the pipeline for the normal online processing from this point on.

Needles to say, while $n_i^Q \leq n^C$, the instances are not flowing out of the buffer. Even after n_i^Q has exceeded n^C, all the instances are retained in the buffer (i.e. $u_i > 0, t_i = 0$) as long as the oldest buffered instances are not outdated. When the oldest instances become outdated, they start flowing out of the buffer (i.e. $u_i > 0$, $t_i > 0$). From this point on, if documents are flowing in at a relatively constant rate, we can assume that $u \doteq u_i \doteq t_i$ or, in other words, $n^Q \doteq n_i^Q$ at each i.

5   Implementation and Testing

We implemented the online document stream visualization pipeline in C# on top of LATINO[4], our software library providing a range of data mining and machine learning algorithms with the emphasis on text mining, link analysis, and data visualization. The only part of the visualization pipeline that is implemented in C++ (and not in C#) is the least-squares solver.

To measure the throughput of the visualization pipeline, we processed the first 30,000 news (i.e. from 20.8.1996 to 4.9.1996) of the Reuters Corpus Volume 1 dataset[5]. Rather than checking if the visualization pipeline is capable of processing the stream at its natural rate (i.e. roughly 1.4 news documents per minute), we measured the maximum possible throughput of the pipeline at constant u (document inflow batch size) and n^Q (buffer capacity). In our experiments, the buffer capacity was set to $n^Q = 5,000$, the document batch size to $u = 10$, the number of clusters and thus representative instances to $n^C = 100$, the k-means convergence criterion to $\varepsilon^{CL} = 10^{-3}$, the stress majorization convergence criterion to $\varepsilon^{SM} = 10^{-3}$, the size of neighborhoods to $n^N = 30$, the size of extended neighborhood to $n^E = 60$, and the least-squares convergence criterion to $\varepsilon^{LS} = 10^{-10}$.

Figure 4 shows the time that packets spent in separate stages of the pipeline (in milliseconds) when streaming the news into the system chronologically. The

---

[4] LATINO (Link Analysis and Text Mining Toolbox) is open-source—mostly under the LGPL license—and is available at http://latino.sourceforge.net/
[5] Available at http://trec.nist.gov/data/reuters/reuters.html

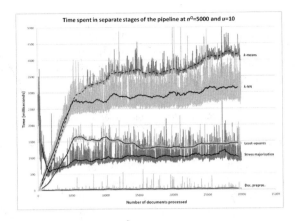

Fig. 4. Time spent in separate stages of the pipeline when streaming the news into the system in chronological order

timing started when a particular packet (i.e. a batch of documents and the corresponding data computed in the preceding stage) entered a stage and stopped when it has been processed[6]. We measured the actual time rather than the processor time to get a good feel for the performance in real-life applications. Our experiments were conducted on a simple laptop computer with an Intel processor running at 2.4 GHz, having 2 GB of memory. The purpose of the experiment was to empirically verify that each stage of the pipeline processes a packet in constant time provided that n^Q is constant. However, the chart in Figure 4 is not very convincing as the time spent in some of the stages seems to increase towards the end of the stream segment (e.g. the k-means clustering algorithm takes less than 3 seconds when 10,000 documents are processed and slightly over 4 seconds when 30,000 documents are processed). Luckily, this phenomenon turned out to be due to some temporal dataset properties. Specifically, for some reason which we do not explore in this work (e.g. "big" events, changes in publishing policies, different news vendors...), the average length of news documents in the buffer has increased over a certain period of time. This resulted in an increase of non-zero components in the corresponding TF-IDF vectors and caused dot product computations to slow down as on average more scalar products were required to compute a dot product. In other words, the positive trends in consecutive timings of the k-means clustering and neighborhoods computation algorithms are coincidental and do *not* imply that the pipeline will eventually overflow.

To prove this, we conducted another experiment in which we randomly shuffled the first 30,000 news documents and thus fed them into the system in random order. Figure 5 shows the time spent in separate stages of the pipeline when streaming the news into the system in random order. From the chart, it is possible to see that each of the pipeline stages is up to the task. After the number of

---

[6] Even if the next pipeline stage was still busy processing the previous packet, the timing was stopped in this experimental setting.

instances in the buffer has reached n^Q, it is possible to clearly observe that the processing times are kept in reasonable bounds that do not increase over time, which implies constant processing time at each time step. The gray series in the chart represent the actual times while the black series represent the moving average over 100 steps (i.e. over 1,000 documents).

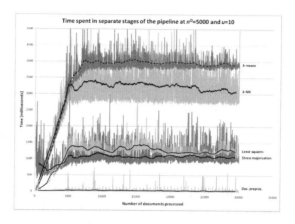

Fig. 5. Time spent in separate stages of the pipeline when streaming the news into the system in random order

In addition to measuring processing times in separate pipeline stages, we computed the delay between packets exiting the pipeline in a real pipeline-processing scenario. We simulated the pipeline processing by taking the separate processing times into account. Note that in order to actually run our algorithm in a true pipeline sense, we would need a machine that is able to process 5 processes in parallel (e.g. a computer with at least 5 cores). Let s_1, s_2, s_{3a}, s_{3b}, and s_4 correspond to separate stages of the pipeline, that is to document preprocessing, k-means clustering, stress majorization, neighborhood computation, and least-squares interpolation, respectively. Note that the stages s_{3a} and s_{3b} both rely only on the preprocessing in s_2 and can thus be executed in parallel. These two stages can be perceived as a single stage, s_3, performing stress majorization and neighborhood computation in parallel. The time a packet spends is s_3 is equal to the longer of the two times spent in s_{3a} and s_{3b}. Figure 6 shows the delay between packets exiting the pipeline. From the chart, it is possible to see that after the buffer has been filled up, the delay between two packets—this corresponds to the delay between two consecutive updates of the visualization—is roughly 4 seconds on average. This means that we are able to process a stream with a rate of at most 2.5 documents per second. Note that this roughly corresponds to 25% of the entire blogosphere rate and should be sufficient for most real-life applications. Note also that each packet, i.e. each visualization update, is delayed for approximately 9.5 seconds on average from the time a document entered the pipeline to the time it exited and was reflected in the visualization. Furthermore, since $n^Q = 5,000$, at 2.5 documents per second, the visualization

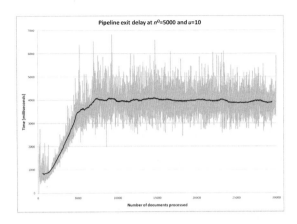

Fig. 6. The delay between packets exiting the pipeline

represents an overview of half an hour worth of documents and is suitable for real-time applications such as public sentiment surveillance in financial market decision-making.

6   Conclusions

In this paper, we presented an online algorithm for document stream visualization through a distance-preserving MDS-like projection onto a 2D canvas. The algorithm can be executed as a 4-stage pipeline, which greatly increases the processing speed. We showed that in a particular setting with limited buffer capacity and constant document batch size, the pipeline can efficiently handle 25% of the entire active blogosphere, which should be sufficient for most real-life applications. Also important to note is that the achieved visualization nicely transitions from one frame to another which enables the user to visually track a point (i.e. a document) gradually moving in the 2D space.

As part of future work, we plan to evaluate the pipeline in several more experimental settings, to better understand how different values of n^Q and u influence the maximum throughput of the pipeline. Furthermore, we aim to optimize the slowest stages of the pipeline. Our experiments indicate that k-means clustering and neighborhoods computation are the slowest stages in the current implementation. Luckily, these two algorithms are trivially parallelizable.

In addition, we plan to evaluate the visualization itself both from the perspective of the distance-preserving projection and in an application scenario. We will compare the quality of the online projection to that of MDS and other offline MDS-like techniques. On the other hand, we plan to employ the presented algorithm in the European project FIRST (Large-scale information extraction and integration infrastructure for supporting financial decision making) starting in October 2010. In FIRST, we will assess the usefulness of the presented visualization technique in the financial market decision-making process.

References

1. Albrecht-Buehler, C., Watson, B., Shamma, D.A.: Visualizing Live Text Streams Using Motion and Temporal Pooling. IEEE Computer Graphics and Applications 25/3, 52–59 (2005)
2. Havre, S., Hetzler, B., Nowell, L.: ThemeRiver: Visualizing Theme Changes over Time. In: Proceedings of InfoVis 2000, pp. 115–123 (2000)
3. Shaparenko, B., Caruana, R., Gehrke, J., Joachims, T.: Identifying Temporal Patterns and Key Players in Document Collections. In: Proceedings of TDM 2005, pp. 165–174 (2005)
4. Krstajić, M., Mansmann, F., Stoffel, A., Atkinson, M., Keim, D.A.: Processing Online News Streams for Large-scale Semantic Analysis. In: Proceedings of DESWeb 2010 (2010)
5. Fortuna, B., Grobelnik, M., Mladenić, D.: Visualization of Text Document Corpus. Informatica, pp. 270–277 (2005)
6. Deerwester, S., Dumais, S., Furnas, G., Landuer, T., Harshman, R.: Indexing by Latent Semantic Analysis. Journal of the American Society of Information Science 41/6, 391–407 (1990)
7. Groenen, P.J.F., van der Velden, M.: Multidimensional Scaling. Econometric Institute Report EI 2004-15, Netherlands, April 6 (2004)
8. Paulovich, F.V., Nonato, L.G., Minghim, R.: Visual Mapping of Text Collections through a Fast High Precision Projection Technique. In: Proceedings of the 10th Conference on Information Visualization, pp. 282–290 (2006)
9. Salton, G.: Developments in Automatic Text Retrieval. Science 253, 974–979 (1991)
10. Hartigan, J.A., Wong, M.A.: Algorithm 136: A k-Means Clustering Algorithm. Applied Statistics 28, 100–108 (1979)
11. Gansner, E.R., Koren, Y., North, S.C.: Graph Drawing by Stress Majorization, pp. 239–250 (2004)
12. Sorkine, O., Cohen-Or, D.: Least-Squares Meshes. In: Proceedings of Shape Modeling International, pp. 191–199 (2004)
13. Paige, C.C., Saunders, M.A.: Algorithm 583: LSQR: Sparse Linear Equations and Least Squares Problems. ACM Transactions on Mathematical Software 8, 195–209 (1982)
14. Rakhlin, A., Caponnetto, A.: Stability of k-Means Clustering. In: Advances in Neural Information Processing Systems, pp. 1121–1128 (2007)

Bridging Conjunctive and Disjunctive Search Spaces for Mining a New Concise and Exact Representation of Correlated Patterns

Nassima Ben Younes, Tarek Hamrouni, and Sadok Ben Yahia

URPAH, Computer Science Department, Faculty of Sciences of Tunis, Tunis, Tunisia
benyounes.nassima@gmail.com,
{tarek.hamrouni,sadok.benyahia}@fst.rnu.tn

Abstract. In the literature, many works were interested in mining frequent patterns. Unfortunately, these patterns do not offer the whole information about the correlation rate amongst the items that constitute a given pattern since they are mainly interested in appearance frequency. In this situation, many correlation measures have been proposed in order to convey information on the dependencies within sets of items. In this work, we adopt the correlation measure *bond*, which provides several interesting properties. Motivated by the fact that the number of correlated patterns is often huge while many of them are redundant, we propose a new exact concise representation of frequent correlated patterns associated to this measure, through the definition of a new closure operator. The proposed representation allows not only to efficiently derive the correlation rate of a given pattern but also to exactly offer its conjunctive, disjunctive and negative supports. To prove the utility of our approach, we undertake an empirical study on several benchmark data sets that are commonly used within the data mining community.

Keywords: Concise representation, Correlated pattern, *bond* measure, Closure operator, Equivalence class, Conjunctive support, Disjunctive support.

1 Introduction and Motivations

In data mining, frequent pattern mining from a data set constitutes an important step within the overall knowledge extraction process. Since its inception, this key task grasped the interest of many researchers since frequent patterns constitute a source of information on the relations between items. Unfortunately, the number of mined patterns from a real-life database is often huge. As a consequence, many concise representations of frequent patterns appeared in the literature. These representations are associated to different quality measures. However, the most used one is the frequency measure (*aka* the conjunctive support or, simply, support) since it sheds light on the simultaneous appearances of items in the data set. Beyond this latter measure, recently some works [8,13] have taken into account another measure, called *disjunctive support*, which conveys information about the complementary occurrences of items. However, the size of these representations remains voluminous and many frequent patterns, having weakly correlated items, are often extracted. Moreover, whenever the minimum support threshold,

B. Pfahringer, G. Holmes, and A. Hoffmann (Eds.): DS 2010, LNAI 6332, pp. 189–204, 2010.

denoted *minsupp*, is set very low, a huge number of patterns will be generated. Additionally, within the mined set of patterns, a large portion of them is redundant or uninformative. In this situation, setting a high value of *minsupp* can solve this problem, however many interesting patterns will be missed. Therefore, in order to overcome this problem and to reduce the size of representations, many correlation measures were proposed in the literature [11,12,15,18,24]. The mined correlated patterns have then been proven to be interesting for various application domains, such as text mining, bioinformatics, market basket study, and medical data analysis, etc.

To choose the appropriate measure w.r.t. a specific aim, there are various criteria which help the user in his choice. In our case, we are interested in the *bond* measure [18]. Indeed, in addition to the information on items correlations conveyed by this measure, it offers valuable information about the conjunctive support of a pattern as well as its disjunctive and negative supports. In spite of its advantages that can be exploited in several application contexts, few studies were dedicated to the *bond* measure. One of the main reasons of this negligence is that the extraction of correlated patterns w.r.t. *bond*, is proved to be more difficult than that of correlated patterns associated to other measures, like *all-confidence*, as mentioned in [15]. In this paper, we will study the behavior of the *bond* measure w.r.t. some key criteria. We then introduce a new exact concise representation of frequent correlated patterns associated to this measure. This representation – based on a new closure operator – relies on a simultaneous exploration of both conjunctive and disjunctive search spaces, whose associated patterns are respectively characterized through the conjunctive and disjunctive supports. Indeed, in a rough manner, this new representation can be considered as a compromise between both exact representations based, respectively, on the frequent closed patterns [19] and the disjunctive closed patterns [8]. Thus, it also offers the main complementary advantages of these representations, such as the direct derivation of the conjunctive and disjunctive supports of a given pattern. Interestingly enough, the proposed representation makes it also possible to find the correlation dependencies between items of a given data set without the costly computation of the inclusion-exclusion identities [5]. To the best of our knowledge, this representation is the first one proposed in the literature associated to the *bond* measure.

The remainder of the paper is organized as follows: Section 2 presents the background used throughout the paper. We also discuss related work in Section 3. Section 4 details the f_{bond} closure operator and its main properties. Moreover, it presents the new concise representation of frequent correlated patterns associated to the *bond* correlation measure. The empirical evidences about the utility of our representation are provided in Section 5. The paper ends with a conclusion of our contributions and sketches forthcoming issues in Section 6.

2 Key Notions

In this section, we briefly sketch the key notions used in the remainder of the paper.

Definition 1. - ***Data set*** - *A data set is a triplet $\mathcal{D} = (\mathcal{T}, \mathcal{I}, \mathcal{R})$ where $\mathcal{T}$ and $\mathcal{I}$ are, respectively, a finite set of transactions and items, and $\mathcal{R} \subseteq \mathcal{T} \times \mathcal{I}$ is a binary relation between the transaction set and the item set. A couple $(t, i) \in \mathcal{R}$ denotes that the transaction $t \in \mathcal{T}$ contains the item $i \in \mathcal{I}$.*

Example 1. In the remainder, we will consider the running data set $\mathcal{D}$ given in Table 1.

Table 1. An example of a data set

	A	B	C	D	E	F
1			×	×		×
2	×	×		×		
3	×	×	×	×	×	×
4	×	×	×		×	
5			×	×		×

A pattern can be characterized by three kinds of supports presented by Definition 2.

Definition 2. - ***Supports of a pattern*** - *Let $\mathcal{D} = (\mathcal{T}, \mathcal{I}, \mathcal{R})$ be a data set and I be a non-empty pattern. We mainly distinguish three kinds of supports related to I:*
- ***Conjunctive support:*** $Supp(\wedge I) = | \{t \in \mathcal{T} \mid (\forall i \in I, (t, i) \in \mathcal{R})\} |$
- ***Disjunctive support:*** $Supp(\vee I) = | \{t \in \mathcal{T} \mid (\exists i \in I, (t, i) \in \mathcal{R})\} |$
- ***Negative support:*** $Supp(\neg I) = | \{t \in \mathcal{T} \mid (\forall i \in I, (t, i) \notin \mathcal{R})\} |$

Example 2. Let us consider the data set of Table 1. We have $Supp(\wedge(BE)) = | \{3, 4\} |$ $= 2.$ [1] $Supp(\vee(BE)) = | \{2, 3, 4\} | = 3$. Moreover, $Supp(\neg(BE)) = | \{1, 5\} | = 2$.

Note that $Supp(\wedge \emptyset) = |\mathcal{T}|$ since the empty set is included in all transactions, while $Supp(\vee \emptyset) = 0$ since the empty set does not contain any item [13]. Moreover, $\forall i \in \mathcal{I}$, $Supp(\wedge i) = Supp(\vee i)$, while in the general case, for $I \subseteq \mathcal{I}$ and $I \neq \emptyset$, $Supp(\wedge I) \leq Supp(\vee I)$. A pattern I is said to be *frequent* if $Supp(\wedge I)$ is greater than or equal to a user-specified minimum support threshold, denoted *minsupp* [1]. The following lemma shows the links that exist between the different supports of a non-empty pattern I. These links are based on the *inclusion-exclusion identities* [5].

Lemma 1. - ***Inclusion-exclusion identities*** - *The inclusion-exclusion identities ensure the links between the conjunctive, disjunctive and negative supports of a non-empty pattern I.*

$$Supp(\wedge I) = \sum_{\emptyset \subset I_1 \subseteq I} (-1)^{|I_1| - 1} Supp(\vee I_1) \qquad (1)$$

$$Supp(\vee I) = \sum_{\emptyset \subset I_1 \subseteq I} (-1)^{|I_1| - 1} Supp(\wedge I_1) \qquad (2)$$

$$Supp(\neg I) = |\mathcal{T}| - Supp(\vee I) \; \text{(The De Morgan's law)} \; (3)$$

An operator is said to be a closure operator if it is *extensive*, *isotone* and *idempotent* [6]. We present patterns that help to delimit the equivalence classes induced by the conjunctive closure operator f_c [19] and the disjunctive closure operator f_d [8], respectively.

Definition 3. *[19] - **Conjunctive closure of a pattern** - The conjunctive closure of a pattern $I \subseteq \mathcal{I}$ is: $f_c(I) = \max_{\subseteq} \{I' \subseteq \mathcal{I} \mid (I \subseteq I')$ and $(Supp(\wedge I') = Supp(\wedge I))\} = I \cup \{i \in \mathcal{I} \backslash I \mid Supp(\wedge I) = Supp(\wedge (I \cup \{i\}))\}.*

---

[1] We use a separator-free form for the sets, *e.g.*, BE stands for the set of items {B, E}.

A minimal element within a conjunctive equivalence class is called *minimal generator* and is defined as follows.

Definition 4. *[19] - **Minimal generator** - A pattern $I \subseteq \mathcal{I}$ is said to be minimal generator if and only if Supp($\wedge I$) < min{Supp($\wedge I\backslash\{i\}$) | $i \in I$}.*

The following definition formally introduces a disjunctive closed pattern.

Definition 5. *[8] - **Disjunctive closure of a pattern** - The disjunctive closure of a pattern $I \subseteq \mathcal{I}$ is: $f_d(I) = \max_{\subset}\{I_1 \subseteq \mathcal{I} | (I \subseteq I_1) \wedge (Supp(\vee I) = Supp(\vee I_1))\} = I \cup \{i \in \mathcal{I}\backslash I | Supp(\vee I) = Supp(\vee (I \cup \{i\}))\}$.*

The antipode of a disjunctive closed pattern within the associated disjunctive equivalence class is called *essential pattern* and is defined as follows.

Definition 6. *[4] - **Essential pattern** - A pattern $I \subseteq \mathcal{I}$ is said to be essential if and only if Supp($\vee I$) > max{Supp($\vee I\backslash\{i\}$) | $i \in I$}.*

Definition 7 and Definition 8 introduce some properties that are interesting for the evaluation of quality measures, while Definition 9 and Definition 10 describe interesting pruning strategies that will be used in the remainder for reducing the number of generated patterns.

Definition 7. *[16] - **Descriptive or statistical measure** - A measure is said to be descriptive if its value is invariant w.r.t. the total number of transactions. Otherwise, it is said to be a statistical measure.*

Definition 8. *[21] - **Symmetric measure** - A measure μ is said to be symmetric if $\forall X$, $Y \subseteq \mathcal{I}$, $\mu(XY) = \mu(YX)$.*

Definition 9. *[17] - **Anti-monotone constraint** - Let $I \subseteq \mathcal{I}$. A constraint Q is said to be anti-monotone if $\forall I_1 \subseteq I$: I satisfies Q implies that I_1 satisfies Q.*

Definition 10. *[24] - **Cross-support patterns** - Given a threshold $t \in]0, 1[$, a pattern $I \subseteq \mathcal{I}$ is a cross-support pattern w.r.t. t if I contains two items x and y such that $\frac{Supp(\wedge x)}{Supp(\wedge y)} < t$.*

3   Related Work

Several works in the literature mainly paid attention to the extraction of frequent patterns. Nevertheless, the conjunctive support, used to estimate their respective frequency, only conveys information on items co-occurrences. Thus, it is not enough for giving the information about other kinds of items relations like their complementary occurrences as well as their mutual dependencies and inherent correlations. In order to convey information on the dependencies within sets of items and, then, to overcome the limits of the use only of the frequency measure, many correlation and similarity measures have been proposed. These latter measures were then applied in different fields like statistics, information retrieval, and data mining, for analyzing the relationships among items. For example, *lift* and χ^2 are typical correlation measures used for mining association rules [3], while *any-confidence*, *all-confidence* and *bond* [18] are used in pattern mining to assess the relationships within sets of patterns. There are also many other interestingness measures and metrics studied and used in a variety of fields and applications in order to select the most interesting patterns w.r.t. a given task. In order to select the right measure for a given application, several key properties should be examined. Recent studies have identified a critical property, null-invariance, for measuring associations among items in large data sets, but many measures do not have this property. Indeed, in [23], the authors re-examine a set of null-invariant, *i.e.*, uninfluenced by the number of null transactions, interestingness measures and they express them as a generalized mathematical mean. However in their work, the authors only considered the application of the studied measures only for patterns of size two. Moreover, other studies are based on the analysis of measures w.r.t. some desirable properties, such as the nice property of anti-monotonicity, like carried out in [14]. In this respect, anti-monotone measures are extensively used to develop efficient algorithms for mining correlated patterns [12,15,18,24]. However, almost all dedicated works to correlated patterns do not address the problem of the huge number of mined patterns while many of them are redundant. To the best of our knowledge, only the work proposed in [12] allows the extraction of a concise representation of frequent correlated patterns based on the *all-confidence* measure. Furthermore, the proposed works only rely on the exploration of the conjunctive search space for the extraction of the correlated patterns and no one was interested in the exploration of the disjunctive search space.

In addition, our work can also be linked with that proposed in [20]. This latter work presents a general framework for setting closure operators associated to some measures through the introduction of the so-called *condensable function*. In comparison to our work, that of [20] does not propose any concise representation for frequent correlated patterns using the condensable measure *bond*. In addition, the authors neither studied the structural properties of this measure nor paid attention to the corresponding link between the patterns associated to this measure and those characterizing the conjunctive search space and the disjunctive one. All these points are addressed in the following.

4 New Concise and Exact Representation of Frequent Correlated Patterns

We concentrate now on the proposed representation of frequent correlated patterns. We firstly introduce a structural characterization of the *bond* measure and, then, detail the associated closure operator on which the representation is based.

4.1 Structural Characterization of the *bond* Measure

We study, in this subsection, different interesting properties of the *bond* measure. In the literature, other equivalent measures to *bond* are used in different application contexts such as *Coherence* [15], *Tanimoto coefficient* [22], and *Jaccard* [10]. With regard to data mining, the *bond* measure is similar to the conjunctive support but w.r.t. a subset of the data rather than the entire data set. Indeed, semantically speaking, this measure conveys the information about the correlation of a pattern I by computing the ratio between the number of co-occurrences of its items and the cardinality of its universe, which is equal to the transaction set containing a non-empty subset of I. It is worth mentioning that, in the previous works dedicated to this measure, the disjunctive support has never been used to express it. Indeed, none of these works highlighted the link between the denominator – the cardinality of the universe of I – and the disjunctive support. Thus, we propose a new expression of *bond* in Definition 11.

Definition 11. - *The bond measure* - *The bond measure of a non-empty pattern $I \subseteq \mathcal{I}$ is defined as follows:*
$$bond(I) = \frac{Supp(\wedge I)}{Supp(\vee I)}$$

The use of the disjunctive support allows to reformulate the expression of the *bond* measure in order to bring out some pruning conditions for the extraction of the patterns fulfilling this measure. Indeed, as shown later, the *bond* measure satisfies several properties that offer interesting pruning strategies allowing to reduce the number of generated pattern during the extraction process. Note that the value of the *bond* measure of the empty set is undefined since its disjunctive support is equal to 0. However, this value is positive because $\lim_{I \to \emptyset} bond\ (I) = \frac{|\mathcal{T}|}{0} = +\infty$. As a result, the empty set will be considered as a correlated pattern for any minimal threshold of the *bond* correlation measure. To the best of our knowledge, none of the literature works was interested in the properties of this measure in the case of the empty set.

The following proposition presents interesting properties verified by *bond*.

Proposition 1. - *Some properties of the bond measure* - *The bond measure is descriptive and symmetric.*

Proof. The numerator of the bond measure represents the conjunctive support of a pattern I, while the denominator represents its disjunctive support. Being the ratio between two descriptive and symmetric measures, bond is also descriptive and symmetric.

Several studies [21,23] have shown that it is desirable to select a descriptive measure that is not influenced by the number of transactions that contain none of pattern items. The symmetric property fulfilled by the *bond* measure makes it possible not to treat all the combinations induced by the precedence order of items within a given pattern. Noteworthily, the anti-monotony property, fulfilled by the *bond* measure as proven in [18], is very interesting. Indeed, all the subsets of a correlated pattern are also necessarily correlated. Then, we can deduce that any pattern having at least one uncorrelated proper subset is necessarily uncorrelated. It will thus be pruned without computing the value of its *bond* measure. In the next proposition, we introduce the relationship between the *bond* measure and the cross-support property.

Proposition 2. - *Cross-support property of the bond measure* - *Any cross-support pattern $I \subseteq \mathcal{I}$, w.r.t. a threshold $t \in\ ]0, 1[$, is guaranteed to have bond$(I) < t$.*

Proof. Let $I \subseteq \mathcal{I}$ and $t \in\ ]0, 1[$. If I is a cross-support pattern w.r.t. the threshold t, then $\exists x$ and $y \in I$ such as $\frac{Supp(\wedge x)}{Supp(\wedge y)} < t$. Let us prove that bond$(I) < t$: bond$(I) =$
$$\frac{Supp(\wedge (I))}{Supp(\vee (I))} \leq \frac{Supp(\wedge (xy))}{Supp(\vee (xy))} \leq \frac{Supp(\wedge (xy))}{Supp(\vee y)} \leq \frac{Supp(\wedge x)}{Supp(\vee y)} = \frac{Supp(\wedge x)}{Supp(\wedge y)} < t.$$

The cross-support property is very important. Indeed, any pattern, containing two items fulfilling the cross-support property w.r.t. a minimal threshold of correlation, is not correlated. Thus, this property avoids the computation of its conjunctive and disjunctive supports, required to evaluate its *bond* value.

The set of frequent correlated patterns associated to *bond* is defined as follows.

Definition 12. - *The set of frequent correlated patterns* - *Considering the support threshold minsupp and the correlation threshold minbond, the set of frequent correlated patterns, denoted $\mathcal{FCP}$, is equal to: $\mathcal{FCP} = \{I \subseteq \mathcal{I} \mid bond(I) \geq minbond$ and $Supp(\wedge I) \geq minsupp\}$.*

The following proposition establishes the relation between the values of the *bond* measure as well as the conjunctive and disjunctive supports of two patterns linked by set inclusion.

Proposition 3. *Let I and I_1 be two patterns such as $I \subseteq I_1 \subseteq \mathcal{I}$. We have $bond(I) = bond(I_1)$ if and only if $Supp(\wedge I) = Supp(\wedge I_1)$ and $Supp(\vee I) = Supp(\vee I_1)$.*

Proof. The bond correlation measure of a pattern is the ratio between its conjunctive and disjunctive supports. So, if there is two patterns I and $I_1 \subseteq \mathcal{I}$, with $I \subseteq I_1$, and if they have equal values of the bond measure, they also have equal values of the conjunctive and disjunctive supports. Indeed, to have $\frac{a}{b} = \frac{c}{d}$ (where a, b, c, and d are four positive integers), three cases are possible: ($a = c$ and $b = d$) or ($a > c$ and $b > d$) or ($a < c$ and $b < d$), such that $a \times d = b \times c$. So, when we add an item i to the pattern I, its conjunctive and disjunctive supports vary inversely proportionally to each other such that $\forall i \in \mathcal{I}, Supp(\wedge I) \geq Supp(\wedge (I \cup \{i\}))$ and $Supp(\vee I) \leq Supp(\vee (I \cup \{i\}))$. Thus, the unique possibility to have $\dfrac{Supp(\wedge I)}{Supp(\vee I)} = \dfrac{Supp(\wedge (I \cup \{i\}))}{Supp(\vee (I \cup \{i\}))}$ occurs when $Supp(\wedge I) = Supp(\wedge (I \cup \{i\}))$ and $Supp(\vee I) = Supp(\vee (I \cup \{i\}))$. In an incremental manner, it can be easily shown that whenever $\dfrac{Supp(\wedge I)}{Supp(\vee I)} = \dfrac{Supp(\wedge (I \cup I_1))}{Supp(\vee (I \cup I_1))}$, $Supp(\wedge I) = Supp(\wedge (I \cup I_1))$ and $Supp(\vee I) = Supp(\vee (I \cup I_1))$.

4.2 Closure Operator Associated to the *bond* Measure

Since many correlated patterns share exactly the same characteristics, an interesting solution in order to reduce the number of mined patterns is to find a closure operator associated to the *bond* measure. Indeed, thanks to the non-injectivity property of the closure operator, correlated patterns having common characteristics will be mapped without information loss into a single element, namely the associated closed correlated pattern. The proposed closure operator is given by the following definition.

Definition 13. - *The f_{bond} operator* - *Let $\mathcal{D} = (\mathcal{T}, \mathcal{I}, \mathcal{R})$ be a data set. Let f_c and f_d be, respectively, the conjunctive closure operator and the disjunctive one. Formally, the f_{bond} operator is defined as follows:*

$$f_{bond} : \mathcal{P}(\mathcal{I}) \rightarrow \mathcal{P}(\mathcal{I})$$
$$I \mapsto f_{bond}(I) = I \cup \{i \in \mathcal{I} \setminus I \mid bond(I) = bond(I \cup \{i\})\}$$
$$= \{i \in \mathcal{I} \mid i \in f_c(I) \cap f_d(I)\}$$

The fact that the application of f_{bond} to a given pattern is exactly equal to the intersection of both its conjunctive and disjunctive closures, associated respectively to the f_c and f_d operators (*cf.* Definition 3 and Definition 5), results from Proposition 3.

Example 3. Consider the data set illustrated by Table 1. We have: since $f_c(AB) = AB$ and $f_d(AB) = ABE$, then $f_{bond}(AB) = AB$. Since $f_c(AC) = ABCE$ and $f_d(AC) = ABCDEF$, then $f_{bond}(AC) = ABCE$.

The next proposition proves that f_{bond} is a closure operator.

Proposition 4. *The f_{bond} operator is a closure operator.*

Proof. Let I, $I' \subseteq \mathcal{I}$ be two patterns. $f_{bond}(I) = f_c(I) \cap f_d(I)$ and $f_{bond}(I') = f_c(I') \cap f_d(I')$. Let us prove that the f_{bond} operator is a closure operator.
    (1) Extensivity: *Let us prove that* $I \subseteq f_{bond}(I)$
$$\begin{cases} f_c \text{ is a closure operator} \Rightarrow I \subseteq f_c(I) \\ f_d \text{ is a closure operator} \Rightarrow I \subseteq f_d(I) \end{cases} \Rightarrow \{ I \subseteq (f_c(I) \cap f_d(I)) \Rightarrow \{ I \subseteq f_{bond}(I).$$

Thus, the f_{bond} operator is extensive.

    (2) Isotony: *Let us prove that* $I \subseteq I' \Rightarrow f_{bond}(I) \subseteq f_{bond}(I')$

$$I' \subseteq I \Rightarrow \begin{cases} f_c(I') \subseteq f_c(I) \\ f_d(I') \subseteq f_d(I) \end{cases} \Rightarrow \{ (f_c(I') \cap f_d(I')) \subseteq (f_c(I) \cap f_d(I))$$
$$\Rightarrow \{ f_{bond}(I') \subseteq f_{bond}(I).$$
Thus, the f_{bond} operator is isotone.

    (3) Idempotency: *Let us prove that* $f_{bond}(f_{bond}(I)) = f_{bond}(I)$

According to *(1)*, we have $f_{bond}(I) \subseteq f_{bond}(f_{bond}(I))$. We will prove by absurdity that $f_{bond}(f_{bond}(I)) = f_{bond}(I)$.
    Suppose that $f_{bond}(I) \subset f_{bond}(f_{bond}(I))$. This is equivalent to $bond(I) \neq bond(f_{bond}(I))$.
    However, this is impossible because $bond(f_{bond}(I)) = bond(I)$ (*cf.* Proposition 3).
    Thus, $f_{bond}(f_{bond}(I)) = f_{bond}(I)$, i.e., the f_{bond} operator is idempotent.
According to *(1)*, *(2)* and *(3)*, the operator f_{bond} is a closure operator.

Definition 14. - *Closed pattern by f_{bond}* - *Let $I \subseteq \mathcal{I}$ be a pattern. The associated closure $f_{bond}(I)$ is equal to the maximal set of items containing I and having the same value of bond as that of I.*

Example 4. Consider our running data set. We have the maximal set of items which have an equal value of the *bond* measure than *AF* is *ABCDEF*. Then, $f_{bond}(AF) = ABCDEF$

The next definition introduces the set of frequent closures while Definition 16 presents the minimal patterns associated to f_{bond}.

Definition 15. *The set $\mathcal{FCCP}$ of frequent closed correlated patterns is equal to: $\mathcal{FCCP} = \{ I \in \mathcal{FCP} \mid bond(I) > bond(I \cup \{i\}), \forall\, i \in \mathcal{I} \backslash I \}$.*

Definition 16. - *Frequent minimal correlated pattern* - Let $I \in \mathcal{FCP}$. The pattern I is said to be minimal if and only if $\forall\, i \in I$, $bond(I) < bond(I \backslash \{i\})$ or, equivalently, $\nexists\, I_1 \subset I$ such that $f_{bond}(I) = f_{bond}(I_1)$.

Example 5. Consider our running data set illustrated by Table 1 for *minsupp* = *1* and *minbond* = *0.30*. The pattern *CE* is a minimal one since $bond(CE) < bond(C)$ and $bond(CE) < bond(E)$. Moreover, the pattern *CE* is correlated and frequent since $bond(CE) = 0.50 > 0.30$ and $Supp(\wedge(CE)) = 2 \geq 1$.

The next proposition links a minimal pattern with the key notions of minimal generator (*cf.* Definition 4) and essential pattern (*cf.* Definition 6) of the conjunctive and disjunctive search spaces, respectively.

Proposition 5. *Every minimal generator (resp. essential pattern) is a minimal pattern.*

Proof. Let $I \subseteq \mathcal{I}$ be a minimal generator (resp. essential pattern). $\forall\, i \in I$, $Supp(\wedge I) < Supp(\wedge(I \backslash \{i\}))$ and $Supp(\vee I) \geq Supp(\vee(I \backslash \{i\}))$ (resp. $Supp(\vee I) > Supp(\vee(I \backslash \{i\}))$ and $Supp(\wedge I) \leq Supp(\wedge(I \backslash \{i\})))$. Thus, in both cases, $\dfrac{Supp(\wedge I)}{Supp(\vee I)} < \dfrac{Supp(\wedge (I \backslash \{i\}))}{Supp(\vee (I \backslash \{i\}))}$. As a result, $bond(I) \neq bond(I \backslash \{i\})$ and, hence, I is a minimal pattern.

It is important to note that a minimal pattern can be neither an essential pattern nor a minimal generator. This is illustrated through the following example.

Example 6. Consider our running data set illustrated by Table 1. According to Example 5, *CE* is a minimal pattern, although it is neither a minimal generator (since $Supp(\wedge (CE)) = Supp(\wedge E)$) nor an essential pattern (since $Supp(\vee (CE)) = Supp(\vee C)$).

In the remainder, we will consider the empty set as a frequent minimal correlated pattern given that the values of its conjunctive support and that of its *bond* measure exceed both *minsupp* and *minbond* thresholds, respectively (the conjunctive support of the empty set is equal to $|\mathcal{T}| \geq$ *minsupp* and the value of its *bond* measure tends to $+\infty$ when I tends to $\emptyset$). Besides, we will consider the closure of the empty set as equal to itself. Let us note that these considerations are important and allow the set of frequent minimal correlated patterns to be flagged as *order ideal* (*aka downward closed set*) [6], without having any effect neither on the supports of the other patterns nor on their closures.

Proposition 6. *The set $\mathcal{FMCP}$ of the frequent minimal correlated pattern is an order ideal. Thus, it fulfills the following properties:*

- If $X \in \mathcal{FMCP}$, then $\forall\, Y \subseteq X$, $Y \in \mathcal{FMCP}$, i.e., the constraint "be a frequent minimal correlated pattern" is anti-monotone.

- If $X \notin \mathcal{FMCP}$, then $\forall\, Y \supseteq X$, $Y \notin \mathcal{FMCP}$, i.e., the constraint "not to be a frequent minimal correlated pattern" is monotone.

Proof. The proof results from the following fact: for $i \in \mathcal{I}$ and for all $X \subseteq Y \subset \mathcal{I}$, if $bond(X \cup \{i\}) = bond(X)$, then $bond(Y \cup \{i\}) = bond(Y)$, i.e., if $(X \cup \{i\})$ is not a minimal pattern, so $(Y \cup \{i\})$ is also not minimal. The constraint "not to be a minimal pattern" is hence monotone, w.r.t. set inclusion. We deduce that the constraint "be a minimal pattern" is anti-monotone. Moreover, the constraint "be a frequent minimal correlated pattern" is anti-monotone since resulting from the conjunction of three *anti-monotone constraints: "to be frequent", "to be correlated", and "to be minimal". Conversely, the constraint "not to be a frequent minimal correlated pattern" is mono-tone. We then deduce that the set $\mathcal{FMCP}$ is an order ideal.*

The closure operator f_{bond} induces an equivalence relation on the power-set of the set of items $\mathcal{I}$, which splits it into disjoint subsets, called f_{bond} *equivalence classes*. In each class, all the elements have the same f_{bond} closure and the same value of *bond*. The minimal patterns of a *bond* equivalence class are the smallest incomparable members, w.r.t. set inclusion, while the f_{bond} closed pattern is the largest one.

   To establish the link with the conjunctive and disjunctive search spaces, an f_{bond} equivalence class as well as conjunctive and disjunctive classes are given in Figure 1. The equivalence class associated to the *bond* measure can then be considered as an intermediary representation of both conjunctive and disjunctive ones.

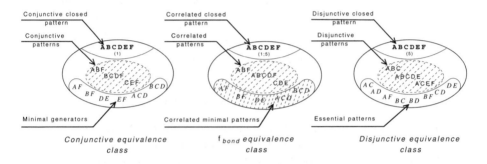

Fig. 1. Structural characterization of the equivalence classes associated respectively (from left to right) to the f_c, f_{bond}, and f_d closure operators w.r.t. the data set given in Table 1.

4.3   New Concise Representation Associated to the *bond* Measure

Based on the f_{bond} closure operator, we can design two representations which cover the same frequent correlated patterns. The first is based on the frequent closed correlated patterns, whereas the second one is based on the frequent minimal correlated patterns. In this work, we focus on the first one, since it is considered more concise thanks to the fact that a f_{bond} equivalence class always contains only one closed pattern, but potentially several minimal patterns. Let us define the new concise representation of frequent correlated patterns based on the frequent closed correlated patterns associated to the *bond* measure.

Definition 17. *The representation $\mathcal{RFCCP}$ based on the set of frequent closed corre-lated patterns associated to f_{bond} is defined as follows:*

$$\mathcal{RFCCP} = \{(I, Supp(\wedge I), Supp(\vee I)) \mid I \in \mathcal{FCCP} \}.$$

Example 7. Consider our running data set illustrated by Table 1. For *minsupp = 2* and *minbond = 0.60*, the representation $\mathcal{RFCCP}$ of the $\mathcal{FCP}$ set is equal to: $\{(\emptyset, 5, 0), (C, 4, 4), (D, 4, 4), (E, 2, 2), (F, 3, 3), (AB, 3, 3), (CF, 3, 4), (DF, 3, 4), (ABE, 2, 3), (CDF, 3, 5)\}$.

The next theorem proves that the proposed representation is an exact one of frequent correlated patterns.

Theorem 1. *The representation $\mathcal{RFCCP}$ constitutes an exact concise representation of the $\mathcal{FCP}$ set.*

Proof. Thanks to a reasoning by recurrence, we will demonstrate that, for an arbitrary pattern $I \subseteq \mathcal{I}$, its f_{bond} closure, $f_{bond}(I)$, belongs to $\mathcal{FCCP}$ if it is frequent correlated. In this regard, let $\mathcal{FMCP}_k$ be the set of frequent minimal correlated patterns of size k and $\mathcal{FCCP}_k$ be the associated set of closures by f_{bond}. The hypothesis is verified for single items i inserted in $\mathcal{FMCP}_1$, and their closures $f_{bond}(i)$ are inserted in $\mathcal{FCCP}_1$ if $Supp(\wedge i) \geq minsupp$ (since $\forall i \in \mathcal{I}, bond(i) = 1 \geq minbond$). Thus, $f_{bond}(i) \in \mathcal{FCCP}$. Now, suppose that $\forall I \subseteq \mathcal{I}$ such as $|I| = n$. We have $f_{bond}(I) \in \mathcal{FCCP}$ if I is frequent correlated. We show that, $\forall I \subseteq \mathcal{I}$ such as $|I| = (n+1)$, we have $f_{bond}(I) \in \mathcal{FCCP}$ if I is frequent correlated. Let I be a pattern of size $(n+1)$. Three situations are possible:
(a) if $I \in \mathcal{FCCP}$, then necessarily $f_{bond}(I) \in \mathcal{FCCP}$ since f_{bond} is idempotent.
(b) if $I \in \mathcal{FMCP}_{n+1}$, then $f_{bond}(I) \in \mathcal{FCCP}_{n+1}$ and, hence, $f_{bond}(I) \in \mathcal{FCCP}$.
(c) if I is neither closed nor minimal – $I \notin \mathcal{FCCP}$ and $I \notin \mathcal{FMCP}_{n+1}$ – then $\exists I_1 \subset I$ such as $|I_1| = n$ and $bond(I) = bond(I_1)$. According to Proposition 3, $f_{bond}(I) = f_{bond}(I_1)$, and I is then frequent correlated. Moreover, using the hypothesis, we have $f_{bond}(I_1) \in \mathcal{FCCP}$ and, hence, $f_{bond}(I) \in \mathcal{FCCP}$.

It is worth noting that maintaining both conjunctive and disjunctive supports for each pattern belonging to the representation allows to avoid the cost of the evaluation of the inclusion-exclusion identities. Indeed, this evaluation can be very expensive, in particular in the case of long correlated patterns to be derived. For example, for a pattern containing 20 items, the evaluation of an inclusion-exclusion identity will involve the computation of the supports of all its non-empty subsets, *i.e.*, 2^{20} - 1 terms (*cf.* Lemma 1, page 191). Such an evaluation will be mandatory if we retain only one support and not both. It will then be carried out in order to derive the non-retained support to compute the value of the *bond* measure for each pattern. Thus, contrarily to the main concise representations of the literature, the regeneration of the whole frequent correlated patterns from the representation $\mathcal{RFCCP}$ can be carried out in a very simple and effective way. Indeed, in an equivalence class associated to the *bond* measure, patterns present the same value of this measure and consequently the same conjunctive, disjunctive and negative supports. Then, to derive the information corresponding to a frequent correlated pattern, it is enough to locate the smallest frequent closed correlated pattern which covers it and which corresponds to its closure by f_{bond}. Thus, we avoid the highly costly evaluation of the inclusion-exclusion identities.

Note however that the closure operator associated to the *bond* measure induces a strong constraint. Indeed, the f_{bond} operator gathers the patterns having the same conjunctive and disjunctive supports (*cf.* Proposition 3). Consequently, the number of patterns belonging to a given equivalence class associated to this operator is in almost all cases lower than those resulting when the conjunctive and the disjunctive closure operators are separately applied. Fortunately, the pruning based on both thresholds *minsupp* and *minbond* drastically reduces the size of our concise representation, as shown in the next section.

5   Experimental Results

In this section, our objective is to show, through extensive experiments, that our concise representation based on frequent closed correlated patterns provides interesting compactness rates compared to the whole set of frequent correlated patterns. All experiments were carried out on a PC equipped with a 2 GHz Intel processor and 4 GB of main memory, running the Linux Ubuntu 9.04 (with 2 GB of swap memory). The experiments were carried out on benchmark data sets[2].

We first show that the complete set of frequent correlated patterns ($\mathcal{FCP}$) is much bigger in comparison with both that of frequent correlated closed patterns ($\mathcal{FCCP}$) and that of frequent minimal correlated patterns ($\mathcal{FMCP}$) especially for low *minsupp* and *minbond* values. In this respect, Figure 2 presents the cardinalities of these sets when *minsupp* varies and *minbond* is fixed, while, in Figure 3, cardinalities are shown when *minbond* varies and *minsupp* is fixed.

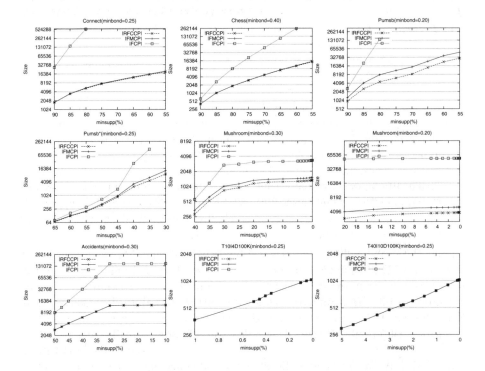

Fig. 2. Number of patterns generated when *minsupp* varies and *minbond* is fixed

The obtained results show that the size of $\mathcal{FCCP}$ is always smaller than that of $\mathcal{FMCP}$ over the entire range of the support and *bond* thresholds. For example, considering the PUMSB data set for *minsupp = 50%* and *minbond = 0.5*: $|\mathcal{FMCP}| = 68, 532$, while $|\mathcal{RFCCP}| = 40, 606$, with a reduction reaching approximately *41%*. These results

---
[2] Available at http://fimi.cs.helsinki.fi/data

are obtained thanks to the closure operator f_{bond} which gathers into disjoint subsets, *i.e.*, f_{bond} *equivalence classes*, patterns that have the same characteristics.

The key role of this operator is all the more visible when we compare the number of the whole set of correlated patterns with that of the proposed representation. In this respect, Figures 2 and 3 show that $\mathcal{FCCP}$ mining generates a much smaller set than that of frequent correlated patterns. Interestingly enough, compression rates increase proportionally with the decrease of the *minsupp* and *minbond* values. It is hence a desirable phenomenon since the number of frequent correlated patterns increases dramatically as far as one of both thresholds decreases. For example, let us consider the PUMSB* data set, and *minbond* fixed at *0.25%*: for *minsupp* = *60%*: $\frac{|\mathcal{FCP}|}{|\mathcal{RFCCP}|} = \frac{167}{124} = 1.34$, while for *minsupp* = *35%*: $\frac{|\mathcal{FCP}|}{|\mathcal{RFCCP}|} = \frac{116, 787}{4, 546} = 25.69 \gg 1.34$. In fact, in general, only single items can fulfill high values of thresholds. In this situation, the set of frequent correlated patterns only contains items which are in most cases equal to their closures. However, when thresholds are set very low, a high number of frequent correlated patterns, which are in general not equal to their respective closures, is extracted.

Noteworthily, the size reduction rates brought by the proposed representation, w.r.t. the size of the $\mathcal{FCP}$ set, are closely related to the chosen *minsupp* and *minbond*

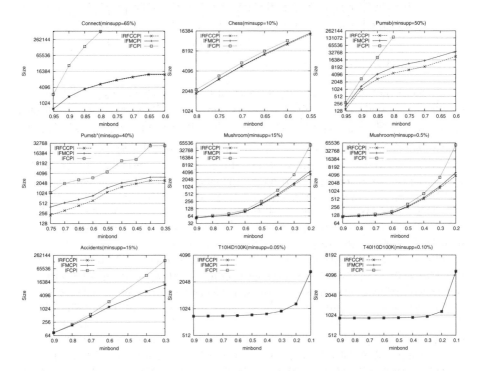

Fig. 3. Number of patterns generated when *minbond* varies and *minsupp* is fixed

values (*cf.* Figure 2 (*resp.* Figure 3) for a variation of the fixed value of *minbond* (*resp. minsupp*) for the MUSHROOM data set). Note that rates vary depending on the data set characteristics and are, hence, more important for some data sets than for others. In this respect, for CONNECT, PUMSB and PUMSB*, the obtained rates are more interesting than for CHESS, MUSHROOM and ACCIDENTS. This is explained by the fact that for the first three data sets, which contain strongly correlated items, the f_{bond} operator produces equivalence classes containing a high number of patterns. Thus, the number of closures is much more reduced compared to that of the whole set of patterns, even for high values of *minsupp* and *minbond*. While for the latter three data sets, items are relatively less correlated than for the three first ones, which decreases the number of patterns having common characteristics, and hence the same closure.

For the two data sets T10I4D100K and T40I10D100K, the size of the representation $\mathcal{RFCCP}$ is almost equal to that of the $\mathcal{FCP}$ set. This is due to the nature of these data sets, which contain a large number of items but only a few of them frequently occur. Moreover, most of them are weakly correlated with each other. This makes the size reduction rates brought by the representation meaningless in such data sets. It is important to note that, these two data sets are the "worst" for the f_{bond} closure operator as well as for the conjunctive and disjunctive closure operators (*cf.* [7,8] for experimental results associated to these two latter operators). In addition, the number of frequent correlated patterns that are extracted from these data sets is relatively reduced for each used value of *minsupp* and *minbond*.

6   Conclusion and Perspectives

In this work, we studied the behavior of the *bond* correlation measure according to some key properties. In addition, we introduced a new closure operator associated to this measure and we thoroughly studied its theoretical properties. Based on this operator, we characterized the elements of the search space associated to the *bond* measure. Then, we introduced a new concise representation of frequent patterns based on the frequent closed correlated patterns. Beyond interesting compactness rates, this representation allows a straightforward computation of the conjunctive, disjunctive and negative supports of a pattern. In nearly all experiments we performed, the obtained results showed that our representation is significantly smaller than the whole set of frequent correlated patterns.

Other avenues for future work mainly address a thorough analysis of the computational time required for mining our representation and, then, for the derivation process of the whole set of frequent patterns. In this respect, efficient algorithms for mining conjunctive closed patterns (like LCM and DCI-CLOSED [2]) and disjunctive closed patterns (like DSSRM [9]) could be adapted for mining frequent closed correlated patterns. Other important tasks consist in applying the proposed approach in real-life applications and extending it by (i) generating association rules starting from correlated frequent patterns, and, (i) extracting unfrequent (*aka* rare) correlated patterns associated to the *bond* measure by selecting the most informative ones.

References

1. Agrawal, R., Srikant, R.: Fast algorithms for mining association rules. In: Proceedings of the 20th International Conference on VLDB 1994, Santiago, Chile, pp. 487–499 (1994)
2. Ben Yahia, S., Hamrouni, T., Mephu Nguifo, E.: Frequent closed itemset based algorithms: A thorough structural and analytical survey. ACM-SIGKDD Explorations 8(1), 93–104 (2006)
3. Brin, S., Motwani, R., Silverstein, C.: Beyond market baskets: generalizing association rules to correlations. In: Proceedings of the ACM SIGMOD International Conference on SIGMOD 1997, Tucson, Arizona, USA, pp. 265–276 (1997)
4. Casali, A., Cicchetti, R., Lakhal, L.: Essential patterns: A perfect cover of frequent patterns. In: Proceedings of the 7th International Conference on DaWaK, Copenhagen, Denmark, pp. 428–437 (2005)
5. Galambos, J., Simonelli, I.: Bonferroni-type inequalities with applications. Springer, Heidelberg (2000)
6. Ganter, B., Wille, R.: Formal Concept Analysis. Springer, Heidelberg (1999)
7. Hamrouni, T.: Mining concise representations of frequent patterns through conjunctive and disjunctive search spaces. Ph.D. Thesis, University of Tunis El Manar (Tunisia) and University of Artois (France) (2009), http://tel.archives-ouvertes.fr/tel-00465733
8. Hamrouni, T., Ben Yahia, S., Mephu Nguifo, E.: Sweeping the disjunctive search space towards mining new exact concise representations of frequent itemsets. Data & Knowledge Engineering 68(10), 1091–1111 (2009)
9. Hamrouni, T., Ben Yahia, S., Mephu Nguifo, E.: Optimized mining of a concise representation for frequent patterns based on disjunctions rather than conjunctions. In: Proceedings of the 23rd International Florida Artificial Intelligence Research Society Conference (FLAIRS 2010), pp. 422–427. AAAI Press, Daytona Beach, Florida, USA (2010)
10. Jaccard, P.: Nouvelles recherches sur la distribution florale. Bulletin de la Société Vaudoise des Sciences Naturelles 44, 223–270 (1908)
11. Ke, Y., Cheng, J., Yu, J.X.: Efficient discovery of frequent correlated subgraph pairs. In: Proceedings of the 9th IEEE International Conference on Data Mining, Miami, Florida, USA, pp. 239–248 (2009)
12. Kim, W.Y., Lee, Y.K., Han, J.: CCMINE: Efficient mining of confidence-closed correlated patterns. In: Proceedings of the 8th International Pacific-Asia Conference on KDD, Sydney, Australia, pp. 569–579 (2004)
13. Kryszkiewicz, M.: Compressed disjunction-free pattern representation versus essential pattern representation. In: Corchado, E., Yin, H. (eds.) IDEAL 2009. LNCS, vol. 5788, pp. 350–358. Springer, Heidelberg (2009)
14. Le Bras, Y., Lenca, P., Lallich, S.: Mining interesting rules without support requirement: a general universal existential upward closure property. Annals of Information Systems 8, 75–98 (2010)
15. Lee, Y.K., Kim, W.Y., Cai, Y.D., Han, J.: CoMine: Efficient mining of correlated patterns. In: Proceedings of the 3rd IEEE International Conference on Data Mining, Melbourne, Florida, USA, pp. 581–584 (2003)
16. Lenca, P., Vaillant, B., Meyer, P., Lallich, S.: Association rule interestingness measures: Experimental and theoretical studies. In: Quality Measures in Data Mining, Studies in Computational Intelligence, vol. 43, pp. 51–76. Springer, Heidelberg (2007)
17. Mannila, H., Toivonen, H.: Levelwise search and borders of theories in knowledge discovery. Data Mining and Knowledge Discovery 1(3), 241–258 (1997)
18. Omiecinski, E.R.: Alternative interest measures for mining associations in databases. IEEE Transactions on Knowledge and Data Engineering 15(1), 57–69 (2003)

19. Pasquier, N., Bastide, Y., Taouil, R., Stumme, G., Lakhal, L.: Generating a condensed representation for association rules. Journal of Intelligent Information Systems 24(1), 25–60 (2005)
20. Soulet, A., Crémilleux, B.: Adequate condensed representations of patterns. Data Mining and Knowledge Discovery 17(1), 94–110 (2008)
21. Tan, P.N., Kumar, V., Srivastava, J.: Selecting the right interestingness measure for association patterns. In: Proceedings of the 8th ACM SIGKDD International Conference on KDD, Edmonton, Alberta, Canada, pp. 32–41 (2002)
22. Tanimoto, T.T.: An elementary mathematical theory of classification and prediction. Technical Report, I.B.M. Corporation Report (1958)
23. Wu, T., Chen, Y., Han, J.: Re-examination of interestingness measures in pattern mining: a unified framework. Data Mining and Knowledge Discovery (2010) doi: 10.1007/s10618-009-0161-2
24. Xiong, H., Tan, P.N., Kumar, V.: Hyperclique pattern discovery. Data Mining and Knowledge Discovery 13(2), 219–242 (2006)

Graph Classification Based on Optimizing Graph Spectra

Nguyen Duy Vinh[1], Akihiro Inokuchi[1,2], and Takashi Washio[1]

[1] The Institute of Scientific and Industrial Research, Osaka University
[2] PRESTO, Japan Science and Technology Agency
{inokuchi,washio}@ar.sanken.osaka-u.ac.jp

Abstract. Kernel methods such as the SVM are becoming increasingly popular due to their high performance in graph classification. In this paper, we propose a novel graph kernel, called SPEC, based on graph spectra and the Interlace Theorem, as well as an algorithm, called OPT-SPEC, to optimize the SPEC kernel used in an SVM for graph classification. The fundamental performance of the method is evaluated using artificial datasets, and its practicality confirmed through experiments using a real-world dataset.

Keywords: Graph Kernel, Interlace Theorem, Graph Spectra.

1 Introduction

A natural way of representing structured data is to use graphs. As an example, the structural formula of a chemical compound is a graph where each vertex corresponds to an atom in the compound, and each edge corresponds to a bond between two atoms therein. By using such graph representations, a new research field has emerged from data mining, namely graph mining, with the objective of mining information from a database consisting of graphs. With the potential to find meaningful information, graph mining has raised great interest, and research in the field has increased rapidly in recent years. Furthermore, since the need for classifying graphs has increased in many real-world applications, *e.g.*, analysis of proteins in bioinformatics and chemical compounds in cheminformatics [11], graph classification has also been widely researched worldwide. The main objective of graph classification is to classify graphs of similar structures into the same classes. This originates from the fact that instances represented by graphs usually have similar properties if their graph representations have high structural similarity.

Kernel methods such as the SVM are becoming increasingly popular due to their high performance in graph classification [10]. Most graph kernels are based on the idea of an object decomposed into substructures and a feature vector

B. Pfahringer, G. Holmes, and A. Hoffmann (Eds.): DS 2010, LNAI 6332, pp. 205–220, 2010.

containing counts of the substructures. Since the dimensionality of feature vectors is typically very high and includes the subgraph isomorphism matching problem that is known to be NP-complete [4], kernels deliberately avoid explicit computations of feature values and employ efficient procedures.

One of the representative graph kernels is the Random Walk Kernel [12,10], which computes $k(g_i, g_j)$ in $O(|g|^3)$ for graphs g_i and g_j, where $|g|$ is the number of vertices in g_i and g_j. The kernel returns a high value if the random walk on the graph generates many sequences with the same labels for vertices and edges, *i.e.*, the graphs are similar to each other. The Neighborhood Hash Kernel (NHK) [6] is another recently proposed kernel that computes $k(g_i, g_j)$ in $O(|g|d)$ for g_i and g_j, where d is the average degree of the vertices. The NHK uses logical operations such as the exclusive-OR on the label set of the connected vertices. The updated labels given by repeating the hash, propagate the label information over the graph and uniquely represent the high order structures around the vertices beyond the vertex or edge level. An SVM with two graph kernels works very well with benchmark data consisting of graphs with common small subgraphs (consisting of 1-6 vertices).

In many real-world applications using graph structured data, large subgraphs have a greater significance than small ones, because they contain more useful structural information. However, existing algorithms employing graph kernels, including the Random Walk Kernel and Neighborhood Hash Kernel, do not achieve good performance when classifying graphs whose classes depend on whether the graphs contain some large common subgraphs. The main reason for this is that the core principle of kernel algorithms is the use of a very small number of neighbor vertices when characterizing each vertex. Our experiments show that these two graph kernels do not work well with this kind of graph, and thus the application thereof is limited.

Based on the background, in this paper we aim to solve a classification problem of graphs where the binary class of each graph is defined by whether it contains some graphs in a large graph set as induced subgraphs. For this purpose, we propose a new graph kernel, called SPEC, and an algorithm, referred to as OPTSPEC (*OPT*imizing graph *SPEC*tra for graph classification) which optimizes the classification performance of SPEC when included in an SVM. The key feature of our algorithm is the use of graph spectra and the Interlace Theorem [7,8] for constructing the SPEC graph kernel. The graph spectrum of a graph is a vector consisting of eigenvalues of a matrix representing the graph, while the Interlace Theorem gives conditions of the largest common subgraph of two graphs by comparing their graph spectra. This theorem provides a very sensitive measure to identify graphs with large common subgraphs. Thus, an SVM using the SPEC kernel, which takes full advantage of this theorem, can efficiently classify graphs based on whether or not they contain some graphs in a large graph set as induced subgraphs. The cost of calculating graph spectra for a graph composed of $|g|$ vertices is $O(|g|^3)$, and therefore, our kernel requires only $O(|g|^3)$ computation time to compute $k(g_i, g_j)$ for graphs g_i and g_j.

2   Problem Definition

A labeled graph g is represented as $g = (V, E, L, l)$, where $V = \{v_1, \cdots, v_z\}$ is a set of vertices, $E = \{(v, v') \mid (v, v') \in V \times V\}$ is a set of edges[1], and L is a set of labels such that $l : V \cup E \to L$. If an edge exists between two vertices, the vertices are said to be adjacent. The number of vertices in a graph g is referred to as the size of the graph, and is denoted as $|g|$. Given two graphs $g = (V, E, L, l)$ and $g' = (V', E', L', l')$, g' is called a subgraph of g, if there exists an injective function $\phi : V' \to V$ that satisfies the following three conditions for $\forall v, v_1, v_2 \in V'$.

1. $(\phi(v_1), \phi(v_2)) \in E$, if $(v_1, v_2) \in E'$,
2. $l'(v) = l(\phi(v))$,
3. $l'((v_1, v_2)) = l((\phi(v_1), \phi(v_2)))$.

In addition, a subgraph g' of g is an "induced subgraph", where $\phi(v_1)$ and $\phi(v_2)$ are adjacent in g, if and only if v_1 and v_2 in $V(g')$ are adjacent in g'.

Classification problem of graphs is defined as follows. Given training examples $\{(g_i, y_i)\}$ $(i = 1, \cdots, n)$, where each example is a pair of a labeled graph g_i and class $y_i \in \{+1, -1\}$ that the graph belongs to, the objective of the learning machine is to learn a function f to predict the classes of test examples correctly. As mentioned in Section 1, in this paper, we focus on the classification of graphs whose class labels are determined by whether or not they contain some large induced subgraphs belonging to a given graph set S. In concrete terms, the class label 1 is assigned to graphs containing some graph in S as an induced subgraph, while the class label -1 is assigned to the other graphs[2].

3   Graph Kernel for Large Graph Classification Problems

In this section, we present our method for constructing the SPEC graph kernel to measure similarity between two graphs efficiently. Based on graph spectra and the Interlace Theorem, the SPEC kernel is very sensitive when working with graphs containing large common subgraphs. As a result of this sensitivity, an SVM employing the SPEC kernel is expected to have high accuracy in the classification of graphs.

3.1   Matrix Representation of Graphs

Matrices are a very useful representation of graphs, because we can obtain useful information about the topological structure of a graph from its matrix representation. In fact, there are various kinds of matrices that can be used to represent

---

[1] Although this paper focuses on undirected graphs only, we discuss in Section 6 that the proposed method is also applicable to directed graphs without loss of generality.

[2] In Section 6, we also discuss the classification of graphs whose class labels are determined by whether or not they contain some large common subgraphs as "general" subgraphs, but not "induced" subgraphs.

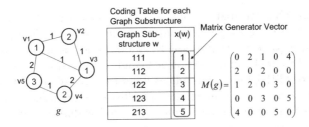

Fig. 1. Coding table and adjacency matrix

a graph, each of which captures different features of the graph. The most basic matrix representation of a graph is the *adjacency matrix*. In this paper, for a graph $g = (V, E, L, l)$ containing $|g|$ vertices, its adjacency matrix representation $M(g)$ is a $|g| \times |g|$ matrix whose elements are given by

$$M(g)_{i,j} = \begin{cases} c(w(i,j)) \text{ if } \{v_i, v_j\} \in E, 1 \leq i, j \leq |g|, \\ 0 \qquad\qquad \text{otherwise}, \end{cases} \tag{1}$$

where $M(g)_{i,j}$ is the (i,j)-th element of $M(g)$ and $w(i,j)$, called the *graph substructure*, is a combination of labels $[l(v_i)\ l(\{v_i, v_j\})\ l(v_j)]$ expressing the structure between two vertices v_i and v_j, which corresponds to an element the adjacency matrix. Moreover, $c(w(i,j))$ is a code characterizing the graph substructure $w(i,j)$ and is represented by a real number. All graph substructures and their corresponding codes are collated into a table called the *Coding table*. Furthermore, we call the vector consisting of all the codes the *matrix generator vector x*. Because we assume that the graph g is an undirected graph, the adjacency matrix of g is symmetric. Moreover, the eigenvalues of $M(g)$ are real.

For example, the adjacency matrix $M(g)$ of the graph g in Fig. 1 is created based on the Coding table and the matrix generator vector $x = (1, 2, 3, 4, 5)^T$. Let us consider the two vertices v_3 with label 1 and v_4 with label 2 in g in our discussion of the construction of $M(g)$. Because there is an edge with label 2 connecting these two vertices, the graph substructure corresponding to these two vertices is $w(3, 4) = [1\ 2\ 2]$. Therefore, since the Coding table assigns the value 3 to the graph substructure $[1\ 2\ 2]$, the two elements $M(g)_{3,4}$ and $M(g)_{4,3}$ of $M(g)$ have the value 3. On the other hand, since there is no edge connecting the two vertices v_1 and v_4 in g, the value 0 is assigned to $M(g)_{1,4}$ and $M(g)_{4,1}$.

3.2   Graph Spectrum and Interlace Theorem

We can calculate the *graph spectrum* of a matrix representing a graph, such as an adjacency matrix. The graph spectrum of a graph is a vector consisting of ordered eigenvalues of the matrix representing the graph. Due to the nature of matrices, a graph spectrum is known to be one of the graph invariants. This arises from the fact that the eigenvalues of a matrix remain constant when its rows and columns of the matrix are exchanged. Using graph spectra, we can compute ranges for the eigenvalues of matrices of common induced subgraphs that may exist in two arbitrary graphs by the following theorem.

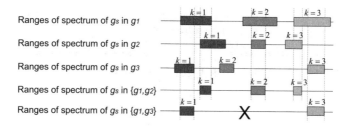

Ranges of spectrum of g_s in g_1

Ranges of spectrum of g_s in g_2

Ranges of spectrum of g_s in g_3

Ranges of spectrum of g_s in $\{g_1, g_2\}$

Ranges of spectrum of g_s in $\{g_1, g_3\}$

Fig. 2. Example of calculating the ranges of a common subgraph spectrum

Theorem 1 (Cauchy's Interlace Theorem [7,8]). *Let* $\lambda_1 \leq \lambda_2 \leq \ldots \leq \lambda_p$ *be eigenvalues of a symmetric matrix* A *of size* $p \times p$, *and let* $\mu_1 \leq \mu_2 \leq \ldots \leq \mu_m$ *be eigenvalues of a symmetric matrix* B *of size* $m \times m$ $(m < p)$. *If* B *is a principal submatrix of* A, *then*

$$\lambda_k \leq \mu_k \leq \lambda_{k+p-m}, \qquad k = 1, \ldots, m. \tag{2}$$

∎

If g_s of size m is an induced subgraph of g of size p, $M(g_s)$ is a principal submatrix of $M(g)$. Therefore, if g_s is an induced subgraph of g, Eq. (2) holds for eigenvalues of the matrices $M(g)$ and $M(g_s)$. In the remainder of this paper, we simply denote the i-th eigenvalue of the graph spectrum of a graph g as $\gamma_i(g)$ $(1 \leq i \leq |g|)$, and the graph spectrum of g as $\gamma(g) = \{\gamma_1(g), \cdots, \gamma_{|g|}(g)\}$.

By utilizing the Interlace Theorem, we can either compute the range of the k-th eigenvalue of the spectrum of common induced subgraphs contained in two given graphs or decide that the graphs do not contain a common induced subgraph of size m. This is illustrated in Fig. 2. Each hatched rectangle in the first, second, and third parts of Fig. 2 represents the range that eigenvalues of a spectrum of an induced subgraph with 3 vertices in g_1, g_2, and g_3, respectively, can take. The ranges of eigenvalues which the graph spectrum of a common induced subgraph g_s of g_1 and g_2 can take are limited to the intersections of the corresponding ranges given by g_1 and g_2, as shown by rectangles in the fourth part of Fig. 2. On the other hand, g_1 and g_3 cannot contain g_s with 3 vertices as a common induced subgraph, because the ranges given by g_1 and g_3 do not have any intersection for the second eigenvalue of the graph spectrum of g_s.

Interlace Theorem is very sensitive to large induced subgraphs of a graph. Let the graph spectrum of a graph g be $\gamma(g) = \{\gamma_1(g), \cdots, \gamma_{|g|}(g)\}$, and an induced subgraph of size m in g be g_s^m. A range that the k-th eigenvalue of the graph spectrum of g_s^m can take is $[\gamma_k(g), \gamma_{k+|g|-m}(g)]$. Thus, width of a range that the k-th eigenvalue of the graph spectrum of $g_s^{m'}$ $(m < m')$ can take is no more than that of g_s^m, because $k + |g| - m' < k + |g| - m$ and $\gamma_{k+|g|-m'}(g) \leq \gamma_{k+|g|-m}(g)$. Therefore, the theorem is very sensitive to large induced subgraphs of a graph.

Given two arbitrary graphs g_i and g_j, one of the problems we intend to solve in this paper is the construction of a graph kernel $k(g_i, g_j)$ that can measure the similarity between the two graphs efficiently, especially when they contain large common induced subgraphs. For this purpose, we employ graph spectra and the

Interlace Theorem described above, since a good kernel is the key to the success of an SVM in the classification of graphs.

3.3 Graph Kernel for Large Graph Classification

First, we describe the kernel function, the kernel matrix characterized by a kernel, and their requirements. Given two graphs g_i and g_j, let $k(g_i, g_j)$ denote a kernel function between graphs g_i and g_j. If g_i and g_j have high similarity, $k(g_i, g_j)$ should be large. Given a graph dataset $\{(g_i, y_i)\}$ $(1 \leq i \leq n)$ and a kernel function $k(\cdot, \cdot)$, we compute each element of a kernel matrix K as

$$K_{i,j} = K_{j,i} = k(g_i, g_j) \quad (1 \leq i, j \leq n).$$

To be applicable to an SVM, the graph kernel must be a PSD (positive semidefinite) kernel [13]. Let k_1 and k_2 be arbitrary PSD kernel functions, $\boldsymbol{x}$ and $\boldsymbol{y}$ be arbitrary vectors, and A be a $|\boldsymbol{x}| \times |\boldsymbol{y}|$ matrix where $|\boldsymbol{x}|$ is the dimensionality of the vector $\boldsymbol{x}$. Moreover, let $\boldsymbol{x} = (\boldsymbol{x}_a^T, \boldsymbol{x}_b^T)^T$, where $\boldsymbol{x}$ is a concatenation of $\boldsymbol{x}_a$ and $\boldsymbol{x}_b$. $k(\boldsymbol{x}, \boldsymbol{y})$ is another PSD kernel if one of the following holds:

$$k(\boldsymbol{x}, \boldsymbol{y}) = \exp(k_1(\boldsymbol{x}, \boldsymbol{y})), \tag{3}$$

$$k(\boldsymbol{x}, \boldsymbol{y}) = \boldsymbol{x}^T A \boldsymbol{y}, \tag{4}$$

$$k(\boldsymbol{x}, \boldsymbol{y}) = k_1(\boldsymbol{x}_a, \boldsymbol{y}_a) + k_2(\boldsymbol{x}_b, \boldsymbol{y}_b), \quad \text{or} \tag{5}$$

$$k(\boldsymbol{x}, \boldsymbol{y}) = k_1(\boldsymbol{x}_a, \boldsymbol{y}_a) \times k_2(\boldsymbol{x}_b, \boldsymbol{y}_b). \tag{6}$$

To construct our graph kernel, we consider two graphs g_i and g_j whose graph spectra are given by

$$\gamma(g_i) = \{\gamma_1(g_i), \cdots, \gamma_{|g_i|}(g_i)\} \text{ and } \gamma(g_j) = \{\gamma_1(g_j), \cdots, \gamma_{|g_j|}(g_j)\},$$

respectively. By the Interlace Theorem, if a common induced subgraph g_s of size m is contained in the graphs g_i and g_j, the range that $\gamma_k(g_s)$ $(1 \leq k \leq m \leq min(|g_i|, |g_j|))$ can take is the intersection of

$$[\gamma_k(g_i), \gamma_{k+|g_i|-m}(g_i)] \text{ and } [\gamma_k(g_j), \gamma_{k+|g_j|-m}(g_j)]. \tag{7}$$

Consider a matrix $\frac{1}{c_{max}} M(g_s)$ of g_s where c_{max} is the maximum absolute value among elements in $M(g_s)$. Since it is a random symmetric matrix with elements of absolute value at most 1, the probability that the k-th eigenvalue of $\frac{1}{c_{max}} M(g_s)$ deviates from its median by more than t is at most $4e^{-t^2/32k^2}$, where $1 \leq k \leq m$ [1]. On the other hand, if the eigenvalues are uniformly spaced, the average interval $\bar{d}$ between two of the eigenvalues is at most $\bar{d} = \frac{2(m-1)}{m-1} = 2$, since m eigenvalues of $\frac{1}{c_{max}} M(g_s)$ whose diagonal elements are 0 as defined by Eq. (1) must exist in $[-(m-1), (m-1)]$. Comparing the interval $\bar{d}$ with the standard deviation $4k$ of the distribution of the eigenvalues [1], the standard deviation is large enough for all k. Since the eigenvalues do not extremely concentrate around the median and are widely distributed, there is a strong possibility that $\gamma_k(g_s)$

exists in the intersection of the rages (7) when the width of the intersection is large. Furthermore, if the possibility that $\gamma_k(g_s)$ exists in the intersection increases for all k, the possibility that g_s is included as an induced subgraph in g_i and g_j also increases according to the Interlace Theorem. Therefore, we define the *SPEC* (graph SPECtra) kernel based on the widths of the intersections of the ranges (7) to measure the similarity between two graphs.

If g_i and g_j contain common induced subgraphs of size m, the two ranges (7) for every k must intersect each other. This can only be satisfied when

$$\gamma_k(g_i) \leq \gamma_{k+|g_j|-m}(g_j) \text{ and } \gamma_k(g_j) \leq \gamma_{k+|g_i|-m}(g_i), \qquad (8)$$

because $\gamma_k(g_i) > \gamma_{k+|g_j|-m}(g_j)$ and $\gamma_k(g_j) > \gamma_{k+|g_i|-m}(g_i)$ cannot be satisfied simultaneously. Eq. (8) is equivalent to

$$(\gamma_{k+|g_j|-m}(g_j) - \gamma_k(g_i))(\gamma_{k+|g_i|-m}(g_i) - \gamma_k(g_j)) \geq 0, \qquad (9)$$

and by taking the exponential of Eq. (9), we obtain the following inequality.

$$h(\cdot) = \exp(\gamma_{k+|g_j|-m}(g_j)\gamma_{k+|g_i|-m}(g_i) + \gamma_k(g_i)\gamma_k(g_j))$$
$$\times \exp(-\gamma_{k+|g_j|-m}(g_j)\gamma_k(g_j))\exp(-\gamma_{k+|g_i|-m}(g_i)\gamma_k(g_i)) \geq 1.$$

$h(\cdot)$ can be further rewritten as $h(\cdot) = exp(\boldsymbol{\lambda}_{mk}'^T \boldsymbol{\theta}_{mk}') \times (\phi_{\lambda_{mk}}\phi_{\theta_{mk}})$, where

$$\boldsymbol{\lambda}_{mk}' = \begin{bmatrix} \gamma_k(g_i) \\ \gamma_{k+|g_i|-m}(g_i) \end{bmatrix}, \; \boldsymbol{\theta}_{mk}' = \begin{bmatrix} \gamma_k(g_j) \\ \gamma_{k+|g_j|-m}(g_j) \end{bmatrix},$$
$$\phi_{\lambda_{mk}} = \exp(-\gamma_{k+|g_i|-m}(g_i)\gamma_k(g_i)), \; \text{and}$$
$$\phi_{\theta_{mk}} = \exp(-\gamma_{k+|g_j|-m}(g_j)\gamma_k(g_j)).$$

We can easily see that the former exponential term in the rhs includes the inner product of $\boldsymbol{\lambda}_{mk}'$ and $\boldsymbol{\theta}_{mk}'$ in its exponent. Since the inner product is a PSD kernel function in the event that A is an identity matrix in Eq. (4), this term is a PSD kernel function based on Eq. (3). Besides, when $|\boldsymbol{x}| = |\boldsymbol{y}| = 1$ and $A = 1$, Eq. (4) mentions that the product of two independent scalars is a PSD kernel. Therefore, the latter product term of two ϕs is also a PSD kernel function based on Eq. (4). These observation shows that $h(\cdot)$ which is a product of the former and the latter terms is a PSD kernel based on Eq. (6). We restate $h(\cdot)$ as

$$k_{mk}'(\boldsymbol{\lambda}_{mk}, \boldsymbol{\theta}_{mk}) = \exp(\boldsymbol{\lambda}_{mk}'^T \boldsymbol{\theta}_{mk}') \times (\phi_{\lambda_{mk}}\phi_{\theta_{mk}}),$$

$$\text{where } \boldsymbol{\lambda}_{mk} = \begin{bmatrix} \gamma_k(g_i) \\ \gamma_{k+|g_i|-m}(g_i) \\ \gamma_k(g_i)\gamma_{k+|g_i|-m}(g_i) \end{bmatrix} \text{ and } \boldsymbol{\theta}_{mk} = \begin{bmatrix} \gamma_k(g_j) \\ \gamma_{k+|g_j|-m}(g_j) \\ \gamma_k(g_j)\gamma_{k+|g_j|-m}(g_j) \end{bmatrix}.$$

Since the ranges (7) must intersect each other for "all k" $(k = 1, \cdots, m)$ when g_i and g_j contain a common subgraph g_s of size m, we take the product of k_{mk}' over all k, and have a new PSD kernel function:

$$k_m(\boldsymbol{\lambda}_m, \boldsymbol{\theta}_m) = \prod_{k=1}^{m} k_{mk}'(\boldsymbol{\lambda}_{mk}, \boldsymbol{\theta}_{mk}),$$

where $\boldsymbol{\lambda}_m = [\boldsymbol{\lambda}_{m1}^T, \boldsymbol{\lambda}_{m2}^T, \cdots, \boldsymbol{\lambda}_{mm}^T]^T$ and $\boldsymbol{\theta}_m = [\boldsymbol{\theta}_{m1}^T, \boldsymbol{\theta}_{m2}^T, \cdots, \boldsymbol{\theta}_{mm}^T]^T$,

based on Eq. (6). Furthermore, since the ranges (7) must intersect each other for "at least one of m" ($m = 1, \cdots, min(|g_i|, |g_j|)$) when g_i and g_j contain an arbitrary common subgraph g_s, we take a summation of k_m over all m which tends to be large when some k_m is large, and provide a new PSD kernel function:

$$k(g_i, g_j) = k(\boldsymbol{\lambda}, \boldsymbol{\theta}) = \sum_{m=1}^{min(|g_i|, |g_j|)} k_m(\boldsymbol{\lambda}_m, \boldsymbol{\theta}_m), \qquad (10)$$

where $\boldsymbol{\lambda} = [\boldsymbol{\lambda}_1^T, \boldsymbol{\lambda}_2^T, \cdots, \boldsymbol{\lambda}_m^T]^T$ and $\boldsymbol{\theta} = [\boldsymbol{\theta}_1^T, \boldsymbol{\theta}_2^T, \cdots, \boldsymbol{\theta}_m^T]^T$,

based on Eq. (5). $k(\boldsymbol{\lambda}, \boldsymbol{\theta})$ is expected to have a high score when measuring the similarity between the graphs, especially where g_i and g_j have large common induced subgraphs. We call this kernel function the *SPEC* kernel, and use it in an SVM to classify graphs.

By summarizing the above discussion, the following lemmas are derived.

Lemma 1. *The SPEC kernel is a PSD kernel.*   ■

Lemma 2. *The SPEC kernel $k(g_i, g_j)$ is computed in $O(|g|^3)$ where $|g|$ is the maximum number of vertices in g_i and g_j.*   ■

Proof. To obtain the eigenvalues from the adjacency matrices of g_i and g_j requires computation time of $O(|g|^3)$. In addition, we require $O(|g|^2)$ computation time to compute Eq. (10) from the eigenvalues. Therefore, the SPEC kernel $k(g_i, g_j)$ can be computed in $O(|g|^3)$.   □

4   Optimizing Graph Spectra for Large Graph Classification

In the previous section, we gave the details of the SPEC kernel, constructed especially for the classification of graphs with large common induced subgraphs. In this section, we propose a new algorithm, called OPTSPEC to obtain high classification accuracy for graphs using the SPEC kernel.

4.1   Basic Idea for Optimizing Graph Spectra

The SPEC kernel $k(g_i, g_j)$ between graphs g_i and g_j is computed from eigenvalues of adjacency matrices $M(g_i)$ and $M(g_j)$ defined by Eq. (1), and the eigenvalues depend on a matrix generator vector $\boldsymbol{x}$ defining elements of $M(g_i)$ and $M(g_j)$. Therefore, $k(g_i, g_j)$ depends on a matrix generator vector $\boldsymbol{x}$. Even if the value of only a single element of $\boldsymbol{x}$ is changed, it leads to a change in the graph spectra of g_i and g_j, and thus the intersection of the ranges (7) is also changed, as well as the value of the kernel function $k(g_i, g_j)$. In other words, by choosing a suitable matrix generator vector $\boldsymbol{x}$, there is a possibility to accurately measure the similarity between two arbitrary graphs using the Interlace Theorem.

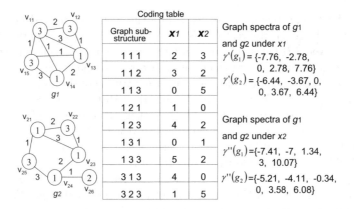

Fig. 3. Example of using different matrix generator vectors x_1 and x_2

Figure 3 shows an example of a difference between two matrix generator vectors in the computation of graph spectra. Given graphs g_1 and g_2, two graph spectra $\gamma'(g_1)$ and $\gamma'(g_2)$ are calculated using the matrix generator vector x_1 shown in the second column of the table in Fig. 3, and two graph spectra $\gamma''(g_1)$ and $\gamma''(g_2)$ are calculated using another vector x_2 shown in the third column of the table in Fig. 3. From this example, we can see that the graph spectra of g_1 and g_2 are completely different. Moreover, while the largest common induced subgraph of g_1 and g_2 is the subgraph consisting of 3 vertices (v_{12}, v_{13}, and v_{14} in g_1 and v_{22}, v_{23}, and v_{24} in g_2), we cannot conclude this fact by using $\gamma'(g_1)$ and $\gamma'(g_2)$ and the Interlace Theorem. The ranges of $\gamma_1'(g_s)$ for a subgraph g_s of size 4 in g_1 and g_2 are [-7.76,-2.78] and [-6.44,-2.78], respectively, and their intersection for $\gamma_1'(g_s)$ is [-6.44,-2.78]. Similarly, the intersections for $\gamma_2'(g_s)$, $\gamma_3'(g_s)$, and $\gamma_4'(g_s)$ are calculated as [-2.78,0], [0,2.78], and [2.78,6.44], respectively. Since the intersections for four elements of $\gamma'(g_s)$ exist, we cannot conclude the correct size of the largest induced subgraph g_s between g_1 and g_2. On the other hand, the ranges of $\gamma_1''(g_s)$ for a subgraph g_s of size 4 in g_1 and g_2 are [-7.41,-7] and [-5.21,-0.34], respectively. These non-intersecting ranges exclude a common induced subgraph of size 4 in g_1 and g_2.

As shown in the above example, if we choose an unsuitable matrix generator vector x, it is hard to correctly know the existence of a common induced subgraph of two arbitrary graphs g_i and g_j and the maximum size of their common induced subgraph. Thus, the importance of a proper selection of x is very clear. In the next subsection, we propose a method for choosing a suitable matrix generator vector x based on an optimization technique.

4.2   Algorithm for Optimizing Graph Spectra for Classification

The data handled in this paper is graphs whose class labels are determined by whether or not they contain some large induced subgraphs belonging to a given graph set S, and the problem we intend to solve in this paper is the construction a classifier for the graphs using an SVM and the SPEC kernel. Although using a suitable matrix generator vector $\boldsymbol{x}$ is expected to achieve high performance of the SVM with the SPEC kernel, obtaining the suitable matrix generator vector is difficult by hand-tuning. We, then, empirically optimize the matrix generator vector $\boldsymbol{x}$ using training examples together with the training of the SVM. If a suitable matrix generator vector is chosen to compute the SPEC kernel, we can correctly measure the similarity between graphs with the kernel, and the number of misclassified training examples is reduced using an SVM with the SPEC kernel. Therefore, we optimize the matrix generator vector $\boldsymbol{x}$ so that the number of examples misclassified by an SVM is reduced using training examples.

We propose the OPTSPEC algorithm for optimizing a matrix generator vector based on the framework of Generalized Multiple Kernel Learning (GMKL) [14]. GMKL minimizes the number of misclassified training examples by alternately learning parameters of an SVM and a parameter $\boldsymbol{x}$ of the kernel function. Given a set of training examples, the OPTSPEC algorithm aims at minimizing the distance between boundaries on support vectors and the number of misclassified training examples in the classification of graphs as follows.

$$\min_{\boldsymbol{x}} T(\boldsymbol{x}) \ s.t. \ \boldsymbol{x} \geq 0 \ \text{(outer loop)} \tag{11}$$

$$where \ T(\boldsymbol{x}) = \min_{\boldsymbol{w}} \frac{1}{2} \boldsymbol{w}^T \boldsymbol{w} + \frac{C}{2} \sum_{i=1}^{n} \max(0, 1 - y_i f(g_i))$$

$$+ r(\boldsymbol{x}), \ \text{(inner loop)} \tag{12}$$

where $\boldsymbol{w}$ is a parameter learned by the SVM, C is a constant value specified by the user, f is a function to be learned as mentioned in Section 2, and r is a regularizer to incorporate a scale parameter within $T(\boldsymbol{x})$ in form of $r(\boldsymbol{x}) = ||\boldsymbol{x}||^2 - 1$. In OPTSPEC, the constraint $\boldsymbol{x} \geq 0$ is relaxed so that the learned kernel is PSD as mentioned in [14]. The optimal kernel is learned by optimizing over $\boldsymbol{x}$ in the outer loop, while the matrix generator vector $\boldsymbol{x}$ remains fixed and the parameter $\boldsymbol{w}$ are learned using an SVM in the inner loop. In the calculation of the outer loop, the matrix generator vector $\boldsymbol{x}$ is updated to another matrix generator vector $\boldsymbol{x} - \alpha(\frac{\partial T}{\partial x_1}, \frac{\partial T}{\partial x_2}, \cdots, \frac{\partial T}{\partial x_{|\boldsymbol{x}|}})^T$ using the Steepest Descent method. In this computation, since $T(\boldsymbol{x})$ which contains the number of misclassified training examples $\frac{1}{2} \sum_{i=1}^{n} \max(0, 1 - y_i f(g_i))$ in its second term is a discrete function, we cannot calculate its differentials. To overcome this difficulty, we employ sensitivity analysis to calculate $\frac{\partial T}{\partial x_i} = \frac{T(\boldsymbol{x}+\tau\Delta\boldsymbol{x}_i)-T(\boldsymbol{x})}{\tau\Delta\boldsymbol{x}_i^T\Delta\boldsymbol{x}_i}$ $(i = 1, \cdots, |\boldsymbol{x}|)$ for each element x_i, one by one, where

$$\Delta\boldsymbol{x}_i^T = \begin{pmatrix} 1 & \cdots & i-1 & i & i+1 & \cdots & |\boldsymbol{x}| \\ 0 & \cdots & 0 & 1 & 0 & \cdots & 0 \end{pmatrix}.$$

The computations in the inner loop using an SVM and the outer loop using the Steepest Descent method are alternately continued, while the number of misclassified examples is reduced.

5   Experiments

The proposed method was implemented in Java. All experiments were done on an Intel Xeon L5240 3 GHz computer with 2 GB memory and running Microsoft Windows 2008 Server. We compared the accuracy of the training and the prediction performance of OPTSPEC with those of SVMs using the Random Walk Kernel and Neighborhood Hash Kernel. In the remainder of this paper, for simplicity, we refer to the SVMs using Random Walk Kernel and Neighborhood Hash Kernel as RWK-SVM and NHK-SVM, respectively. We varied parameters $\lambda = \{0.9, 0.8, \cdots, 0.2, 0.1, 0.01, 0.001\}$ which represents the termination probability for the random walk kernel and $R = \{1, 2, \cdots, 9, 10, 20, \cdots, 90, 100, 150, 200\}$ which represents maximum order of neighborhood hash for Neighborhood Hash Kernel. For RWK-SVM and NHK-SVM, we show the best prediction accuracy for various λ and R in the next subsections. For learning from the kernel matrices generated by the above graph kernels, we used the LIBSVM package[3] using 10-fold cross validation. The performance of the proposed method was evaluated using artificial and real-world data.

Table 1. Default parameters for the data generation program

	Avg. size of graphs	Proportion of size of induced subgraphs	Prob. of edge existence	# of vertex and edge labels
Default values	$\|g\| = 100$	$p_V = 0.7$	$p = 5\%$	$\|L\| = 3$

5.1   Experiments on Artificial Datasets

We generated artificial datasets of graphs using the four parameters listed in Table 1. For each dataset, 50 graphs, each with an average of $|g|$ vertices, were generated. Two vertices in a graph were connected with probability p of the existence of an edge, and one of $|L|$ labels was assigned to each vertex or edge in the graph. In parallel with the dataset generation and using the same parameters p and $|L|$, three graphs g_{s1}, g_{s2}, and g_{s3} with an average of $p_V \times |g|$ vertices were also generated for embedding in the 50 graphs as common induced subgraphs. g_{s1} was randomly embedded in half of the 50 graphs. The embedding process was then repeated using g_{s2} and g_{s3}. Finally, the class label 1 was assigned to the graphs containing g_{s3}, which was the last to be embedded, while the class label -1 was assigned to the other graphs. Even under this tough condition in which graphs have high similarity, i.e., they contain parts of g_{s1} and/or g_{s2} as common induced subgraphs, a good classifier should be able to classify graphs labeled according to whether or not they contain g_{s3} correctly.

First, we varied only p_V to generate various datasets with the other parameters set to their default values. The proportion of subgraphs embedded in each dataset to the average size of the graphs was varied from 0.1 to 0.9. The values in Table 2 denote the average and standard deviation of the accuracy of the three

---

[3] http://www.csie.ntu.edu.tw/~cjlin/libsvm/

Table 2. Results for various p_V: accuracy (standard deviation)

p_V	NHK-SVM		RWK-SVM		OPTSPEC		BoostOPTSPEC	
	Training	Test	Training	Test	Training	Test	Training	Test
0.1	95% (4%)	58% (17%)	52% (1%)	34% (13%)	68% (10%)	60% (16%)	82% (10%)	56% (28%)
0.3	74% (6%)	30% (24%)	52% (2%)	28% (13%)	67% (10%)	57% (17%)	79% (10%)	54% (26%)
0.5	80% (6%)	50% (22%)	52% (2%)	30% (13%)	83% (9%)	72% (19%)	93% (4%)	70% (25%)
0.7	99% (2%)	54% (22%)	52% (2%)	30% (13%)	86% (20%)	74% (31%)	96% (4%)	86% (14%)
0.9	100% (0%)	48% (24%)	51% (2%)	32% (13%)	98% (6%)	96% (8%)	100% (0%)	92% (10%)

Table 3. Results for various $|g|$: accuracy (standard deviation)

$|g|$	NHK-SVM		RWK-SVM		OPTSPEC		Boost OPTSPEC	
	Training	Test	Training	Test	Training	Test	Training	Test
36	99% (2%)	70% (22%)	89% (3%)	86% (13%)	92% (6%)	88% (22%)	99% (1%)	94% (14%)
60	90% (3%)	60% (18%)	52% (2%)	28% (13%)	90% (10%)	64% (25%)	98% (2%)	82% (15%)
100	99% (2%)	54% (22%)	52% (2%)	52% (2%)	86% (20%)	74% (31%)	96% (4%)	86% (14%)
180	91% (2%)	68% (20%)	51% (1%)	38% (6%)	85% (15%)	72% (27%)	98% (3%)	80% (16%)
360	100% (0%)	38% (28%)	–	–	93% (3%)	90% (14%)	–	–

classifiers in the experiments. RWK-SVM did not perform well with either the training or test datasets. Although the accuracy of NHK-SVM is high for the training datasets, it was unable to classify graphs correctly in the test datasets. On the other hand, the accuracy of OPTSPEC is high for the test dataset, especially with a high p_V, compared with both NHK-SVM and RWK-SVM. Since the accuracy of OPTSPEC is constantly higher than 50%, we combined our OPTSPEC algorithm with AdaBoost [3] to create a strong learner. The experimental results of OPTSPEC combined with AdaBoost are shown in the last two columns of Table 2. By combining OPTSPEC with AdaBoost, the classification performance of OPTSPEC was enhanced particularly for the training data.

In the remaining sections, we set p_V to 0.7 as its default value to emphasize the classification performance of OPTSPEC for graphs labeled by whether or not they contain g_{s3}. Table 3 gives the experimental results for datasets generated by varying the values of $|g|$ and keeping the other parameters set to their default values. The average size $|g|$ of the graphs in the datasets was varied between 36 and 360. In the table, "–" indicates that results were not obtained due to intractable computation times exceeding 3 hours. The accuracy of OPTSPEC is comparable with that of NHK-SVM and RWK-SVM for small $|g|$. However, as $|g|$ increases, so the accuracy of NHK-SVM and RWK-SVM decreases, while the accuracy of OPTSPEC remains high. This arises from the fact that NHK-SVM and RWK-SVM make use of only small structures in the graphs to calculate their kernel functions, and therefore, they are unable to perform well with graphs containing large common induced subgraphs. On the other hand, in OPTSPEC, the possibility of a large common graph is identified in the computation of SPEC and then the matrix generator vector $\boldsymbol{x}$ is optimized based on GMKL.

Tables 4 and 5 give the experimental results for datasets generated by varying values of $|L|$ and p, respectively, while keeping the other parameters set to their default values. The number of labels $|L|$ in the graphs was varied between 1 and 5 in Table 4, while the probability p of the existence of an edge between

Table 4. Results for various $|L|$: accuracy (standard deviation)

$|L|$	NHK-SVM		RWK-SVM		OPTSPEC		Boost OPTSPEC	
	Training	Test	Training	Test	Training	Test	Training	Test
1	76% (3%)	40% (16%)	52% (1%)	28% (10%)	83% (18%)	55% (19%)	82% (10%)	56% (28%)
2	82% (5%)	44% (17%)	52% (2%)	32% (13%)	83% (7%)	74% (25%)	95% (3%)	80% (19%)
3	99% (3%)	54% (22%)	52% (2%)	52% (2%)	84% (11%)	70% (22%)	96% (4%)	86% (14%)
4	84% (3%)	38% (24%)	53% (1%)	26% (13%)	86% (20%)	74% (31%)	99% (3)%	96% (8%)
5	95% (2%)	70% (22%)	52% (2%)	30% (16%)	98% (3%)	98% (6%)	100% (0%)	96% (8%)

Table 5. Results for various p: accuracy (standard deviation)

p	NHK-SVM		RWK-SVM		OPTSPEC		Boost OPTSPEC	
	Training	Test	Training	Test	Training	Test	Training	Test
2.5	88% (3%)	46% (22%)	52% (1%)	32% (10%)	93% (4%)	80% (16%)	99% (2%)	86% (14%)
5	99% (2%)	54% (22%)	52% (2%)	52% (2%)	84% (11%)	70% (22%)	96% (4%)	86% (14%)
10	89% (3%)	66% (28%)	52% (1%)	36% (8%)	92% (4%)	92% (10%)	99% (2%)	92% (14%)
20	76% (5%)	44% (23%)	52% (1%)	34% (13%)	94% (5%)	80% (16%)	98% (4%)	92% (14%)

two vertices was varied between 2.5 and 20 in Table 5. Table 4 shows that the accuracies of OPTSPEC and Boost OPTSPEC remains high for both training and test datasets as the number of labels in the graphs increases. This originates from the fact that the size of the matrix generator vector $\boldsymbol{x}$ becomes large when the number of labels in the graphs increases, which leads to high probability of discovering a suitable vector $\boldsymbol{x}$. Table 5 shows that the accuracies of OPTSPEC and Boost OPTSPEC are high. This is because most of the elements in the adjacency matrices become non-zero when p increases, which also leads to high probability of discovering a suitable vector $\boldsymbol{x}$. On the other hand, similar to the previous experiments, accuracy results for RWK-SVM are low, and the accuracy of NHK-SVM for the test data is also low compared with that of OPTSPEC, while the accuracy of NHK-SVM being high for the training datasets.

5.2   Experiment with Real-World Graphs

To assess the practicability of our proposed method, we experimented on the email-exchange history data of the Enron company [2]. We transformed the mail exchange history per week data to one graph, and, having preprocessed the data as described below, obtained a dataset consisting of 123 graphs for 123 weeks. Each person in the company is represented by a single vertex labeled with his or her position in the company, for example "CEO", "Director", "Employee", "Lawyer", "Manager", "President", "Trader" and "Vice President". An edge connecting two vertices is included if the corresponding individuals exchanged emails for a week. The maximum size of the graphs in the dataset is 70, and the average edge existence probability is 4.9%. Since our aim is to evaluate the proposed method using large graphs, we chose 50 large graphs from the 123, corresponding to 50 continuous weeks.

Because it is very difficult to understand the common graphs contained in the graphs, we transformed the dataset by randomly choosing two graphs g_{s1} and g_{s2} from the 50 graphs of Enron data, and embedded these graphs within the

Table 6. Results for the Enron Dataset: accuracy (standard deviation)

Dataset	NHK-SVM		RWK-SVM		OPTSPEC		Boost OPTSPEC	
	Training	Test	Training	Test	Training	Test	Training	Test
D_1	99% (2%)	60% (18%)	52% (1%)	28% (16%)	85% (14%)	72% (20%)	98% (3%)	92% (14%)
D_2	100% (0%)	62% (21%)	52% (1%)	30% (13%)	91% (7%)	88% (13%)	99% (1%)	92% (14%)
D_3	100% (0%)	68% (20%)	51% (0%)	40% (0%)	96% (9%)	92% (10%)	99% (4%)	98% (6%)

other graphs. This embedding process is similar to the generation of artificial datasets described in the previous subsection. Accordingly, the graphs containing g_{s2} were given the class label 1, while the remaining graphs were given the class label -1. Since the embedded graphs were chosen from the Enron data, they also express the email exchanges for one week, and have an identical character to the other graphs in the dataset. Therefore, the characters of the embedded graphs do not change much. To obtain the exact performance of the proposed method, we chose three pairs of g_{s1} and g_{s2}, and embedded the three pairs in the graphs to create three datasets, respectively. We denote the datasets as D_1, D_2, and D_3, respectively. The numbers of vertices and edges of the embedded graphs in each dataset are as follows:

- D_1: g_{s1}: 25 vertices, 25 edges. g_{s2}: 28 vertices, 26 edges.
- D_2: g_{s1}: 25 vertices, 25 edges. g_{s2}: 22 vertices, 16 edges.
- D_3: g_{s1}: 22 vertices, 16 edges. g_{s2}: 27 vertices, 26 edges.

Table 6 gives the accuracy of the four methods with the three datasets. OPT-SPEC and Boost OPTSPEC outperformed NHK-SVM and RWK-SVM on all three datasets. We confirmed that Boost OPTSPEC performed very well in improving the accuracy of OPTSPEC with the real-world data. In contrast, although NHK-SVM classified the training data with high accuracy, *i.e.*, over 85%, it did not predict the test graphs very well.

6   Discussion

Based on the various experiments, Boost OPTSPEC achieved the best performance for datasets with a variety of different specifications. For graphs whose class label is decided by whether or not they contain some large induced subgraphs in a given set, Boost OPTSPEC predicted the class of the test graphs with very high accuracy. By comparing the accuracy of Boost OPTSPEC and OPTSPEC, it is clear that the boosting method effectively increases the performance of OPTSPEC, thus enabling a powerful classifier for classifying graphs containing large common induced subgraphs. In contrast, although NHK-SVM showed high accuracy in the training process, the classifier was unable to perform the prediction well in almost all the experiments. RWK-SVM and NHK-SVM classifiers performed well only in the experiments with small graphs. The reason for this is that the two classifiers make use of very small structures in the graphs to measure the similarity between them, and therefore do not perform well in cases where the graphs contain large common induced subgraphs.

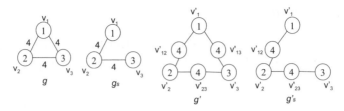

Fig. 4. Conversion of a graph g to another graph g'

In this paper, we focused on the common induced subgraphs within the graphs. Our SPEC kernel can be applied to classify graphs whose classes are determined by whether or not the graphs contain some specified common graph as general subgraphs using the following matrix representation. Given a graph $g = (V, E, L, l)$, the graph is converted to another graph $g' = (V', E', L', l')$, where $V' = V \cup E$, $E' \subseteq V \times E$, $L' = L$ and an edge e' between $v \in V$ and $e \in E$ exists in g' if v is directly linked to another vertex by e in g.

$$l'(v') = \begin{cases} l(v) & \text{if } v' \text{ corresponds to a vertex } v \text{ in } g, \\ l(e) & \text{otherwise if } v' \text{ corresponds to an edge } e \text{ in } g. \end{cases}$$

Since a general subgraph in g corresponds to an induced subgraph in g', the graph g' is represented by an adjacency matrix of size $(|V| + |E|) \times (|V| + |E|)$ using Eq. (1). Computing a kernel matrix using this matrix representation requires $O((|V| + |E|)^3)$ computation time. For example, the graph g with 3 vertices and 3 edges in Fig. 4 is converted to the another graph g' with 6 vertices. In this conversion as shown in Fig. 4, the graph g_s which is a general subgraph of g becomes an induced subgraph g'_s of g'. Therefore, the Interlace Theorem holds between adjacency matrices $M(g'_s)$ and $M(g')$, and the SPEC kernel can be computed from the adjacency matrices.

On the other hand, when given data are directed graphs, an adjacency matrix $M(g)$ of each directed graph g is converted to $M'(g) = \begin{pmatrix} 0 & M(g) \\ M(g)^T & 0 \end{pmatrix}$ to create a symmetric matrix. Therefore, the Interlace Theorem holds between adjacency matrices $M'(g_s)$ and $M'(g)$ where g_s is an induced subgraph of g, and the SPEC kernel can be computed from the adjacency matrices. In this case, computing the kernel matrix is $O((2|V|)^3) = O(|V|^3)$. Therefore, the proposed method can be applied to directed graphs whose classes are determined by whether or not the graph contains some common graphs as general subgraphs.

7   Conclusion

In this paper, we proposed a novel graph kernel named as SPEC based on graph spectra and the Interlace Theorem. We also proposed the OPTSPEC algorithm for optimizing the SPEC kernel used in an SVM for graph classification. We developed a graph classification program and confirmed the performance and practicability of the proposed method through computational experiments using artificial and real-world datasets.

Acknowledgment

We would like to thank Mr. Shohei Hido of IBM Research and Prof. Kouzo Ohara of Aoyama Gakuin University for their help and advice.

References

1. Alon, N., Krivelevich, M., Vu, V.H.: On the Concentration of Eigenvalues of Random Symmetric Matrices. Israel Journal of Mathematics 131(1), 259–267 (2001)
2. Enron Email Dataset, http://www.cs.cmu.edu/~enron/
3. Freund, Y., Schapire, R.E.: A Decision-Theoretic Generalization of On-Line Learning and an Application to Boosting. Journal of Computer and System Sciences 55(1), 119–139 (1997)
4. Garey, M.R., Johnson, D.S.: Computers and Intractability: A Guide to the Theory of NP-Completeness. W.H. Freeman, New York (1979)
5. Gärtner, T., Lloyd, J.W., Flach, P.A.: Kernels and Distances for Structured Data. Machine Learning 57(3), 205–232 (2004)
6. Hido, S., Kashima, H.: A Linear-Time Graph Kernel. In: Proc. of Int'l Conf. on Data Mining, pp. 179–188 (2009)
7. Hwang, S.: Cauchy's Interlace Theorem for Eigenvalues of Hermitian Matrices. American Mathematical Monthly 111, 157–159 (2004)
8. Ikebe, Y., Inagaki, T., Miyamoto, S.: The monotonicity theorem, Cauchy's interlace theorem, and the Courant-Fischer theorem. American Mathematical Monthly 94, 352–354 (1987)
9. Kashima, H., Inokuchi, A.: Kernels for graph classification. In: Proc. of ICDM Workshop on Active Mining (2002)
10. Kashima, H., Tsuda, K., Inokuchi, A.: Marginalized Kernels Between Labeled Graphs. In: Proc. of Int'l Conf. on Machine Learning, pp. 321–328 (2003)
11. Schölkopf, B., Tsuda, K., Vert, J.: Kernel Methods in Computational Biology. The MIT Press, Cambridge (2004)
12. Schölkopf, B., Smola, J.: Learning with kernels. MIT Press, Cambridge (2002)
13. Shawe-Taylor, J., Cristianini, N.: Kernel Methods for Pattern Analysis. Cambridge University Press, Cambridge (2004)
14. Varma, M., Rakesh Babu, B.: More Generality in Efficient Multiple Kernel Learning. In: Proc. of Int'l Conf. on Machine Learning, vol. 134 (2009)
15. Vishwanathan, S.V.N., Borgwardt, K.M., Schraudolph, N.N.: Fast Computation of Graph Kernels. In: Proc. of Annual Conf. on Neural Information Processing Systems, pp. 1449–1456 (2006)

Algorithm for Detecting Significant Locations from Raw GPS Data

Nobuharu Kami[1], Nobuyuki Enomoto[2],
Teruyuki Baba[1], and Takashi Yoshikawa[1]

[1] System Platforms Research Laboratories, NEC Corporation, Kanagawa, Japan
[2] Appliance Business Development Div., NEC BIGLOBE, Ltd., Tokyo, Japan

Abstract. We present a fast algorithm for probabilistically extracting significant locations from raw GPS data based on data point density. Extracting significant locations from raw GPS data is the first essential step of algorithms designed for location-aware applications. Assuming that a location is significant if users spend a certain time around that area, most current algorithms compare spatial/temporal variables, such as stay duration and a roaming diameter, with given fixed thresholds to extract significant locations. However, the appropriate threshold values are not clearly known *in priori* and algorithms with fixed thresholds are inherently error-prone, especially under high noise levels. Moreover, for N data points, they are generally $O(N^2)$ algorithms since distance computation is required. We developed a fast algorithm for selective data point sampling around significant locations based on density information by constructing random histograms using locality sensitive hashing. Evaluations show competitive performance in detecting significant locations even under high noise levels.

1 Introduction

The widespread use of GPS-enabled mobile devices, such as smart phones, enables easy collection of location data and accelerates development of a variety of location-aware applications. A typical application is to visualize a geographical trajectory of activities including traveling, shopping, and sporting, and many utility tools for displaying trajectories on an online map [1] are available so users can edit and open their traces to the public.

Perhaps, the most essential first step in editing raw GPS data would be to extract points of interest that represent significant locations such as shopping centers, restaurants and famous sightseeing spots. In fact, most websites show not only a plain trajectory but also a set of reference points so that readers can grab a summary of the activities. It is obviously time consuming if users have to memorize and manually set those points and find the corresponding locations in a set of many data points. One of the major goals of this paper is to support people using such online applications that display automatically extracted reference points along with the trajectory from uploaded GPS data.

Many algorithms designed for understanding user behavior by mining GPS data often automatically extract significant locations. Assuming that a location

B. Pfahringer, G. Holmes, and A. Hoffmann (Eds.): DS 2010, LNAI 6332, pp. 221–235, 2010.

is significant if one remains there for a long time, most of the current algorithms try to distinguish between "staying" and "moving" segments by comparing spatial/temporal variables, such as stay duration and roaming diameter, or velocity and acceleration, with fixed threshold values. In doing so, significant locations are extracted by locating stay locations. However, any fixed-threshold-based approach is in practice error-prone, and finding an optimal choice of threshold values is often difficult since the appropriate upper/lower bound to distinguish the staying segments from the moving segments depends on GPS data and is not known *in priori*. Threshold values optimized for one data set does not necessarily work for another. The problem becomes more serious when spatial noise level is high, which often happens to actual GPS data. In practice, we often have to tune parameters for each case and set an excessive margin to be on the safe side at the cost of computation time and accuracy. However, we also have to be careful not to choose excessively large threshold values that may even degrade the detection quality due to crosstalk between neighboring significant locations. Furthermore, since those algorithms are basically designed for offline processing, computation time is not well considered. Most of the algorithms are $O(N^2)$ algorithms for N data points since they require direct distance computation between those data points. Obviously, it is preferable to analyze as many datasets as possible for high-quality extraction; therefore, quick computation is important even for offline processing.

To tackle these issues, we take the approach of probabilistically detecting peak locations in density distribution of data points for extracting significant locations. This is based on the observation that noise is generally distributed around the center; therefore, peak locations in density distribution are often close to the stay location, independent of the width of noise distribution. We can also use density information for scoring the detected locations in order of importance because density at one location generally reflects the number of data points there, and thereby how long one stays.

For detecting peak locations, we developed a probabilistic algorithm for selectively sampling data points from high-density regions. By doing so, the obtained subset of the original GPS data is a set of sampled data points where high-density regions are spatially well separated. It is easy to divide such datasets into clusters with low computation cost such that each cluster contains data points sampled from each high-density region. Once we obtain a set of clusters, the final task is to identify the most representative point in each cluster as a reference point and rank it by giving a scoring metric in order of density reflected by importance.

We propose a density-dependent random sampling algorithm for selectively sampling from only high-density regions. The core idea is to construct random histograms using locality sensitive hashing (LSH). Our algorithm is inspired by the randomized algorithm developed in [2] for constructing random histograms representing feature sets of multimedia objects. We take advantage of random histograms for density-dependent random sampling. In the rest of this paper, we describe the algorithm design in Section 2, evaluation in Section 3, and related work in Section 4 followed by conclusion in the final section.

2   Algorithm Design

The goal of our algorithm is to return a set of *waypoints* as output in response to input GPS data X. An *waypoint* is a reference point that designates each significant location and contains information regarding both the geographical location and the importance. This section describes the mechanism of the algorithm in detail.

2.1   Design Overview

Given a GPS data set, high-density locations often indicate significant locations because they imply that one stays for a long time at those locations, and spatial noise distribution is often strongly centered at the actual location. This observation implies that detecting high density points and scoring them by their density is good for extracting *waypoints*.

An ordinary histogram, which is constructed by dividing the space into small bins (cells), is strongly affected by a *binning process* that determines the appropriate size and boundary value for each bin. To prevent the *binning process* being dependent on an input data set, we instead map GPS data points into an auxiliary space using LSH [4][5], a function for mapping two data points to the same value with probability that reflects the similarity (distance) between them. If we take the value to which each data point is mapped by LSH as the *label* of bins, we can construct a histogram whose high-frequency bin contains the data points sampled from a high-density region since data points around the center of this region are close.

Once we obtain a subset of data points selectively sampled from high-frequency bins, it is easy to cluster them such that each cluster contains data points coming only from a single stay location. We call this cluster a *waypoint region*. The final task after we obtain a set of *waypoint regions* is to extract a *waypoint* for each *waypoint region* by identifying the most representative point in each *waypoint region* and compute a scoring metric reflecting density information so we can rank them in order of importance.

Figure 1 illustrates an overview of operations flow of the algorithm. The flow is comprised of the following three steps.

Density-dependent random sampling: samples data points with probability reflecting the density information by creating histograms using LSH such that all the high-density regions of the sampled data points are geographically sparse and well separated.

Waypoint region construction: clusters the sampled data points and constructs a set of *waypoint regions* such that each *waypoint region* satisfies a given clustering policy such as maximum cluster size and stay location resolution (minimum distance between a pair of closest stay locations).

Extracting and scoring waypoints: extracts *waypoints* by identifying the most representative point among all data points in each *waypoint region* and computes a scoring metric reflecting density information at those points for the purpose of ranking.

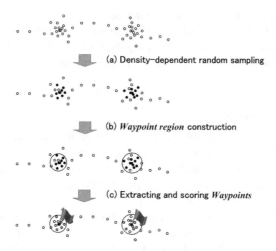

Fig. 1. Overview of operations flow. Given raw GPS data, proposed algorithm (a) performs random sampling from high-density regions, (b) constructs *waypoint regions* by clustering sampled data points, and (c) extracts and scores *waypoint* for each cluster

2.2   Density-Dependent Random Sampling

We first define the *labeling* function that maps each data point $x \in X$ to a *label* space L, a set of *labels* of histogram bins, and develop a method of constructing *random histograms* using this function. Note that we simplify input GPS data X as a sequence of periodically recorded location vectors, $X = \{x \in \mathbb{R}^D\}$. The dimension is $D = 3$ in general but when temporal information such as temporal distance is necessary, we can extend the dimension as $x' = (x, ct)$ with c being a scaling coefficient so we can handle all entries in the same way.

LSH sketch to base C. LSH is a probabilistic method of hashing objects such that two similar objects will likely collide into the same bucket in response to the degree of similarity. Let $\mathcal{F}$ be an LSH family for L_2 distance (refer to [6] for details of LSH family definition). A hashing function, $f \in \mathcal{F}$, is implemented by taking advantage of the property of *p-stable* distribution [5], and LSH sketch [2] [3] takes only the least significant bit of the hash value represented in the binary numeral system. We extend the base of this LSH sketch to the general value C, which is a positive integer greater than or equal to 2:

$$f(x) = \left\lfloor \frac{a \cdot x + r}{W} \right\rfloor \ mod \ C \,, \tag{1}$$

where r is a real number drawn from a uniform distribution $U[0, W)$, a is a D-dimensional vector whose entries are independently drawn from a normal distribution, and W is a parameter called *window size*. Note that when $C = 2$, it reduces to a binary LSH sketch in [2].

Slight modification of the argument on the original LSH described in [5] results in an analytical form of collision probability for two vectors with distance d:

$$p(d) = \mathbf{Pr}\left[\, f(x) = f(y) \mid |x - y| = d \,\right]$$
$$= \int_0^W \frac{1}{d} \sum_{k \in \mathbb{Z}} \phi\left(\frac{kCW + t}{d}\right)\left(1 - \frac{t}{W}\right) dt \,, \tag{2}$$

where $\mathbb{Z}$ is a set of integers and $\phi(t)$ denotes the probability density function of the absolute value of the normal distribution. Figure 2(a) plots the collision probability $p(d)$ of two vectors with distance d for $B = 1$ and $C = 2, 3, 5$. We can see that it almost linearly decreases in response to an increase in d until d reaches a certain value where it then shows a plateau. We call $f(x)$ an *atomic label* of vector x, and concatenation of B independent *atomic labels* constructs a *label* of vector x as $\langle f_1(x), \cdots, f_B(x) \rangle$. The collision probability of two vectors with the same *label* is given by $[p(d)]^B$, which is illustrated in Figure 2(b) for $B = 1, 3, 5$ and $C = 2$. The collision probability quickly converges to $P_{res} \sim C^{-B}$. This residual probability P_{res} means that no matter how far apart two vectors are located, they could collide with probability P_{res}. Therefore, by controlling (B, C, W), we can adjust the shape of the probability that two arbitrary vectors with distance d collide (have a same *label*). The benefit of introducing parameter C is to provide more powerful controllability and reduce computation time by adjusting both bit length B and the base number C. Without C, the desired value of P_{res} is obtained only by increasing B, which also increases computation time since it increases the number of loops for *label* computation in a program code.

Random histograms of GPS data. Let $h(x) = \langle f_1(x), \cdots, f_B(x) \rangle$ be a *labeling* function that maps X to a *label* space $L \subseteq \{0, 1, \cdots, C^B - 1\}$, where a *label* l is expressed in the decimal numeral system using the formula $l = \sum_{b=1}^B f_b(x) C^{b-1}$. If we interpret L as a set of bin *labels*, computation of $h(x)$ for $x \in X$ determines to which bin x is registered. Let $\mathcal{H} = \{h = \langle f_1, \cdots, f_B \rangle | f_i \in \mathcal{F}\}$ be a set of *labeling* functions and $\Lambda_{h \in \mathcal{H}, l}(X) = \{x \in X | l = h(x)\}$ be a set of data points mapped to l by h chosen at random from $\mathcal{H}$. Then, we can construct a *table* (a set of bins) of X, $\Lambda_h(X) = \{\Lambda_{h,l}(X)\}_{l \in L}$.

We can also define the *frequency distribution* (called *random histogram*) of X over L as a vector point in $|L|$-dimensional vector space, $\varphi_h(X) = \sum_{i=0}^{|L|} \lambda_{l_i} \mathbf{e}_i$, where $\lambda_{l_i} = |\Lambda_{h,l_i}(X)|$ is the frequency of the bin labeled by l_i, and $\{\mathbf{e}_i\}_{i=1,\cdots,|L|}$ is a standard basis of the $|L|$-dimensional vector space. Without loss of generality, we can assume that $L = \{l_1, \cdots, l_{|L|}\}$ is in descending order of λ_l such that for any $i < j$, we get $\lambda_{l_i} \geq \lambda_{l_j}$. Then, given the positive integer Q, we can define a sampling operation $\mathscr{S}_Q[\Lambda_h(X)]$ in such a way that it returns a set of bins with frequency being among Q-highests in Λ_h, *i.e.*, $\{\Lambda_{h,l_i}\}_{i=1,\cdots,Q}$. Since histogram construction is a probabilistic operation, sometimes we may fail to sample a large enough number of data points from a high-density region; therefore, we need to augment the sampling quality. To this end, we maintain a *supertable*, $\Xi_H(X) = \langle \mathscr{S}_Q[\Lambda_{h_1}(X)], \cdots, \mathscr{S}_Q[\Lambda_{h_N}(X)] \rangle$, by concatenating N independent *tables*, each of which is constructed by a *labeling* function chosen at random from $\mathcal{H}$. Note that although a large enough value of Q and N ensures good accuracy, choosing these optimal values, N in particular, requires careful consideration because values that are too large directly affect computation time.

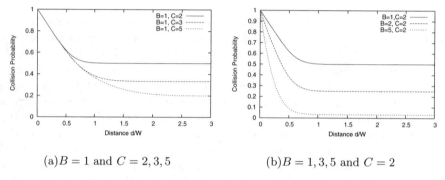

(a)$B = 1$ and $C = 2, 3, 5$ (b)$B = 1, 3, 5$ and $C = 2$

Fig. 2. Collision probability of two vectors with distance d

2.3   Waypoint Region Construction Using Cluster Analysis

Unless we take W that is too large, most of the data points $x \in \Lambda_{h,l_i}(X)$ for $i = 1, \cdots, Q$ are distributed around a single stay location, and the most centered point among them would be a good waypoint for the corresponding significant location. Yet, there is still small probability that the bin contains some irrelevant data points that disturb the extraction quality.

To exclude those "noise" points and group the data points such that each group contains data points coming only from a single stay location, we execute a cluster analysis on the sampled data set. To this end, we use *Ward*'s method [7], a common hierarchal clustering method that works well on our sampled dataset because the dense regions are already well separated by density-dependent random sampling. To appropriately terminate hierarchical clustering, which keeps merging two closest clusters in a one-by-one manner until a given goal condition is satisfied, we need to pre-define the goal condition such that the desired *waypoint regions* are obtained. We describe the goal condition Θ using two parameters; *resolution* ζ of clusters, which designates the shortest permissible distance between two most adjacent clusters, and *radius* ρ of clusters, the maximum allowable distance from a centroid of a cluster for a data point to be grouped in that cluster. The distance between a pair of clusters is defined by the distance between their centroids, which reflects the distance between two corresponding significant locations, and ρ should be determined by the *radius* indicating the possible drift distance from an actual stay location due to a user roaming or noise. Note that ρ can be loosely set using a typical quantity of the standard deviation of noise as we will see in Section 3.

One major drawback in using this cluster analysis is long computation time due to distance computation between all pairs of data points. Therefore, we first separate the sampled dataset into two groups of datasets, *High* and *Low* such that a *High* dataset contains at most M data points registered to bins with high frequency and *Low* contains all others. M is the largest number of data points permissible for quick clustering. After obtaining the datasets, we perform cluster analysis only over the *High* dataset and construct "nuclei" and then individually register data points in the *Low* dataset to the nearest nuclei to obtain *waypoint regions* $R = \{R_i\}$, where R_i is each resulting cluster comprised of a set of data points in X. Initial nucleation over a limited number of data points is aimed to eliminate the dependency on the sequence of registering data points to the result, especially when density distribution of the input data set shows poor structure with few broad peaks.

We should also note that the clustering described here is optional and can be skipped if computation time is a major concern. It is rare that we obtain bins containing irrelevant data points unless two significant locations are too close or the density distribution of the input dataset is too broad and vague in structure. If the purpose is only to eliminate coincidentally registered data points far from a relevant stay location, skipping the hierarchical clustering (*i.e.* $M = 0$) and only eliminating data points that violate the clustering policy Θ will often return a good result. This is quicker because it only requires distance computation between a to-be-added data point and a centroid of each cluster for finding the closest cluster to register.

2.4   Extracting and Scoring Waypoints

Given a set of *waypoint regions* R, and the positive integer K, the final step of the algorithm is to extract a set of K *waypoints*, $\Omega = \{\omega_i\}_{i=1,\cdots,K}$, by identifying the most representative point in each *waypoint region* R_i and rank them by associating a scoring metric that reflects the density at each location. $\omega_i = (r_i, s_i)$ contains information regarding location, $r_i \in \mathbb{R}^D$, and scoring metric, $s_i = (s_{i,1}, s_{i,2}) \in \mathbb{R} \times \mathbb{R}$. Suppose we are to extract a *waypoint* $\omega = (r, s)$ for a given *waypoint region* R_a. Then, r should be the location of the data point in R_a that minimizes the sum of the distance to all data points, *i.e.*,
$$r = \arg \min_{x \in R_a} \sum_{y \in R_a \setminus \{x\}} |x - y|^2,$$
which reduces to the data point that is closest to $\bar{x} = \frac{1}{|R_a|} \sum_{x \in R_a} x$.

On the other hand, scoring metric s of the waypoint ω should reflect the density at r. Observing that the number of data points in R_a and sample variance of distance between r and other data points in R_a reflect the density at r, we define the scoring metrics as $s_1 = |R_a|$ and $s_2 = \frac{1}{|R_a|-2} \sum_{x \in R_a \setminus \{r\}} |x - r|^2$, where as s_1 is large and s_2 is small, the density at the location is higher and thereby ω is more important. Note that we define that the first metric s_1 has priority over the second metric s_2. By computing a *waypoint* $\omega_i = (r_i, s_i)$ for all *waypoint regions* R_i, we can sort a set of *waypoints* in descending order of importance using the scoring metrics $s_{i,1}$ and $s_{i,2}$, and obtain K-most important *waypoints*, $\Omega = \{\omega_i\}_{i=1,\cdots,K}$.

3 Evaluation

3.1 Evaluation Using Artificially Generated Datasets

Using an artificially generated dataset whose noise level is under control, we evaluate the performance and noise tolerance in parameter setting of the proposed algorithm $\mathcal{A}_{DDRS}$ and compare it to a typical fixed-threshold-based algorithm.

Dataset generation algorithm. The artificially generated dataset X_a is a history of periodically recorded locations that contain K_{X_a} "staying" periods alternating with $K_{X_a} + 1$ "moving" periods. In generating X_a, we first prepare for K_{X_a} time slots $\{\tau_i\}$ whose duration τ_i is drawn from the Poisson distribution with average τ_s and randomly allocate these slots without any overlap in X_a with the total number of time steps being N_{X_a}. Each time slot τ_i indicates the staying period in which one stays at a single location and other parts of the dataset represent the moving period. In the moving period, the location vector $\psi(t)$ is updated by the randomly generated step vector $\delta \Delta \psi_m(t)$ such that $\psi(t+1) = \psi(t) + \delta \Delta \psi_m(t)$. The step width δ is drawn from the Poisson distribution with average δ_0, and $\Delta \psi_m(t)$ is a unit vector whose direction is determined by random rotation whose angle is drawn from the uniform distribution $U[-\theta_{max}, \theta_{max}]$, where θ_{max} is the maximum possible angle between the previous and next steps. When the system enters into the staying period at time t_0, $\psi(t)$ keeps being updated by the formula $\psi(t) = \psi(t_0) + \sigma \Delta \psi_s(t)$ until the system consumes the time slot allocated for the staying period and re-enters the moving period. The second term $\sigma \Delta \psi_s(t)$, where σ represents the noise level and $\Delta \psi_s(t)$ is drawn from the two dimensional normal distribution, indicates the spatial noise introduced to an actual stay location due to a weak signal, e.g., one staying inside a building.

Performance measures. Let $\Omega_r = \{r_i\}_{i=1,\cdots,K}$ be a set of *waypoint* locations extracted by an algorithm and $\Psi = \{\psi_i\}_{i=1,\cdots,K_{X_a}}$ be a set of actual stay locations, *i.e.*, an answer for Ω_r. Then we can define two quality measures for Ω_r; *distance* $\delta(\Omega_r, \Psi)$ and *detection ratio* $\varrho(\Omega_r, \Psi)$. $\delta(\Omega_r, \Psi)$ quantifies the distance between Ω_r and Ψ, *i.e.*, a set distance. Since each *waypoint* should correspond to each actual stay location, we should use a set distance for one-to-one matching defined by,

$$\delta(\Omega_r, \Psi) = min \; \frac{1}{\Delta} \sum_{r_i \in \Omega_r} \sum_{\psi_i \in \Psi} a_{r_i, \psi_i} |r_i - \psi_i| \tag{3}$$

$$s.t. \quad \forall \psi_i \in \Psi, \quad \sum_{r_i \in \Omega_r} a_{r_i, \psi_i} \leq 1 \; ,$$

$$\forall r_i \in \Omega_r, \qquad \sum_{\psi_i \in \Psi} a_{r_i, \psi_i} \leq 1 \,,$$

$$\forall r_i \in \Omega_r, \forall \psi_i \in \Psi, \quad a_{r_i, \psi_i} \in \{0, 1\} \,,$$

$$\Delta = min\left\{|\Omega_r|, |\Psi|\right\} = \sum_{r_i \in \Omega_r} \sum_{\psi_i \in \Psi} a_{r_i, \psi_i} \,.$$

The coefficient $a_{r_i, \psi_i} = 1$ means that a member r_i is matched to a member ψ_i. Note that the number of matchings is $\Delta = min\{|\Omega_r|, |\Psi|\}$, whereas each member can be used at most once. We take the average in (3), to prevent a situation in which the number of extracted *waypoints* is too small due to poor extraction ability, which reduces to the shorter distance, *i.e.*, a good matching. ϱ is another quality measure for indicating how many actual stay locations are detected. Since Δ indicates the number of matchings, ϱ is simply defined as $\varrho = \Delta/|\Psi|$. Using $\delta(\Omega_r, \Psi)$ and $\varrho(\Omega_r, \Psi)$, tolerance $\pi_p = [\pi_{p,l}, \pi_{p,u}]$ in a given parameter p is defined by a range in p that achieves $\delta(\Omega_r, \Psi) \leq \sigma$ and $\varrho = 1$, where σ is the spatial noise level defined above. This definition states that as long as $p \in \pi_p$, we can find any actual stay location with an average distance being at most σ from each corresponding *waypoint*.

Noise tolerance. Given N_{X_a}, K_{X_a}, τ_s, δ_0, and θ_{max}, we prepare for input datasets with various noise levels by controlling σ. For comparison, we also implemented a fixed-threshold-based algorithm $\mathcal{A}_{FT}$, similar to the one described in [10], which is simple and intuitive but shows considerably good performance at least under low noise levels. $\mathcal{A}_{FT}$ is a deterministic algorithm that has two parameters; *roaming distance* l_{th} and *stay duration* t_{th}. l_{th} represents the maximum distance that determines the region where one can stray to be counted as a "stay", and t_{th} is the minimum duration one must stay in the segment to be qualified as a "stay". By comparing these two parameters to the given dataset X_a, we can detect segments corresponding to *waypoint regions* with a longer duration than t_{th} and a diameter of the staying region less than l_{th}, and the *waypoint* in each segment is extracted in the same way.

Obviously, l_{th} strongly affects the detection quality, and these parameter settings are difficult, especially when the spatial noise level is high and unknown. A large enough l_{th} can cover all data points coming from each stay location but data points coming from two geographically close stay locations may be mixed. On the other hand, a too small l_{th} can detect no stay segment when t_{th} is large, and if t_{th} is too small, it will detect too many stay segments, most of which are of little importance. In addition, large l_{th} and t_{th} leads to an increase in the number of data points and thereby increases the computation time since all pairs of those data points must be compared to l_{th}. Therefore, the tolerance in the spatial parameter setting will become small when noise level is high, and careful parameter tuning is required for maximizing detection quality. The parameters in $\mathcal{A}_{DDRS}$ corresponding to l_{th} are $\Theta = (\zeta, \rho)$, which determine to which *waypoint region* each sampled data point should be registered. Although both l_{th} and Θ require an estimate of the stretch in spatial noise distribution, finding good Θ in $\mathcal{A}_{DDRS}$ is not as difficult as finding the appropriate l_{th} in $\mathcal{A}_{FT}$. Since $\mathcal{A}_{DDRS}$

samples data points selectively from around each stay location, small ζ and ρ work even under high noise levels, and the crosstalk between data points coming from two neighboring stay locations is also suppressed. In short, tolerance in ζ or ρ is wide even against strong spatial noise.

To observe how well $\mathcal{A}_{DDRS}$ detects *waypoints* and how much $\mathcal{A}_{DDRS}$ eases the difficulty in parameter setting, we compared our algorithm $\mathcal{A}_{DDRS}$ to $\mathcal{A}_{FT}$ using datasets with noise levels $\sigma = \{100, 200, 300, 400, 500, 600\}$ and measured the tolerance in setting parameters ρ for $\mathcal{A}_{DDRS}$ and l_{th} for $\mathcal{A}_{FT}$. Note that we only controlled ρ by setting $\zeta = 2\rho$ for simplicity and all other parameters were configured at the loosely optimized point. Table 1 summarizes the parameter values used for the evaluation. Figures 3(a) and 3(b) plot both the average values of $\delta(\Omega_{r,alg.}, \Psi)$ and $\varrho(\Omega_{r,alg.}, \Psi)$ over ten independently generated X_a for $\sigma = 200$ and $\sigma = 500$, respectively. Ψ indicates a set of actual stay locations in X_a, and $\Omega_{r,alg.}$ indicates the output of each algorithm $\mathcal{A}_{alg.}$ for $alg. = DDRS$ or FT. Note that we also executed ten independent trials for evaluating $\mathcal{A}_{DDRS}$ for each dataset since it is a probabilistic algorithm and requires taking the average $\delta(\Omega_{r,alg.}, \Psi)$ and $\varrho(\Omega_{r,alg.}, \Psi)$ over the trials for fair comparison. Note that each X_a is generated in such a way that it contains at least one pair of neighboring stay locations whose distance is upper-bounded by around $2\sigma \sim 3\sigma$ to limit the upper bound of tolerance π_p for $p = l_{th}$ and ρ.

Basically, the optimal point ρ^* and l_{th}^*, which is around the center of the tolerance π_p for $p = \rho$ and l_{th}, respectively, increase in response to the increase in σ, but the upper/lower bound of the tolerance in both parameters shows a different response. For all σ, small l_{th} degrades both $\delta(\Omega_{r,FT}, \Psi)$ and $\varrho(\Omega_{r,FT}, \Psi)$, and the lower bound $\pi_{p,l}$ for $p = l_{th}$ is relatively large. This is because l_{th} that is too small detects only few stay locations as indicated by $\varrho(\Omega_{r,FT}, \Psi) \sim 0.1$, where $\delta(\Omega_{r,FT}, \Psi)$ for such l_{th} is too large to be displayed in the plot. On the other hand, $\mathcal{A}_{DDRS}$ shows good tolerance even for small ρ since it can sample many data points even from regions with such a small radius. On the opposite side of the spectrum, the upper bound of both parameters is basically determined by the crosstalk between data points coming from the most adjacent stay locations. However, $\mathcal{A}_{DDRS}$ shows moderately better performance (a gentler slope) because $\mathcal{A}_{DDRS}$ samples few data points located around the "edge" of high-density regions and suppresses crosstalk. Furthermore, $\mathcal{A}_{DDRS}$ took long computation time when l_{th} is large, whereas $\mathcal{A}_{DDRS}$ did not show noticeable difference in response to an increase in ρ. This is because $\mathcal{A}_{FT}$ needs to compute distance among many data points when l_{th} is large. Figure 4 illustrates profiles of the tolerance and dynamic range DR against σ ranging $[100, 600]$, where DR is a ratio of tolerance to the optimal point: $DR = \pi_p/p^*$. $\mathcal{A}_{DDRS}$ shows a wide dynamic range due to small $\pi_{p,l}$ even for large σ. A wide dynamic range is important in parameter setting because it allows parameter setting to work for a variety of input data sets. In fact, Figure 4 shows that there is a parameter band $[max\{\pi_{\rho,l}(\sigma)\}, min\{\pi_{\rho.u}(\sigma)\}]$, in which ρ works for all σ, whereas there is no l_{th} that is universally valid for all σ. Since it is, in practice, often difficult to make a good estimate about the noise level contained in actual GPS datasets,

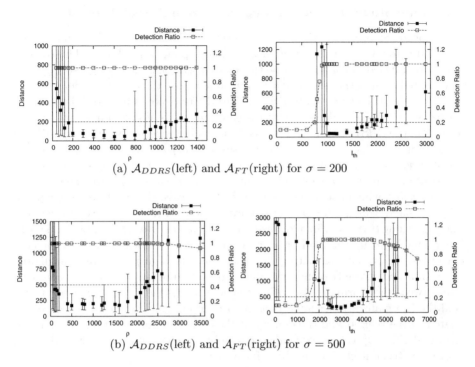

(a) $\mathcal{A}_{DDRS}$(left) and $\mathcal{A}_{FT}$(right) for $\sigma = 200$

(b) $\mathcal{A}_{DDRS}$(left) and $\mathcal{A}_{FT}$(right) for $\sigma = 500$

Fig. 3. Performance comparison of distance $\delta(\Omega_{r,alg.}, \Psi)$ and $\varrho(\Omega_{r,alg.}, \Psi)$ between $alg. = \mathcal{A}_{DDRS}$ and $alg. = \mathcal{A}_{FT}$ for (a)$\sigma = 200$ and (b)$\sigma = 500$. Points designate average over ten different input datasets whereas error bars range from min to max value. Note that tolerance is defined by region below dashed line representing reference distance.

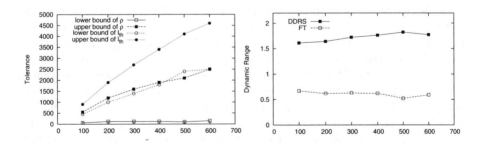

Fig. 4. Tolerance π_p(left) and Dynamic range DR(right) for $p = \rho$ in $\mathcal{A}_{DDRS}$ and $p = l_{th}$ in $\mathcal{A}_{FT}$ for various σ

fixed-threshold-based algorithms require trial-and-error repetitions until an optimal parameter setting for each data can be found.

3.2   Case Study: Extracting *waypoints* from Actual Travel Data

We applied the proposed algorithm to actual GPS data recorded when a person traveled throughout Miyako island, in Okinawa prefecture, Japan. The traveler basically traveled by car and stopped at several locations, such as sightseeing spots distributed around the island. We examined how well the algorithm detected the spots that the traveler actually visited. The GPS data contained 1617 data points, each of which was recorded at about 15-second intervals. Figure 5(a) shows the trajectory of the GPS data and the fourteen top-ranked *waypoints* extracted with the algorithm. The algorithm successfully returned *waypoints* that corresponded well to all major locations that the traveler actually visited; sightseeing spots, shops, a gas station, a hotel, an airport, *etc.* Figure 5(b) illustrates an enlargement around one *waypoint*, which encompasses *Higashi-Hennazaki*, the most eastern cape on Miyako island, which is famous for panorama views of the ocean. We can see from the sojourning trajectory in a localized region that the traveler stopped and spent some time enjoying the landscape, and the *waypoint* is located around the center of the region. For extracting the *waypoints*, the algorithm took a negligibly short time and showed excellent responsiveness. Figure 5(c) illustrates a comparison between the (normalized) scoring metric (the first metric s_1 of each *waypoint*) and the (normalized) measured density at each location. By observing excellent matching between s_1 and the density at the corresponding location, we can confirm that the proposed algorithm successfully samples data points such that the scoring metric reflects the density information.

4   Related Work

The recent widespread use of GPS-enabled devices encourages many location-aware services, such as GeoLife project [8] to use algorithms for automatically extracting significant locations for understanding users' activity patterns. For example, Ashbrook et al. [9] designed a fixed-threshold-based algorithm for detecting significant locations from GPS data and using a set of those locations for behavior prediction. Hariharan et al. [10], Liao et al. [11] [12], and Zheng et al. [13] [14] [15] also developed a similar fixed-threshold-based algorithm for detecting segments of GPS data and identifying the most representative point in each segment. Although there are many variations, all these algorithms have the basic principle of detecting locations where one stays at least for a certain time in a limited region with a certain diameter using spatial/temporal thresholds. However, fixed-threshold-based algorithms do not generally work well under high noise levels, and the optimal parameter setting tends to be difficult.

Agamennoni et al. [16] developed a different algorithm for extracting significant locations by introducing a score associated with each location using velocity information and linking the top-scored locations to create clusters that designate significant locations. Although this algorithm shows good noise tolerance, it still uses the velocity threshold to compute the score. Any fixed-threshold-based algorithm is inherently error-prone where one cannot make a good guess about

Table 1. Parameters used for evaluation

$N_{X_a} = 2000$	# of data points for X_a
$K_{X_a} = 10$	# of stay locations for X_a
$\delta_0 = 50\ m$	average step width for X_a
$\tau_s = 30\ min$	average stay duration for X_a
$\theta_{max} = 5°$	maximum angle between the previous and next step direction
$\zeta = 2\rho$	cluster *resolution* for a given ρ (cluster *radius*)
$N = 10$	# of tables concatenation
$B = 5$	big length
$C = 21$	base number
$W = \rho$	window size
$Q = 10$	# of bins for sampling
$K = 15$	# of extracted *waypoints* for display
$t_{th} = 10\ min$	stay duration for $\mathcal{A}_{FT}$

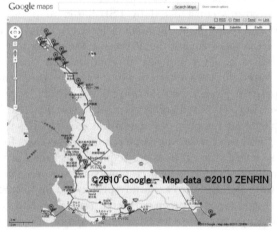

(a) An overview of GPS data trajectory and extracted *waypoints*.

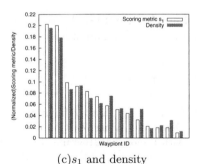

(b)[Enlargement] Higashi-Hennnazaki   (c)s_1 and density

Fig. 5. Trajectory and extracted *waypoints* for actual travel data in *Miyako* island. (a) overview of GPS trajectory and extracted *waypoints*. (b) Higashi-Hennnazaki, a *waypoint*(enlargement). (c) Comparison between first scoring metric of each *waypoint* and actually measured density (defined by number of data points in circle with radius of 40 meters around measured point). All maps are displayed using Google Maps[1].

the optimal quantity in the control variable. Furthermore, most studies do not focus on computation time, even though a number of data points are generally involved. Unlike the work described above, we took a probabilistic approach and explored the way to make parameter setting easy even under high noise levels, and made loose optimization work even when we did not precisely know the optimal threshold value in the control variables.

5   Conclusion

We proposed an algorithm that automatically extracts *waypoints*, points of reference designating significant locations, from raw GPS data. In extracting *waypoints*, the proposed algorithm probabilistically detects high-density regions using random histograms constructed by LSH-based mapping for computing a *label* of bins. Since it samples data points selectively from high-density regions, it shows competitive performance in extracting *waypoints* from even input data with high spatial noise level and also in computation time. Evaluations with artificially generated data with various noise levels revealed that our algorithm possesses competitive *waypoint* extraction ability as well as very wide tolerance in parameter setting compared to the typical fixed-threshold-based algorithm. This result implies that the proposed algorithm greatly eases the difficulty in parameter setting, and we can use the same parameter settings for input data with a variety of noise levels. Since it does not require direct distance computation between data points, it shows excellent responsiveness even against an increase in the number of data points. The case study performed for actual travel data also shows excellent consistency between extracted *waypoints* and actually visited locations. Also, the location of each extracted *waypoint* agrees with the center of high-density regions, and the scoring metric reflects actual density. We believe that the proposed algorithm works well for many location-aware applications.

Acknowledgments. A part of this work is supported by National Institute of Information and Communication Technology (NICT), Japan.

References

1. Google Maps, http://maps.google.com/
2. Dong, W., Wang, Z., Charikar, M., Li, K.: Efficiently Matching Sets of Features with Random Histograms. In: Proceedings of the 16th ACM International Conference on Multimedia, pp. 179–188 (2008)
3. Dong, W., Charikar, M., Li, K.: Asymmetric Distance Estimation with Sketches for Similarity Search in High-Dimensional Spaces. In: Proceedings of the 31st Annual International ACM SIGIR Conference on Research and Development on Information Retrieval, pp. 123–130 (2008)
4. Indyk, P., Motwani, R.: Approximate nearest neighbors: towards removing the curse of dimensionality. In: Proceedings of the Thirtieth Annual ACM Symposium on Theory of Computing, pp. 604–613 (1998)

5. Datar, M., Immorlica, N., Indyk, P., Mirrokni, V.S.: Locality-sensitive hashing scheme based on p-stable distributions. In: Proceedings of the Twentieth Annual Symposium on Computational Geometry, pp. 253–262 (2004)
6. Charikar, M.S.: Similarity estimation techniques from rounding algorithms. In: STOC 2002: Proceedings of the Thiry-Fourth Annual ACM Symposium on Theory of Computing, pp. 380–388 (2002)
7. Ward, J.H.: Hierarchical grouping to optimize an objective function. Journal of the American Statistical Association 58(301), 236–244 (1963)
8. GeoLife Project, http://research.microsoft.com/en-us/projects/geolife/
9. Ashbrook, D., Starner, T.: Using GPS to learn significant locations and predict movement across multiple users. In: Personal and Ubiquitous Computing, vol. 7, pp. 275–286 (2003)
10. Hariharan, R., Toyama, K.: Project Lachesis: Parsing and Modeling Location Histories. In: Egenhofer, M.J., Freksa, C., Miller, H.J. (eds.) GIScience 2004. LNCS, vol. 3234, pp. 106–124. Springer, Heidelberg (2004)
11. Liao, L., Fox, D., Kautz, H.: Location-based Activity Recognition using Relational Markov networks. In: Proc. of the International Joint Conference on Artificial Intelligence (2005)
12. Liao, L., Patterson, D.J., Fox, D., Kautz, H.: Building Personal Maps from GPS Data. In: Annals of the New York Academy of Sciences, vol. 1093, pp. 249–265 (2006)
13. Zheng, Y., Liu, L., Wang, L., Xie, X.: Learning Transportation Mode from Raw GPS Data for Geographic Applications on the Web. In: Proceeding of the 17th International Conference on World Wide Web, pp. 247–256 (2008)
14. Zheng, Y., Li, Q., Chen, Y., Xie, X., Ma, W.: Understanding mobility based on GPS data. In: Proceedings of the 10th International Conference on Ubiquitous Computing, vol. 21, pp. 312–321 (2008)
15. Zheng, Y., Chen, Y., Xie, X., Ma, W.: Mining Interesting Locations and Travel Sequences From GPS Trajectories. In: Proceedings of the 18th International Conference on World Wide Web, vol. 21, pp. 791–800 (2009)
16. Agamennoni, G., Nieto, J.I., Nebot, E.: Mining GPS data for extracting signifcant places. In: Proceedings of the 2009 IEEE International Conference on Robotics and Automation (ICRA), pp. 855–862 (2009)

Discovery of Conservation Laws via Matrix Search

Oliver Schulte and Mark S. Drew[*]

School of Computing Science, Simon Fraser University,
Burnaby, B.C., Canada V5A 1S6
{oschulte,mark}@cs.sfu.ca

Abstract. One of the main goals of Discovery Science is the development and analysis of methods for automatic knowledge discovery in the natural sciences. A central area of natural science research concerns reactions: how entities in a scientific domain interact to generate new entities. Classic AI research due to Valdés-Pérez, Żytkow, Langley and Simon has shown that many scientific discovery tasks that concern reaction models can be formalized as a matrix search. In this paper we present a method for finding conservation laws, based on two criteria for selecting a conservation law matrix: (1) maximal strictness: rule out as many unobserved reactions as possible, and (2) parsimony: minimize the L1-norm of the matrix. We provide an efficient and scalable minimization method for the joint optimization of criteria (1) and (2). For empirical evaluation, we applied the algorithm to known particle accelerator data of the type that are produced by the Large Hadron Collider in Geneva. It matches the important Standard Model of particles that physicists have constructed through decades of research: the program rediscovers Standard Model conservation laws and the corresponding particle families of baryon, muon, electron and tau number. The algorithm also discovers the correct molecular structure of a set of chemical substances.

1 Introduction: Reaction Data and Conservation Laws

As scientific experiments amass larger and larger data sets, sometimes in the millions of data points, scientific data mining and automated model construction become increasingly important. One of the goals of Discovery Science is the development and analysis of methods that support automatic knowledge discovery in the sciences. The field of automated scientific discovery has developed many algorithms that construct models for scientific data, in domains ranging from physics to biology to linguistics [1,2,3]. From a cognitive science point of view, automated scientific discovery examines principles of learning and inductive inference that arise in scientific practice and provides computational models of scientific reasoning [3], [1].

---

[*] This work was supported by Discovery Grants from NSERC (Natural Sciences and Engineering Research Council of Canada) to each author. We thank the anonymous referees for helpful comments.

B. Pfahringer, G. Holmes, and A. Hoffmann (Eds.): DS 2010, LNAI 6332, pp. 236–250, 2010.

One of the key problems in a scientific domain is to understand its *dynamics*, in particular how entities react with each other to produce new entities [2,4,5,6,7,8]. Classic AI research established a general computational framework for such problems. Many discovery problems involving reaction data can be formulated as a matrix multiplication equation of the form

$$RQ = Y,$$

where R is a matrix representing reaction data for a set of known entities, the vector Y defines constraints on the model given the data, and Q is a matrix to be discovered [2,9]. One interpretation of the Q matrix is that it defines a hidden or *latent feature vector* for each entity involved in the observed reactions; these hidden features explain the observed interactions. So the matrix equation framework is an instance of using matrix models for discovering latent features. The framework models problems from particle physics, molecular chemistry and genetics [2]. An important special case are **conservation matrices** which satisfy a matrix equation of the form $RQ = 0$. Intuitively, a conservation equation says that the sum of a conserved quantity among the reactants is the same as its sum among the products of a reaction. This paper describes a new procedure for finding conservation matrices.

Approach. There are infinitely many matrices that satisfy the conservation equation $RQ = 0$, so a model selection criterion is required. Valdés-Pérez proposed using the L1-norm to select conservation law matrices, which is the sum of the absolute values of the matrix entries [8,7]. The L1-norm is often used as a *parsimony* metric for a matrix [10]. Seeking parsimonious explanations of data is a fundamental principle of scientific discovery, widely applied in machine learning [11, Ch.28.3]. Schulte [9] recently introduced a new criterion for selecting a hidden feature matrix Q: The matrix should be *maximally strict*, meaning that Q should be consistent with the observed reaction phenomena, but inconsistent with as many unobserved reactions as possible. The maximum strictness criterion formalizes a basic principle of scientific discovery: it is not only important to explain the processes that do occur in nature, but also why some processes *fail* to occur [9]. In this paper we combine the two criteria and consider *maximally simple maximally strict* (MSMS) matrices that have minimal L1-norm among maximally strict matrices. The main algorithmic contribution of this paper is an efficient new optimization scheme for this criterion that scales linearly with the number of observed data points.

Evaluation. In principle, the theory and algorithms in this paper apply to matrix search in any scientific domain. Here we focus on high-energy particle physics (HEP) as the application domain, for several reasons. (1) The problem of analyzing particle accelerator data is topical as a new set of data is expected from the record-breaking energy settings of the Large Hadron Collider (LHC) in Geneva. (2) An easily accessible source of particle accelerator data is the Review of Particle Physics [12], an authoritative annual publication that collects the current knowledge of the field. (3) Most of the previous work on discovering conservation laws has analyzed particle physics data [6,8,7,9].

In particle physics, we compare our algorithm with the centrally important *Standard Model* of particles [13,14]. The main concept of the Standard Model is to view quarks as fundamental building blocks for all other entities in nature. Since Gell-Mann introduced the quark model in his Nobel-prize winning work, physicists have used it as a basis to develop, over decades of research, the Standard Model, which is consistent with virtually all known observations in particle physics. One of the goals of the LHC is to probe new phenomena that test the Standard Model and may require an extension or modification. A key component of the Standard Model are conservation laws, in particular the conservation of Electric Charge, and of the Baryon, Tau, Electron and Muon Numbers. Applying our program to data from particle accelerators, the combination of laws + particle families found by the program is equivalent to the combination of laws + particle families in the Standard Model: both classify reactions as possible and impossible in the same way. The algorithm agrees with the Standard Model on the particle families corresponding to Baryon, Tau, Electron and Muon families, in the sense that MSMS conservation matrices define these particle families.

We also apply our procedure to the chemistry reaction data set used in the evaluation of the DALTON*system [1]. While this is a small data set, it illustrates the generality of the matrix equation framework and of our conservation law discovery procedure by applying both in a second domain. The procedure correctly recovers the molecular structure of a set of chemical substances given reactions among them. Our code and datasets are available online at http://www.cs.sfu.ca/~oschulte/particles/conserve_zipfile.zip.

This paper addresses the problem of finding theories to explain reaction data that have been accepted by the scientific community. A challenging and practically important extension is reconstructing raw sensory data with a data reaction matrix that separates the true experimental signal from background noise [9, Sec.1]. Matrix reconstruction methods often employ a minimization search with an objective function that measures reconstruction quality; our work suggests that incorporating the MSMS criterion may well improve reconstruction quality.

Contributions. The main contributions of this paper may be summarized as follows.

1. The new MSMS criterion for selecting a set of conserved quantities given an input set of observed reactions: the conserved quantities should be as simple as possible, while ruling out as many unobserved reactions as possible.
2. An efficient minimization routine for finding an MSMS conservation law matrix that scales linearly with the number of observed reactions (data points).
3. A comparison of the output of the algorithm on particle accelerator data with the fundamental Standard Model of particles.

Paper Organization. We begin by reviewing previous concepts and results from the matrix search framework. Then we define the MSMS selection criterion, and describe a scalable local search algorithm for MSMS optimization. The output of our procedure is compared with the Standard Model on actual particle accelerator data, and with the known molecular structure of chemical substances on chemical reaction data.

Related Work. We review related work within the matrix search framework. For discussions of this framework, please see [2,9]. Valdés-Pérez and Erdmann used the L1-norm to select conservation matrices for particle physics data [8,7]; their work is the most advanced in this problem. In contrast to the current paper, it assumes that both observed and unobserved particle reactions are explicitly specified, and it does not use the maximal strictness criterion. In empirical evaluation, they found that their method failed to find more than a single conservation law, and they proved analytically that this is difficult if not impossible to avoid on their approach. Schulte introduced and applied the maximal strictness criterion to develop an algorithm for inferring the existence of hidden or unobserved particles [9]. His paper does not consider the parsimony of conservation matrices. A combined system might first find hidden particles, and then apply the MSMS criterion to find parsimonious laws that include the hidden particles. In effect this is the setting of the experiments of this paper, where knowledge of the hidden particles in the Standard Model is part of the input. To our knowledge the connection between groupings of entities, like particle families, and parsimonious conservation laws is an entirely new topic in scientific discovery.

2   Selecting Maximally Simple Maximally Strict Conservation Laws

We review the matrix framework for representing reaction data and conservation laws, and illustrate it in particle physics and molecular chemistry. Then we define the new matrix selection criterion that is the focus of this paper. At any given time, we have a set $r_1, .., r_m$ of reactions that scientists accept as experimentally established so far. The arrow notation is standard for displaying reactions where reacting entities appear on the left of the arrow and the products of the reaction on the right. For example, the expression $e_1 + e_2 \rightarrow e_3 + e_4$ denotes that two entities e_1, e_2 react to produce another two entities e_3, e_4. For a computational approach, we represent reactions as vectors, following Valdés-Pérez *et al.* [2]. Fix an enumeration of the known entities numbered as $e_1, \ldots, e_n$. In a given reaction r, we may count the number of occurrences of an entity e among the reagents, and among the products; subtracting the second from the first yields the **net occurrence**. For each reaction r, let r be the n-dimensional **reaction vector** whose i-th entry is the net occurrence of entity e_i in r. In what follows we simply refer to reaction vectors as reactions. The conserved quantities of interest in this paper are integers, so a quantity can be represented as an n-dimensional vector with integer entries; in what follows we simply refer to quantity vectors as **quantities** or **quantum numbers**. A quantity q is conserved in reaction r if and only if q is orthogonal to r. We combine m observed reactions involving

Table 1. The representation of reactions and conserved quantities as n-dimensional Vectors. The dimension n is the total number of particles, so $n = 7$ for this table.

Particle	1	2	3	4	5	6	7
Process/Quantum Number	p	π^0	μ^-	e^+	e^-	ν_μ	$\bar{\nu}_e$
$\mu^- \rightarrow e^- + \nu_\mu + \bar{\nu}_e$	0	0	1	0	-1	-1	-1
$p \rightarrow e^+ + \pi^0$	1	-1	0	-1	0	0	0
$p + p \rightarrow p + p + \pi^0$	0	-1	0	0	0	0	0
Baryon Number	1	0	0	0	0	0	0
Electric Charge	1	0	-1	1	-1	0	0

n known entities to form a **reaction data matrix** $R_{m \times n}$ whose rows are the observed reaction vectors. Similarly, combining q quantities assigned to n entities produces a **quantity matrix** $Q_{n \times q}$ whose columns are the quantity vectors. In the context of discovering conserved quantities, we also refer to quantity matrices as **conservation law matrices** or simply conservation matrices. The conservation equation $RQ = 0$ holds iff each quantity in Q is conserved in each reaction in R; in this case we say that Q is **consistent** with all reactions in R.

2.1 Example 1: Reactions and Conservation Laws in Particle Physics

Table 1 illustrates the representation of reactions and quantum numbers as vectors. Table 2 shows the main conservation laws posited by the Standard Model. The table specifies the values assigned to some of the most important particles for the five conserved quantities Electric Charge, Baryon Number, Tau Number, Electron Number, and Muon Number. For future reference, we use their initial letters to refer to these collectively with the abbreviation **CBTEM**. The table shows $n = 22$ particles; our complete study uses $n = 193$.

Particle Families. Particle physicists use *particle ontology* to construct conservation law models from data in a semantically meaningful way [14]. They use the hidden feature vectors (quantum numbers) to group particles together as follows: Each of the q numbers is said to correspond to a *particle family*, and a particle is a member of a given family if it has a nonzero value for the corresponding number. For instance, the physical quantity electric charge corresponds to a particle family that contains all charged particles (e.g., it contains the electron with charge -1, and the proton with charge $+1$), and does not contain all electrically neutral particles (e.g., it does not contain the neutron with charge 0). As Table 2 illustrates, the four **BTEM** families are disjoint, in the sense that they do not share particles. For instance, the neutron n carries Baryon Number 1, and carries 0 of the three other families **TEM**. It is desirable to find conservation models with disjoint particle families, for two reasons. (1) In that case we can interpret the conservation of a quantity as stating that particles from one family can cannot turn into particles from another family, which makes the conservation model more intelligible and intuitively plausible. (2) The inferred

Table 2. Some common particles and conserved quantities assigned to them in the Standard Model of particle physics. The table shows a conservation law matrix.

	Particle	Charge (C)	Baryon# (B)	Tau# (T)	Electron# (E)	Muon#(M)
1	Σ^-	-1	1	0	0	0
2	$\overline{\Sigma}^+$	1	-1	0	0	0
3	n	0	1	0	0	0
4	$\overline{n}$	0	-1	0	0	0
5	p	1	1	0	0	0
6	$\overline{p}$	-1	-1	0	0	0
7	π^+	1	0	0	0	0
8	π^-	-1	0	0	0	0
9	π^0	0	0	0	0	0
10	γ	0	0	0	0	0
11	τ^-	-1	0	1	0	0
12	τ^+	1	0	-1	0	0
13	ν_τ	0	0	1	0	0
14	$\overline{\nu}_\tau$	0	0	-1	0	0
15	μ^-	-1	0	0	0	1
16	μ^+	1	0	0	0	-1
17	ν_μ	0	0	0	0	1
18	$\overline{\nu}_\mu$	0	0	0	0	-1
19	e^-	-1	0	0	1	0
20	e^+	1	0	0	-1	0
21	ν_e	0	0	0	1	0
22	$\overline{\nu}_e$	0	0	0	-1	0

particle families can be checked against particle groupings discovered through other approaches, which provide a cross-check on the model [15,13].

2.2   Example 2: Chemical Reactions and Molecular Structure

The problem of discovering molecular structure from chemical reactions can also be cast as a matrix search problem. Our discussion follows the presentation of the DALTON*system by Langley et al. [1, Ch.8]. Consider chemistry research in a scenario where n chemical substances $s_1, s_2, .., s_n$ are known. In the model of the DALTON*system, Langley et al. take the known substances to be Hydrogen, Nitrogen, Oxygen, Ammonia and Water. In what follows, we assume that the reaction data indicate that various proportions of these substances react to form proportions of other substances. For example, 200ml of Hydrogen combine with 100ml of Oxygen to produce 200ml of Water vapour, 400ml of Hydrogen combine with 200ml of Oxygen to produce 400ml of Water, etc. In arrow notation, we can express this finding with the formula

$$2\,Hydrogen + 1\,Oxygen \rightarrow 2\,Water.$$

Table 3. The Representation of Chemical Reactions as n-dimensional vectors. The dimension n is the total number of substances. The entries in the vector specify the proportions in which the substances react.

Substance / Reaction	1 Hydrogen	2 Nitrogen	3 Oxygen	4 Ammonia	5 Water
2 Hydrogen + 1 Oxygen → 2 Water $= 2s_1 + s_2 \rightarrow 2s_5$	2	0	1	0	-2
3 Hydrogen + 1 Nitrogen → 2 Ammonia $= 3s_1 + s_2 \rightarrow 2s_4$	3	1	0	-2	0

Table 4. The correct structural matrix for our five example substances in terms of the three elements H, N, O. An entry in the matrix specifies how many atoms of each element a molecule of a given substance contains.

Element / Substance	H	N	O
1 Hydrogen	2	0	0
2 Nitrogen	0	2	0
3 Oxygen	0	0	2
4 Ammonia	3	1	0
5 Water	2	0	1

Labelling the five substances $s_1, s_2, ..., s_5$, this kind of reaction data can be represented as vectors, as with particle reactions. Table 3 shows the vector representation for the two chemical reactions discussed by Langley *et al.* [1].

According to Dalton's atomic hypothesis [1], the fixed proportions observed in reactions can be explained by the fact that chemical substances are composed of atoms of chemical elements in a fixed ratio. A chemical element is a substance that cannot be broken down into simpler substances by ordinary chemical reactions. A **structure matrix** S is an $s \times q$ matrix with integer entries ≥ 0 such that entry $S_{i,j} = a$ indicates that substance s_i contains a atoms of element e_j. Table 4 shows the true structure matrix for our example substances and elements. For example, the 4-th row in the matrix indicates that Ammonia molecules are composed of $3H$ atoms and $1N$ atom, corresponding to the modern formula H_3N for Ammonia. An elementary substance is different from the element itself, for example Oxygen from O, because substances may consist of molecules of elements, as the substance Oxygen consists of O_2 molecules. The connection with conservation laws is that *chemical reactions conserve the total number of atoms of each element*. This means that given a reaction data matrix R whose rows represent observed reactions, a structural matrix S should satisfy the conservation equation $RS = 0$.

2.3   Selecting Conservation Law Matrices

The criterion of selecting a maximally strict maximally simple (MSMS) conservation law matrix combines the two main selection criteria investigated in previous research. The construction of conservation laws searches for a solution Q of the matrix equation $RQ = 0$ (here we use Q generically for quantity and structure matrices). Valdés-Pérez and Erdmann [8] proposed selecting a solution that minimizes the L1-norm $|Q|$ that sums the absolute value of matrix entries:

$$|Q_{n \times q}| = \sum_{i=1}^{n} \sum_{j=1}^{q} |Q_{ij}|.$$

The L1-norm is often used as a measure of simplicity or parsimony, for example in regularization approaches to selecting covariance matrices (e.g., [10]). This norm tends to select sparse matrices with many 0 entries. Another selection principle was introduced by Schulte [9]: To select a conservation matrix Q that *rules out as many unobserved reactions as possible*. Formally, a matrix Q is **maximally strict** for a reaction matrix R if $RQ = 0$ and any other matrix Q' with $RQ' = 0$ is consistent with all reactions that are consistent with Q (i.e., if $\mathbf{r}Q = 0$, then $\mathbf{r}Q' = 0$). Each maximally strict conservation matrix Q classifies reactions in the same way: a reaction is possible—conserves all quantities in Q—if and only if it is a linear combination of observed reactions (rows in R). The next proposition provides an efficient algorithm for computing a maximally strict matrix. The **nullspace** of a matrix M is the set of vectors $\mathbf{v}$ mapped to 0 by M (i.e., $M\mathbf{v} = 0$).

Proposition 1 (Schulte 2009 [9]). *Let R be a reaction matrix. A conservation matrix Q is maximally strict for R $\iff$ the space of linear combinations of the columns of Q is the nullspace of R.*

The proposition implies that to find a maximally strict conservation matrix, it suffices to find a basis for the nullspace of the reaction data. A **basis** for a linear space V is a maximum-size linearly independent set of vectors from V. Using the L1-norm to select among maximally strict conservation matrices leads to the new criterion investigated in this paper.

Definition 1. *A conservation matrix Q is **maximally strict maximally simple** (MSMS) for R if Q minimizes the L1-norm $|Q|$, subject to the constraint that Q is maximally strict for R.*

3 A Scalable Optimization Algorithm for Finding Maximally Simple Maximally Strict Conservation Laws

Our goal is to find an integer basis Q for the nullspace of a given reaction matrix R such that the L1-norm of Q is minimal. Valdés-Pérez and Erdman [8] managed to cast L1-minimization as a linear programming problem, but this does not work with the nonlinear nullspace constraint, and also assumes that the user explicitly specifies a set of "bad" reactions that the matrix Q must rule out. A summary of our method is displayed as Algorithm 1. We now discuss and motivate the algorithm design, then analyze its runtime complexity. In the following fix a reaction data matrix $R_{m \times n}$ that combines m reactions involving n entities.

Algorithm 1. Minimization Scheme for Finding a Maximally Simple Maximally Strict Conservation Law Matrix

1. Given a set of input reactions R find an orthonormal basis V for the nullspace of R. The basis V is an $n \times q$ matrix.
2. Let any linear combination of V be given by $Q = VX$, with X an $q \times q$ set of coefficients.
   Initialize X to $X_0 = I$, where I is the identity matrix of dimension q.
   Define $\mathcal{I}_1(X) = |VX|$, the L1-norm of the matrix VX.
   Define $\mathcal{I}_2(X) = \sum(X^T X - I)^2$.
3. Minimize $\mathcal{I}_1 + \alpha\mathcal{I}_2$ over X, with α constant, subject to the following constraint:
   (a) To derive an integer version $\widetilde{Q}$, we assign $Q = VX$; $\widehat{\mathbf{q}}_k = \mathbf{q}_k/max(\mathbf{q}_k)$, $k = 1..q$;
   $$\widehat{Q}\left(\widehat{Q} < \varepsilon\right) = 0; \ \widetilde{Q} = sgn(\widehat{Q}).$$
   (b) $\widetilde{Q}$ must have full rank: $rank(\widetilde{Q}) = q$.

Search Space. The following design operates in a search space with small matrices and facilitates the constraint check.

1. Compute a basis $V_{n \times q}$ for the nullspace of the input reaction matrix R. This is a standard linear algebra problem with efficient solutions, and automatically determines the dimensionality q of the set of quantum numbers as the rank of the nullspace of R.
2. Now any solution Q can be written as $Q_{n \times q} = V_{n \times q} X_{q \times q}$ where X is a square full-rank matrix. In other words, the search space comprises the invertible change-of-basis matrices X that change basis vectors from Q to V. The solution Q is maximally strict if and only if X has full rank. Change of basis matrices are much smaller than conservation matrices, because typically $q \ll n$. In the particle physics domain, $n = 193$ and $q = 5$.

Objective Function. Since our basic goal is to minimize the L1-norm of a solution Q, a natural objective function for a candidate X is

$$\mathcal{I}_1(X) = |VX|,$$

the L1-norm of the matrix VX. However, this drives the search towards sparse matrices X with 0 rows/columns that do not have the full rank q. To avoid the reduction in the rank of Q, we add a second optimization contribution

$$\mathcal{I}_2 = \sum(X^T X - I)^2. \tag{1}$$

This score penalizes matrices with blank rows or columns. Also, if we start with an orthonormal basis V, the score (1) is maximized by matrices X such that the columns in $Q = VX$ are orthogonal to each other and have length 1. Our final objective function is a weighted combination of these two scores:

$$\min_X \ (\mathcal{I}_1 + \alpha\mathcal{I}_2) \tag{2}$$

with free parameter α.

From Continuous to Integer Values. Carrying out the minimization search in the space of continuous matrices creates a much faster algorithm than integer programming. We use the following method to discretize a given set of continuous quantum numbers. The method first decides which values should be set to 0, and then maps the non-zero values to an integer.

Scaling. For each column q of Q, we divide by the maximum absolute value $max(q)$, obtaining a new set of scaled (real-valued) quantum numbers $\widehat{Q}$:

$$Q \rightarrow \widehat{Q} \mid \widehat{q} = q/max(q).$$

Pruning. We then set to zero any element of $\widehat{Q}$ with absolute value less than a small ε. We chose $\varepsilon = 0.01$ as a simple default value.

Discretization. In each column, multiply the non-zero entries by the least common denominator to obtain integer entries (i.e., find the least integer multiplier such that after multiplication the entries are effectively integers).

Example. Applying the local search procedure to the chemistry input reactions from Table 3, leads to a minimum matrix X such that

$$S = VX = \begin{pmatrix} 2/3 & 0 & 0 \\ 0 & 1 & 0 \\ 0 & 0 & 1 \\ 1 & 1/2 & 0 \\ 2/3 & 0 & 1/2 \end{pmatrix}.$$

Multiplying the first column by 3 and the second and third by 2, yields the correct structure matrix shown in Table 4.

Complexity Analysis and Scalability. The number of known entities n defines the dimension of the data vectors; it is a constant in most application domains. In our particle data set (described below), $n = 193$, which is a realistic number for particle physics. The crucial growth factor for complexity analysis is the number m of reactions or data points. For a given input matrix $R_{m \times n}$, the initial computation of the nullspace basis V can be done via a singular value decomposition (SVD) of R. A general upper bound on the complexity of finding an SVD is $O(mn^2)$ [16, Lecture 31], which is linear in the number of data points m. Computing a nullspace basis is especially fast for reaction matrices as they are very sparse, because only a small number of entities participate in any given reaction. For instance, in the particle physics domain, the reaction data do not feature more than 6 entities per reaction out of about 200 total entities, so about 97% of the entries in a reaction matrix will be zeros. The computation of the nullspace basis can be viewed as preprocessing the reaction data to compress it into a matrix $V_{n \times q}$ whose dimension does not depend on the number of data points m.

The basis matrix $V_{n \times q}$ is the input to the minimization routine, where q is the dimension of the nullspace of R. This dimension is bounded is bounded by the dimension of the entire space n, so $q < n$ and the size of the matrix V is less than n^2. In practice, we expect to find relatively few conserved quantities (5 quantities in the physics domain for about 200 particles), so we may consider $q \ll n$ to be a constant. In sum, the data preprocessing step scales linearly with the number of data points, and the search space for the minimization routine comprises matrices of essentially constant dimensions.

4   Implementation and Evaluation

We discuss the implementation of the minimization algorithm and the dataset on which it was evaluated. The dataset is the same as that used by Schulte in the study of finding hidden particles [9]. We report the results of applying the minimization routine of Algorithm 1. Our Matlab code and data are available online at http://www.cs.sfu.ca/~oschulte/particles/conserve_zipfile.zip.

Implementation. The objective function and constraints from Algorithm 1 are implemented using the `fmincon` function in Matlab. Optimization is carried out over float values for X, with the continuous objective function (2). A non-linear rank constraint is applied on the quantum number answer set $\widetilde{Q}$. The threshold for rounding down a float to 0 was $\varepsilon = 0.01$. The Matlab function `null` computes an orthonormal nullspace basis for the input data via SVD.

Selection of Particles and Reactions. The selection is based on the particle data published in the Review of Particle Physics [12]. The Review of Particle Physics is an authoritative annual publication that collects the current knowledge of the field. The Review lists the currently known particles and a number of important reactions that are known to occur. Our particle database contains an entry for each particle listed in the Review, for a total of 193 particles. The reaction dataset D includes 205 observed reactions. This includes a maximum probability decay for each of the 182 particles with a decay mode listed. The additional reactions are important processes listed in textbooks (see [9]).

4.1   Experimental Design and Measurements

We carried out several experiments on particle physics and chemistry data. Our two main experiments compare the quantities and particle families introduced by the MSMS algorithm with the Standard Model matrix S.

1. Apply the algorithm with no further background knowledge.
2. Apply it with the quantum number electric charge **C** as given in the Standard Model.

In the context of particle physics, it is plausible to take electric charge as given by background knowledge, for two reasons: (1) Unlike the quantities **BTEM**, charge is directly measurable in particle accelerators using electric fields. So it is realistic to treat charge as observed and not as a hidden feature of particles. (2) The conservation of electric charge is one of the classical laws of physics that had been established over a century before particle physics research began [14]. To implement adding $\mathbf{C}$ as background knowledge, we added it to the data D and applied the minimization procedure to $D + \mathbf{C}$ as input; if V is a basis for the nullspace of $D + \mathbf{C}$, then $V + \mathbf{C}$ is a basis for the nullspace of D.

We ran the minimization routine for each of the two settings with a number of values of the parameter α; we report the results for the settings $\alpha = 0, 10, 20$ which are representative. If $\alpha = 0$, the program minimizes the L1-norm directly. For both the Standard matrix S and the program's output Q we report the following measures. (1) The runtimes. (2) The values of the objective function $\mathcal{I}$ defined in Equation (2) and of the L1-norm. When the program found a valid maximally strict solution, we recorded also (3) the number of particle families recovered by the program, out of the 4 particle families defined by the quantities $\mathbf{BTEM}$ in the Standard Model.

4.2   Results on Standard Model Laws and Families

Table 5 shows a summary of results for Experiment 1, and Table 6 a summary for Experiment 2. We discuss first the quality of the solutions found, and then the processing speed.

Solution Quality. Our discussion distinguishes two questions: (i) Does the MSMS criterion match the Standard Model quantities, that is, do the conserved quantities in the Standard Model optimize the MSMS criterion on the available particle accelerator data? The answer to this question does not depend on the parameter α of Algorithm 1. (ii) Does Algorithm 1 manage to find an MSMS optimum?

Table 5. Summary of results for the dataset without charge given. The matrix Q is the output produced by the MSMS Algorithm 1. The matrix S is the Standard Model matrix. The objective function of Algorithm 1 is denoted by $\mathcal{I}$.

α	Families Recovered	Runtime (sec)	$\mathcal{I}(Q)$	$\mathcal{I}(S)$	$L1(Q)$	$L1(S)$	difference Q vs. S
20	4/4	16.44	22.67	22.31	22.21	21.96	$\mathbf{C}$ replaced by linear combination
10	4/4	15.74	22.20	22.31	21.96	21.96	$\mathbf{C}$ replaced by linear combination
0	n/a	6.95	15.92	22.31	15.92	21.96	invalid local minimum

Table 6. The same measurements as in Table 5 with electric charge $\mathbf{C}$ fixed as part of the input

α	Families Recovered	Runtime (sec)	$\mathcal{I}(Q)$	$\mathcal{I}(S)$	$L1(Q)$	$L1(S)$	difference Q vs. S
20	2/4	7.68	16.65	15.55	16.63	15.52	$\mathbf{E}, \mathbf{M}$ replaced by linear combination
10	4/4	8.40	15.55	15.55	15.52	15.52	exact match
0	n/a	10.68	11.52	15.55	11.52	15.52	invalid local minimum

This does depend on the parameter settings. The optimization algorithm is fast and allows running the local minimization scheme with different parameter values to find a global minimum. However, our experiments suggest a consistently successful default value ($\alpha = 10$).

(1) We verified that the Standard Model quantities **CBTEM** are maximally strict and maximally simple for the observed reaction matrix R, both with and without charge given.

(2) In Experiment 1 (Table 5) we observed that *all* computed solutions recover the quantities **BTEM** exactly (up to sign). The values of the objective function are close to the L1-norms; the function of the $\mathcal{I}_2$ component is thus likely to guide the initial stages of the search.

(3) The MSMS criterion does not uniquely determine charge because it is possible to replace the quantity **C** by a linear combination of **C** with one of the other quantities without raising the L1-norm. In Experiment 2, the quantity electric charge **C** was taken as given. The program recovered the **BTEM** families exactly for the setting with $\alpha = 10$. With $\alpha = 20$, the program recovered two of the families, **B** and **T**, but replaced **E** and **M** with suboptimal linear combinations of **E** and **M**.

(4) The $\mathcal{I}_2$ component is essential for enabling the program to find a local minimum that satisfies the full-rank constraint: With $\alpha = 0$ the minimization routine settles into a local minimum with a small L1-norm whose rank is too low. This is consistent with the observation of Valdés-Pérez and Erdmann that minimizing the L1-norm with no further constraints produces just one quantum number [8,7]. A value of α that is too large can cause failure to find an objective-function global minimum. When charge is part of the input, this leads to a failure to minimize the L1-norm and to recover the correct particle families (Table 6).

Processing Speed. The measurements were taken on a Quad processor with 2.66 GHz and 8 Gbytes RAM. Overall, the runtimes are small. Computing an SVD with 205 reactions and 193 particles takes about 0.05 seconds. In addition to our theoretical analysis, the speed of SVD on our data set supports our expectation that it will be fast even for data sets with 1000 times more reactions than ours. The minimization operation also ran very fast (17 sec in the worst setting), which shows that the optimization is highly feasible even for relatively large numbers of entities ($n = 193$ in our dataset).

Recovering Particle Families: A Theoretical Explanation. The ability of the MSMS criterion to recover the correct particle families is surprising because the method receives data only about particle dynamics (reactions), not about particle ontology. Schulte and Drew [17,18] provide a theoretical explanation of this phenomenon: It can be proven using linear algebra that if there is some maximally strict conservation law matrix with disjoint corresponding particle families, then the particle families are uniquely determined by the reaction data. Moreover, the conservation matrix corresponding to these particle families is the unique MSMS optimizer (up to changes of sign).

We note that all results are robust with respect to adding more data points consistent with the Standard Model, because the **CBTEM** quantities are maximally strict for our data set D already, hence they remain maximally strict for any larger data set consistent with the Standard Model.

Learning Molecular Structure. Applying the minimization scheme to the chemistry reaction data of Table 3 recovers the correct structure matrix of Table 4. The α optimization parameter was set to 10, and the runtime was about 2 sec. While this dataset is small, it shows the applicability of our procedure in another scientific domain that was previously studied by other researchers.

Summary. Our results show that the MSMS criterion formalizes adequately the *goals* that scientists seek to achieve in selecting conservation theories: MSMS theories explain why unobserved reactions do not occur [9, Sec.4], they minimize the magnitude of conserved quantities, and by the theorem of Schulte and Drew [17,18], they connect conservation laws with disjoint particle families. In contrast, our algorithmic *method* for finding MSMS theories was derived from efficiency considerations and does not match how physicists have gone about finding conserved quantities: they started with plausible particle families, derived conservation laws, then checked them against the data [18]. This amounts to using domain knowledge to solve a computationally challenging problem. Our minimization method could be used to check results derived from domain-specific intuitions, or applied when domain knowledge is not available.

5   Conclusion and Future Work

We applied the classic matrix search framework of Raúl Valdés-Pérez *et al.* [2] to two key problems in the analysis of particle reaction data: Finding conserved quantities and particle families. Our approach is based on a new selection criterion for conservation law theories: to select maximally strict maximally simple models. Maximally strict models rule out as many unobserved reactions as possible, and maximally simple models minimize the L1-norm, the sum of the absolute values of the matrix entries. We described an efficient MSMS optimization procedure, that scales linearly with the number of datapoints (= observed reactions). An analysis of particle accelerator data shows that the fundamental Standard Model of particles is maximally strict and maximally simple. This means that the MSMS criterion makes exactly the same predictions as the Standard Model about which interactions among particles are possible, and it rediscovers four of the standard particle families given our reaction data set (or any extension of it that is consistent with the Standard Model). The MSMS criterion correctly recovers the chemical structure of compounds on the data described by Langley *et al.* [1, Ch.8]. In future work we plan to apply the algorithm to other particle data sets, such as those that will come from the Large Hadron Collider. On new data that have been analyzed less exhaustively it may well be possible for our algorithm to find new conservation theories, or at least to support their discovery.

References

1. Langley, P., Simon, H., Bradshaw, G., Zytkow, J.: Scientific Discovery: Computational Explorations of the Creative Processes. MIT Press, Cambridge (1987)
2. Valdés-Pérez, R., Żytkow, J.M., Simon, H.A.: Scientific model-building as search in matrix spaces. In: AAAI, pp. 472–478 (1993)
3. Valdés-Pérez, R.: Computer science research on scientific discovery. Knowledge Engineering Review 11, 57–66 (1996)
4. Rose, D., Langley, P.: Chemical discovery as belief revision. Machine Learning 1, 423–452 (1986)
5. Valdés-Pérez, R.: Conjecturing hidden entities by means of simplicity and conservation laws: machine discovery in chemistry. Artificial Intelligence 65, 247–280 (1994)
6. Kocabas, S.: Conflict resolution as discovery in particle physics. Machine Learning 6, 277–309 (1991)
7. Valdés-Pérez, R.: Algebraic reasoning about reactions: Discovery of conserved properties in particle physics. Machine Learning 17, 47–67 (1994)
8. Valdés-Pérez, R., Erdmann, M.: Systematic induction and parsimony of phenomenological conservation laws. Computer Physics Communications 83, 171–180 (1994)
9. Schulte, O.: Simultaneous discovery of conservation laws and hidden particles with smith matrix decomposition. In: IJCAI 2009, pp. 1481–1487 (2009)
10. Schmidt, M., Niculescu-Mizil, A., Murphy, K.: Learning graphical model structure using L1-regularization path. In: AAAI (2007)
11. MacKay, D.J.C.: Information Theory, Inference and Learning Algorithms. Cambridge University Press, Cambridge (2003)
12. Eidelman, S., et al. (Particle Data Group): Review of Particle Physics. Physics Letters B 592, 1+ (2008)
13. Cottingham, W., Greenwood, D.: An introduction to the standard model of particle physics, 2nd edn. Cambridge University Press, Cambridge (2007)
14. Ne'eman, Y., Kirsh, Y.: The Particle Hunters. Cambridge University Press, Cambridge (1983)
15. Gell-Mann, M., Ne'eman, Y.: The eightfold way. W.A. Benjamin, New York (1964)
16. Bau, D., Trefethen, L.N.: Numerical linear algebra. SIAM, Philadelphia (1997)
17. Schulte, O., Drew, M.S.: An algorithmic proof that the family conservation laws are optimal for the current reaction data. Technical Report 2006-03, School of Computing Science, Simon Fraser University (2006)
18. Schulte, O.: The co-discovery of conservation laws and particle families. Studies in the History and Philosophy of Modern Physics 39(2), 288–314 (2008)

Gaussian Clusters and Noise: An Approach Based on the Minimum Description Length Principle[*]

Panu Luosto[1,3], Jyrki Kivinen[1,3], and Heikki Mannila[2,3]

[1] Department of Computer Science, University of Helsinki, Finland
[2] Department of Information and Computer Science, Aalto University,
Helsinki, Finland
{Panu.Luosto,Jyrki.Kivinen}@cs.helsinki.fi,
Heikki.Mannila@aaltouniversity.fi

Abstract. We introduce a well-grounded minimum description length (MDL) based quality measure for a clustering consisting of either spherical or axis-aligned normally distributed clusters and a cluster with a uniform distribution in an axis-aligned rectangular box. The uniform component extends the practical usability of the model e.g. in the presence of noise, and using the MDL principle for the model selection makes comparing the quality of clusterings with a different number of clusters possible. We also introduce a novel search heuristic for finding the best clustering with an unknown number of clusters. The heuristic is based on the idea of moving points from the Gaussian clusters to the uniform one and using MDL for determining the optimal amount of noise. Tests with synthetic data having a clear cluster structure imply that the search method is effective in finding the intuitively correct clustering.

1 Introduction

Finding hard clusters with underlying normal distributions from data is one of the most fundamental clustering problems. By hard clustering we mean a partitioning of the data so that every element belongs to exactly one cluster. The famous k-means problem can be seen as a special case of this type, even if its probabilistic interpretation is somewhat artificial: the objective is to maximize the likelihood of a mixture model of k spherical normal distributions with equal weights and variances in the limiting case when the variances approach zero. In a more general setting, there are at least two important challenges in applying Gaussian mixture models for clustering in practice. Firstly, real world data seldom fit a pure model very well, and one might want to refine the model for additional robustness. Secondly, one does not usually want to fix the number of clusters arbitrarily but to find the k that fits the data at hand best.

---

[*] Supported by Academy of Finland grant 118653 (Algodan) and the PASCAL Network of Excellence.

B. Pfahringer, G. Holmes, and A. Hoffmann (Eds.): DS 2010, LNAI 6332, pp. 251–265, 2010.

We increase the robustness of the model by adding a component with a uniform distribution in an arbitrary box with axis-aligned edges. An obvious motivation for this are situations where there is uniform background noise in the data. We assume that the true domain of the data is unknown in advance, and we also find no reason to restrict our model so that the set in which the uniform distribution gets positive values would always include all the data. In contrast, our model is capable of adapting to situations where the noise is coming from a separate source and Gaussian clusters may locate also outside the domain of the noise. However, the clustering method that we present in the next section, is not suitable for that kind of data. We discuss this in Sect. 7.

There is a strong tradition of determining the most appropriate model complexity, in this case the number of clusters, with criteria like Bayesian information criterion, Akaike information criterion, minimum message length and different forms of minimum description length (MDL). See discussion about the differences of the methods in [3]. The modern form of MDL, normalized maximum likelihood (NML), has many important optimality properties, but it is not directly usable for encoding of the data in our case for reasons that are explained in detail in the Sects. 5 and 6. As our main contribution we introduce a practical and well-grounded quality measure for model selection. Our quality criterion has a clear coding based interpretation, and according to the MDL principle, we avoid arbitrary assumptions about the data. We also deal with the problems that arise from the singularities of the code length function. As models we use spherical and axis-aligned normal distributions, for which the code lengths are easy to derive, but not general normal distributions with arbitrary covariance matrices.

It is naturally possible to use our code length function for the selection of the best clustering from a set of candidates with different number of clusters, no matter which search method has been used. For example, the expectation maximization (EM) algorithm [2] using a model with one uniform cluster would be adequate for the purpose. As an alternative, we propose a simple heuristic in which only normally distributed clusters are searched for first and the noise cluster is determined then by removing points from the Gaussian clusters. Synthetic data is used in experiments whose objective is to test if the code length and the method work as expected when the intuitively correct clustering is known.

2   Search Method

Even if the main contribution of this paper lies in the code length function of the clustering, we present our search method first. It requires a quality measure of the clustering that takes the complexity of the model into account. This differs from the typical situation, where the best clustering with a certain number of clusters maximizes the likelihood of the data, and the MDL principle (or a similar criterion) is used for choosing the best clustering out of a set of clusterings with different number of clusters. In our case, the complexity of the model changes within a single run of the method, which makes a maximum likelihood based approach unfeasible.

First, we search for k Gaussian clusters with a greedy version of the common EM algorithm, which is somewhat simpler and faster than the standard EM algorithm with soft cluster assignments. The value k should be larger than the actual number of clusters we expect to find in the data. In a variant of the method we also include a uniform component in the model of the EM algorithm. The greedy EM algorithm gives us a mixture model in which every point belongs to exactly one cluster. We estimate the weights of the clusters as well as their parameters, and order the points according to ascending density in the model. Then, using this ordering of the points, which we do not change any more, we move points one by one from the Gaussian clusters into the uniform one. After each move, we update the parameters of the Gaussian cluster that the point is taken from and the uniform cluster that the point is moved into, but we do not alter any other parameters or assignments of points to clusters. Notice that the geometry of the uniform cluster is not fixed but the cluster may grow because of the new point. After each move we also calculate the MDL of the corresponding clustering. Naturally, every Gaussian cluster becomes empty at some point of the method. At the end we have as many code lengths as there are points, corresponding to clusterings with 0 to k Gaussian clusters. The clustering with the shortest code length is returned.

The quality or the code length of the final clustering is dependent on the initial number of Gaussian clusters, and we benefit from running the algorithm with several different values of k, even if tests with synthetic data imply that the method is not sensitive to the choice of k. Because using random seeding at the EM phase of the method is advisable, the method should be run several times with each k too.

We now give the pseudocode of the method. Steps 1 and 2 describe the seeding phase. The D^2 seeding of Arthur and Vassilvitskii [1] is a random seeding algorithm that tends to pick initial centres lying far apart form each other.

Let $x^n = (x_1, x_2, \ldots, x_n) \in (\mathbb{R}^d)^n$ be a data sequence and let $k \in \{1, 2, \ldots, n\}$ be the initial number of clusters. We consider the covariance matrices of normal distributions that are either of the type $\sigma^2 I_d$ (spherical model) or of the type $\mathrm{diag}(\sigma_1^2, \sigma_2^2, \ldots, \sigma_d^2)$ (axis-aligned model). Let $x \sqsubset y$ denote that x is a subsequence of y, and let $|x|$ denote the length of the sequence x. If the model in the EM algorithm has a uniform component, let the proportion of points belonging to the uniform cluster in the beginning be β.

1. Choose k initial cluster centres $c_1, c_2, \ldots, c_k$ using the D^2 seeding.
2. For each $i \in \{1, 2, \ldots, k\}$, set the cluster $S_i = (x_{i1}, x_{i2}, \ldots, x_{im_i}) \sqsubset x^n$ so that for all $j \in \{1, 2, \ldots, m_i\}$, it holds $\|x_{ij} - c_i\| = \min\{\|x_{ij} - c_h\| \mid h \in \{1, 2, \ldots, k\}\}$. Every element of x^n must belong to exactly one cluster. If there are multiple possible cluster assignments here, the ties should be broken randomly. If the model has a uniform component, the points determining the smallest enclosing box of x^n are assigned to the uniform cluster as well as a uniformly at random picked subsequence of other points so that the proportion of the points in the uniform cluster equals β.

3. Run the greedy EM algorithm until the maximum likelihood converges.
4. Sort the points according to the ascending density in the model found. Let the sorted sequence be $(y_1, y_2, \ldots, y_n)$.
5. From now on, the model always includes a uniform component. If necessary, add an empty uniform cluster. Calculate the MDL of the clustering according to the model.
6. For each $i \in \{1, 2, \ldots, n\}$:
   (a) Move y_i to the uniform cluster.
   (b) Recalculate the parameters of the uniform cluster and the cluster y_i used to belong to; if the original cluster of y_i became empty, decrement the number of clusters in the model by one.
   (c) Calculate and store the MDL of the new clustering.
7. Return the clustering that had the smallest MDL.

The order in which the points are moved to the uniform cluster does not change after the step 4 in the previous method. A natural variant of the method would be to update also the densities and the order of the points that are not yet in the uniform cluster after each move. We do not consider this computationally more demanding version in this paper.

3   Minimum Description Length Principle

In this section, we describe briefly some basic concepts of the MDL principle, including the NML. The MDL principle [3,11] was first introduced by Rissanen in [7] and then developed e.g. in [9,10]. Informally, the best clustering is in our case the one that enables the most effective compression of the data and the classification of the points into clusters according to the model classes used. By a model class we mean a parametric collection of probability distributions. A model class is only a technical means for encoding, the MDL principle does not assume that the data is a sample from a probabilistic source. The most effective way to encode a sequence of data $x^n \in (\mathbb{R}^d)^n$ according to a model class would be using the maximum likelihood parameters. The problem is that the receiver cannot know the right parameters in advance. In a two-part code, the maximum likelihood parameters are encoded in the first part of the message, and the data is encoded according to the maximum likelihood distribution in the second part.

Even if the two-part encoding scheme is easily understandable, it is not the most effective one. When the parameters are continuous, two-part coding also always includes the problem how the parameter space should be discretized optimally: if more bits are used for encoding of the parameters, the second part of the code becomes shorter. Modern MDL favours one-part NML code because of its optimality properties. If the NML code can be defined given a model class, it is the best worst-case code. NML code also comes closest to the unreachable optimum, the maximum likelihood code, in the probabilistic sense when the mean is taken with respect to the worst possible data generating distribution. In case of a discrete probability distribution over a finite set X, the NML distribution according to a certain model class is

$$P_{\text{NML}}(x) = \frac{P(x; \hat{\theta}(x))}{\sum_{y \in X} P(y; \hat{\theta}(y))} , \tag{1}$$

where $\hat{\theta}(x)$ denotes the maximum likelihood parameters of x. Denoting the base-2-logarithm as log, the corresponding code length of x is then $-\log P_{\text{NML}}(x)$. The quantity $\log \sum_{y \in X} P(y; \hat{\theta}(y))$ is called the parametric complexity.

In this paper, we consider continuous distributions, and for deriving the NML distribution we replace the sum in (1) by an integral. The normalizing integral diverges unfortunately in many interesting cases, including those that are relevant for our data encoding. The NML with an infinite complexity is a problematic subject, because there is no simple way to define the best possible code in that case [3]. In our approach, we derive first a NML code length for the case when the data is restricted, then remove the restricting parameters by using very flat priors for them. We circumvent the calculation of difficult integrals by considering a limiting case where the conditional NML density grows unbounded. We do not discuss the optimality properties of our codes more closely. Intuitively, they appear quite effective however.

Following a common practice, we call also negative logarithms of probability densities code lengths, even if the term is not to be taken literally. Comparing negative logarithms of densities is equivalent with comparing actual code lengths in the limit when the coding precision approaches infinity. In practical situations, data are rational numbers and minimizing the negative logarithm of the density is not exactly equal to finding the most effective way to encode the data. This does not usually cause problems, but if the density can grow unbounded in the neighbourhood of some point, the results may be surprising, especially when the data is represented with greater precision than we find trustworthy. Therefore it might be reasonable to fine-tune the model in order to limit the density. That is what we do at the end of Sects. 5 and 6.

4   Outline of the Code Length Calculation

Let $x^n = (x_1, x_2, \ldots, x_n) \in (\mathbb{R}^d)^n$ be a sequence. We calculate the MDL for a clustering of x^n as the sum of the code length for the classification of the points and the code length of the data given the classification. The classification is encoded as a sequence of integers that indicate the cluster memberships. We use the number 0 for the uniform cluster and numbers $1, 2, \ldots, k$ for the Gaussian clusters. In our model class the sequence consist of n independently and identically-distributed categorical random variables, and we use NML code for encoding. The numbers from 1 to k are only used as labels for the Gaussian clusters, for example $(1, 0, 1, 2)$ and $(2, 0, 2, 1)$ denote the same classification. In order to have just one presentation for each classification, we use a canonical numbering of the clusters. The classification sequences are interpreted as numbers with the radix $k + 1$, and the smallest possible number is used to denote a classification (in our example 1012). Let $\mathbf{n} = (n_0, n_1, \ldots, n_k)$ be the cluster sizes according to the canonical numbering scheme. The maximum likelihood of the classification is then

$$P_{\mathrm{ML}}(\mathbf{n}) = k! \prod_{i=0}^{k} \left(\frac{n_i}{n}\right)^{n_i},$$

where we define $0^0 \equiv 1$. Calculating the normalizing sum

$$C(k + 1, n) = \sum_{\substack{m_0, m_1, \ldots, m_k \in \{0, 1, \ldots, n\}: \\ m_0 + m_1 + \cdots + m_k = n}} \frac{n!}{m_0! \, m_1! \ldots m_k!} \prod_{i=0}^{k} \left(\frac{m_i}{n}\right)^{m_i} \qquad (2)$$

efficiently is untrivial [5]. We use instead a very accurate Szpankowski approximation [12,5] for the natural logarithm of $C(k + 1, n)$. According to it

$$\ln C(k, n) = \frac{k - 1}{2} \ln \frac{n}{2} + \ln \frac{\sqrt{\pi}}{\Gamma(k/2)} + \frac{\sqrt{2} \, k \, \Gamma(k/2)}{3\Gamma((k - 1)/2)} \frac{1}{\sqrt{n}}$$
$$+ \left(\frac{3 + k(k - 2)(2k + 1)}{36} - \frac{\Gamma^2(k/2) \, k^2}{9\Gamma^2((k - 1)/2)}\right) \frac{1}{n} + \mathcal{O}(n^{-3/2}).$$

The code length of the classification of the points is $-\log P_{\mathrm{ML}}(\mathbf{n}) + \log C(k+1, n)$.

   In our coding context, the number of Gaussian clusters is not known to the receiver in advance and has to be encoded in beginning of the message. But we assume according to the MDL philosophy that all the possible values of $k \in \{0, 1, \ldots, n\}$ are equally likely, hence the encoding of k yields always the same number of bits. Therefore, we can ignore the code length that comes from the encoding of the k while comparing code lengths of different clusterings.

   We encode the subsequences of the clusters independently and concatenate their code in the canonical cluster ordering after the classification part of the code. In the decoding phase, we first decode the canonical representation of the classification, then the subsequences of the individual clusters. The original data sequence can be reconstructed from the subsequences using the classification. In Sects. 5 and 6, we derive the code lengths for the uniform and Gaussian components in detail. To be precise, we give only the densities whose negative logarithms are called code lengths.

5   Code Length for the Uniform Cluster

Before proceeding to the derivation of the code length for the uniform cluster, we introduce a very flat density function that is used as a prior for parameters in Sects. 5 and 6. The choice of the prior is an important part of the design of the code length functions. In [8], Rissanen gives a density function for the reals in the interval $[1, \infty[$. We generalize it without changing its asymptotic properties by adding a parameter b that defines how strongly the probability mass is concentrated to the vicinity of the origin. Denote x^y as $x \uparrow y$ for typographical reasons. Let $x \uparrow\uparrow 0 = 1$ and let $x \uparrow\uparrow y = \underbrace{x \uparrow x \uparrow \ldots \uparrow x}_{y \text{ copies}}$ for $x > 0$, $y \in \mathbb{N}$. Now let $b = \underbrace{2 \uparrow 2 \uparrow \ldots \uparrow 2 \uparrow \delta}_{k-1 \text{ copies of '2}\uparrow\text{'s}}$ where $k \in \mathbb{N}$ and $\delta \in [1, 2]$. For $x \in \mathbb{R}_+$, we define the density

$$f_{\mathbb{R}_+}(x; b) = \frac{1 - \ln 2}{(\ln 2)^k} \frac{1}{1 + \log \delta \, (\ln 2 - 1)} \frac{1}{(x + b) \, h(x + b)} \tag{3}$$

where

$$h(x) = \begin{cases} 1 & \text{if } \log x \leq 1 \\ \log x \, h(\log x) & \text{otherwise.} \end{cases}$$

It is straightforward to verify that (3) is indeed a density function. For the whole real line we simply use the function $f_{\mathbb{R}}(x; b) = f_{\mathbb{R}_+}(|x|; b)/2$.

Next, we derive a code length for the uniform component. Let the length of the sequence x^n be $n \in \{2, 3, \ldots\}$ for the time being. We denote the centre of the smallest enclosing ball of x^n as $c(x^n) = (\min(x^n) + \max(x^n))/2$ and the radius of that ball as $r(x^n) = (\max(x^n) - \min(x^n))/2$. Our model class consists of uniform distributions in rectangular boxes having axis-aligned edges. It suffices to consider the one-dimensional case because in the model class the coordinates are independent, and we get the density of a point by taking the product of the densities of the coordinates. Formally, the one-dimensional model class is the set of densities $\{f(\cdot; c, r_0) \mid c \in \mathbb{R}, r_0 > 0\}$, where

$$f(x^n; c, r_0) = \begin{cases} (2r_0)^{-n} & \text{if } x^n \in [c - r_0, c + r_0]^n \text{ and } r(x^n) > 0 \\ 0 & \text{otherwise.} \end{cases}$$

At the end of the section, we define code lengths also for sequences $x^n \in \mathbb{R}^n$ having $r(x^n) = 0$.

We consider first a case in which the data is restricted to a certain set and the NML can be defined. Let $c_0 \in \mathbb{R}$ and let $\delta, r_1, r_2 > 0$. Assume that $r_1 < r_2$. Let the set of sequences to be considered be $A = \{x^n \in \mathbb{R}^n \mid c(x^n) \in [c_0 - \delta, c_0 + \delta], r(x^n) \in [r_1, r_2]\}$. The maximum likelihood function for the sequences in A is $g_{\mathrm{ML}}(x^n) = (2r(x^n))^{-n}$, and the corresponding normalizing integral is

$$C(c_0, \delta, r_1, r_2) \quad = \int_{x^n \in A} g_{\mathrm{ML}}(x^n)\, dx^n \tag{4}$$

$$= n(n-1) \iint_{\substack{x_1, x_2 \in \mathbb{R}: \\ (x_1+x_2)/2\, \in [c_0-\delta, c_0+\delta], \\ (x_2-x_1)/2\, \in [r_1, r_2]}} \int_{x_1}^{x_2} \cdots$$

$$\cdots \int_{x_1}^{x_2} \frac{1}{(x_2 - x_1)^n}\, dx_n\, dx_{n-1} \ldots dx_2\, dx_1$$

$$= 2n(n-1) \int_{c_0-\delta}^{c_0+\delta} \int_{r_1}^{r_2} \frac{1}{4r^2}\, dr\, dc \tag{5}$$

$$= n(n-1)\, \delta \left(\frac{1}{r_1} - \frac{1}{r_2} \right).$$

There was a coordinate change $(x_1, x_2) = (c - r, c + r)$ at (5) in the previous integration. Dividing the maximum likelihood by the normalizing integral yields the NML density function

$$f_{\mathrm{NML}}(x^n; c_0, \delta, r_1, r_2) = \frac{1}{(2r(x^n))^n} \frac{1}{n(n-1)} \frac{r_1 r_2}{r_2 - r_1} \frac{1}{\delta} \tag{6}$$

if $x^n \in A$.

The normalizing integral (4) diverges if we let $A = \mathbb{R}^n$, which corresponds to the situation $c_0 = 0$, $\delta \to \infty$, $r_1 = 0$ and $r_2 \to \infty$. The next step is therefore to replace c_0, δ, r_1 and r_2 with more general parameters that allow us to define a non-zero density for all $x^n \in \mathbb{R}$ having $r(x^n) > 0$. We assume that r_1 is independent of δ and c_0. Consider the parameters r_1 and r_2 first. Let $t > 1$ and $r_2(r_1) = tr_1$. Requiring that $r(x^n) \in [r_1, r_2(r_1)] = [r_1, tr_1]$, we replace the coefficient $(r_1 r_2)/(r_2 - r_1) = (tr_1)/(t - 1)$ in (6) with the integral

$$\int_{r(x^n)/t}^{r(x^n)} p_{r_1}(r) \frac{tr}{t - 1}\, dr,$$

where p_{r_1} is a continuous prior of the parameter r_1. Letting t approach 1 from above, we get

$$\lim_{t \to 1+} \int_{r(x^n)/t}^{r(x^n)} p_{r_1}(r) \frac{tr}{t - 1}\, dr$$

$$= \lim_{t \to 1+} \left(r(x^n) - \frac{r(x^n)}{t} \right) p_{r_1}(r(x^n)) \frac{t\, r(x^n)}{t - 1}$$

$$= r(x^n)^2\, p_{r_1}(r(x^n)).$$

Next, we get rid of the coefficient $1/\delta$ and the dependence on c_0 in (6). Let $\delta > 0$ and let p_{c_0} be a continuous prior density function of the parameter c_0. The integration goes over all such values of c_0 that $c(x^n) \in [c_0 - \delta, c_0 + \delta]$. In a similar fashion as above, we substitute $1/\delta$ with the limiting function

$$\lim_{\delta \to 0+} \int_{c(x^n)-\delta}^{c(x^n)+\delta} p_{c_0}(c) \frac{1}{\delta}\, dc = 2p_{c_0}(c(x^n)). \tag{7}$$

The final density function is thus

$$f(x^n; p_{r_1}, p_{c_0}) = \frac{1}{(2r(x^n))^{n-2}} \frac{p_{r_1}(r(x^n)) \, p_{c_0}(c(x^n))}{2n(n-1)}$$ (8)

if $x^n \in \mathbb{R}^n$ and $r(x^n) > 0$.

For practical purposes, at least one problem has to be solved: how to encode sequences consisting of equal points. In an axis-aligned box model the problem arises when the coordinates of the points are equal in some dimension. We cannot continuously extend (8) for the case $r(x^n) = 0$, because $f(x^n; p_{r_1}, p_{c_0})$ grows unbounded when $r(x^n) \to 0$. Our solution is to choose a special prior so that $f(x^n; p_{r_1}, p_{c_0})$ has a constant value when $r(x^n) \in \,]0, \epsilon]$. We give the prior in a slightly restricted case. Let $b = \underbrace{2 \uparrow 2 \uparrow \ldots \uparrow 2 \uparrow \delta}_{k-1 \text{ copies of '2↑'s}}$ and let $\epsilon = \underbrace{2 \uparrow 2 \uparrow \ldots \uparrow 2 \uparrow \alpha - b}_{k-1 \text{ copies of '2↑'s}}$

where $k \in \mathbb{N}$ and $\alpha, \delta \in [1, 2]$, $\alpha > \delta$. A continuous density giving fulfilling the previous requirements is

$$p_{r_1}(r_1) = \begin{cases} c \, f_{\mathbb{R}_+}(\epsilon; b) \, \epsilon^{2-n} \, r_1^{n-2} & \text{if } r_1 \in [0, \epsilon[\\ c \, f_{\mathbb{R}_+}(r_1; b) & \text{if } r_1 \geq \epsilon, \end{cases}$$

where $f_{\mathbb{R}_+}$ is a density defined in (3) and c is a constant for normalization. Because

$$\int_0^\epsilon f_{\mathbb{R}_+}(x; b) \, dx = \int_0^\epsilon \frac{1 - \ln 2}{(\ln 2)^k} \frac{1}{(x + b) \, h(x + b)} \, dx$$

$$= \frac{1 - \ln 2}{(\ln 2)^k} \int_b^{b+\epsilon} \frac{1}{y \, h(y)} \, dy$$

$$= \frac{1 - \ln 2}{(\ln 2)^k} \bigg/_{y=b}^{b+\epsilon} (\ln 2)^k \log^{(k)} y$$

$$= (1 - \ln 2)(\log \alpha - \log \delta)$$

and

$$\int_0^\epsilon f_{\mathbb{R}_+}(\epsilon; b) \, \epsilon^{2-n} \, r^{n-2} \, dr = \frac{f_{\mathbb{R}_+}(\epsilon; b)}{n - 1} \epsilon,$$

we get

$$c = \left(1 - (1 - \ln 2)(\log \alpha - \log \delta) + \frac{f_{\mathbb{R}_+}(\epsilon; b)}{n - 1} \epsilon\right)^{-1}.$$

We still need another density, if the length of the sequence is 1. A natural choice is $f((x); p_{r_1}, p_{c_0}) = f_{\mathbb{R}}(x)$.

6   Code Lengths for Spherical and Axis-Aligned Gaussian Clusters

In this section, we derive a code length function according to the model class consisting of spherical normal distributions. Based on that, we get the code length according to the model class with axis-aligned normal distributions in a trivial way. Deriving the NML at (10) resembles the one-dimensional case, see [4] pp. 195–213 with considerations about restricting the parameters.[1]

Let $x^n = (x_1, x_2, \ldots, x_n) \in (\mathbb{R}^d)^n$ where $n \in \{2, 3, \ldots\}$. Let φ_{μ,σ^2} denote the density function of a normal distribution with the mean μ and the covariance matrix $\sigma^2 I_d$. Let $\hat{\mu} = \hat{\mu}(x^n) = (1/n) \sum_{i=1}^{n} x_i$ be the maximum likelihood mean and $\hat{\sigma}^2 = \hat{\sigma}^2(x^n) = \sum_{i=1}^{n} \|x_i - \hat{\mu}\|^2/(dn)$ the ML variance. Now, if x^n is an i.i.d sample from a $\mathcal{N}(\mu, \sigma^2 I_d)$ source, then $\hat{\mu} \sim \mathcal{N}(\mu, (\sigma^2/n)I_d)$. The maximum likelihood estimate of the variance can be written as

$$\hat{\sigma}^2(x^n) = \frac{\sigma^2}{dn} \sum_{i=1}^{d} \underbrace{\frac{1}{\sigma^2} \sum_{j=1}^{n} (x_j(i) - \hat{\mu}(j))^2}_{\sim \chi^2(n-1)}$$

where $x_j(i)$ and $\hat{\mu}(i)$ denote the ith coordinate of x_j and $\hat{\mu}$ respectively. It can be seen that $\hat{\sigma}^2(x^n)$ is proportional to a sum of d independent $\chi^2(n-1)$ variables, and that $(dn\hat{\sigma}(x^n))/\sigma^2 \sim \chi^2(dn-d)$. Additionally, $\hat{\mu}$ and $\hat{\sigma}^2$ are independent.

The density of x^n can be therefore factorized as

$$\varphi_{\mu,\sigma^2}(x^n) = \varphi_{\mu,\sigma^2}(x^n \mid \hat{\mu}, \hat{\sigma}^2) \, \varphi_{\mu,\sigma^2/n}(\hat{\mu}) \, f_{\chi^2, dn-d}\left(\frac{dn}{\sigma^2}\hat{\sigma}^2\right) \frac{dn}{\sigma^2}$$

where $f_{\chi^2, dn-d}$ is the density function of a chi-square distribution with $dn - d$ degrees of freedom, and the product of the two last coefficients is the density of $\hat{\sigma}^2$. We need the maximum likelihood factorization, which is

$$\begin{aligned}
\varphi_{\hat{\mu},\hat{\sigma}^2}(x^n) &= \varphi_{\hat{\mu},\hat{\sigma}^2}(x^n \mid \hat{\mu}, \hat{\sigma}^2) \, \varphi_{\hat{\mu},\hat{\sigma}^2/n}(\hat{\mu}) \, f_{\chi^2, dn-d}(dn) \frac{dn}{\hat{\sigma}^2} \\
&= \varphi_{\hat{\mu},\hat{\sigma}^2}(x^n \mid \hat{\mu}, \hat{\sigma}^2) \, C_{d,n} \cdot (\hat{\sigma}^2)^{-(d/2)-1}
\end{aligned}$$

where

$$C_{d,n} = \left(\frac{dn}{2e}\right)^{(dn)/2} (d\pi)^{-d/2} \left(\Gamma\left(\frac{dn-d}{2}\right)\right)^{-1}.$$

Assume that $\hat{\mu} \in B(\mu_0, r)$ and $\hat{\sigma}^2 \in [\sigma_1^2, \sigma_2^2]$, where $r > 0$ and $\sigma_1^2, \sigma_2^2 > 0$. Let $\Theta(\mu, \sigma^2) = \{y^n \in (\mathbb{R}^d)^n \mid \hat{\mu}(y^n) = \mu, \hat{\sigma}^2(x^n) = \sigma^2\}$. We get the normalizing integral

$$I(\mu_0, r, \sigma_1^2, \sigma_2^2) \tag{9}$$

$$= \int_{\mu \in B(\mu_0, r)} \int_{\sigma_1^2}^{\sigma_2^2} \int_{x^n \in \Theta(\mu,\sigma^2)} \varphi_{\mu,\sigma^2}(x^n \mid \mu, \sigma^2) \, C_{d,n}(\sigma^2)^{-(d/2)-1} \, dx^n \, d\sigma^2 \, d\mu$$

$$= C_{d,n} \, V_d(r) \frac{2}{d} \left(\frac{1}{\sigma_1^d} - \frac{1}{\sigma_2^d}\right).$$

---

[1] There are some minor mistakes in [4]: on p. 203 at (8.13) the arguments of the exponential function have been written falsely; on p. 206 on the 6th line the result should be $k_n 4R/\sigma_0$.

Dividing the maximum likelihood by $I(\mu_0, r, \sigma_1^2, \sigma_2^2)$ yields the normalized maximum likelihood

$$f_{\text{NML}}(x^n; \mu_0, r, \sigma_1^2, \sigma_2^2) = \varphi_{\hat{\mu}, \hat{\sigma}^2}(x^n) \frac{d}{2\,C_{d,n}} \frac{1}{V_d(r)} \frac{\sigma_1^d \sigma_2^d}{\sigma_2^d - \sigma_1^d}. \tag{10}$$

The normalizing integral (9) diverges if taken over $(\mathbb{R}^d)^n$. Similarly as in the previous subsections, we use the continuous priors f_{μ_0} and f_{σ_1} for μ_0 and σ_1, respectively, and we let $\sigma_2(\sigma_1) = t\sigma_1$. We denote the square root of the sample variance as $\hat{\sigma}(x^n)$. Noting that

$$\lim_{r \to 0+} \int_{\mu \in B(\hat{\mu}(x^n), r)} \frac{1}{V_d(r)} f_{\mu_0}(\mu)\, d\mu = f_{\mu_0}(\hat{\mu}(x^n))$$

and that

$$\lim_{t \to 1+} \int_{\hat{\sigma}(x^n)/t}^{\hat{\sigma}(x^n)} \frac{\sigma^d(t\sigma)^d}{(t\sigma)^d - \sigma^d} f_{\sigma_1}(\sigma)\, d\sigma = \hat{\sigma}(x^n)^{d+1} f_{\sigma_1}(\hat{\sigma}(x^n)) \frac{1}{d},$$

we have the final density

$$f(x^n; g_{\mu_0}, g_{\sigma_1})$$
$$= \frac{1}{2} (\pi d)^{(d-dn)/2} n^{-dn/2} \Gamma\left(\frac{dn-d}{2}\right) \hat{\sigma}(x^n)^{d-dn+1} f_{\mu_0}(\hat{\mu}(x^n)) f_{\sigma_1}(\hat{\sigma}(x^n)).$$

Let Σ be any diagonal covariance matrix in the model $\mathcal{N}(\mu, \Sigma)$. Because the coordinates are independent, a density for this model can be formed by setting $d = 1$ in the spherical model, which yields a one-dimensional model, and taking a product of the coordinate densities. In the same way as at the end of Sect. 5, we avoid singularities by a special prior

$$f_{\sigma_1}(\sigma) = \begin{cases} c\, f_{\mathbb{R}_+}(\epsilon; b)\, \epsilon^{d(1-n)+1}\, \sigma^{d(n-1)-1} & \text{if } \sigma \in [0, \epsilon[\\ c\, f_{\mathbb{R}_+}(\sigma; b) & \text{if } \sigma \geq \epsilon \end{cases}$$

where

$$c = \left(1 - (1 - \ln 2)(\log \alpha - \log \delta) + \frac{f_{\mathbb{R}_+}(\epsilon; b)}{d(n-1)} \epsilon\right)^{-1}.$$

7   Experiments

Data. Empirical tests were made using synthetic data with a clear cluster structure. However, the uniform component is very dominant in the 2-dimensional data sets with a large number of clusters, making the discovery of the clustering challenging.

There were two main categories of 2, 5 and 20-dimensional synthetic data in the experiments. In the category *1*, all the the Gaussian clusters were surrounded by uniform background noise in an axis-aligned box. In the category *2*, the Gaussian clusters could also locate outside the box containing uniform noise. The main categories were each divided into two subcategories in which the Gaussian clusters were either spherical (S) or axis-aligned (A). The standard deviations of the normal distributions were drawn uniformly at random from the interval $[1,3]$ in all the cases. The number of Gaussian clusters varied from 1 to 20. For each category (*1S*, *1A*, *2S* and *2A*), dimension and number of Gaussian clusters, 5 data sets were generated using different parameters of the generating distributions.

We describe first the point generation in the category *1*. The dimensions of the box for the uniform background noise were drawn uniformly at random from the interval $[50, 200]$. The location of the closest corner of the box to the origin was drawn uniformly from the set $[-50, 50]^d$ where d is the number of dimensions. Then the means of the Gaussian distributions were drawn uniformly out of the box, ensuring however, that they do not lie too close to the border of the box or to each other. To be precise, in the case of spherical Gaussians (*1S*), the distance to the border was at least 3 times the standard deviation of the Gaussian cluster, and the distance between two means at least 3 times the sum of the standard deviations of the generating distributions. The category with axis-aligned Gaussians (*1A*) was similar but for each cluster the maximum of the standard deviations on the diagonal of the covariance matrix was used.

The sizes of the Gaussian clusters for point generation were drawn uniformly at random from the set $\{50, \ldots, 150\}$. The size of the uniform cluster was always 2 times the sum of the sizes of the Gaussian clusters. After the points had been drawn at random from the distributions, they were classified so that each point was assigned to the distribution in which the density of the points was largest. Finally, the data set was accepted only if each resulting cluster had at least 30 points.

The point generation in the category *2* was otherwise similar to that in the category *1*, but the box that is determined at the beginning was not used for the uniform cluster, but the means of the Gaussian clusters were drawn from that box instead. A smaller box B was generated for the uniform distribution by drawing a coefficient from the interval $[0.2, 1.0]$ for each dimension, scaling the dimensions of the original box with the coefficients and placing the box B in a random location inside the original box.

Experiments and results. Three methods were tested. The first two corresponded to the description in Sect. 2 and they used either a pure Gaussian mixture or a Gaussian mixture with a uniform component in the EM phase. We refer to these methods as GEM-plus and GUEM-plus, respectively. For comparison, we also ran the greedy EM algorithm with a Gaussian mixture and a uniform component as such, using our code length for choosing the best clustering from several candidates corresponding to different parameter settings and test runs of the EM algorithm. We call this method GUEM. We denote the original number of Gaussian clusters of a data set as k and the number of Gaussian clusters in the beginning of the method as m. With the GEM-plus method, we ran our method for each data set with the values $m \in \{1, 2, \ldots, 50\}$. For GUEM-plus and GUEM we used the values $m \in \{1, 2, \ldots, 30\}$ and proportions of uniform points in the beginning $\beta \in \{0.1, 0.3, 0.5, 0.7, 0.9\}$. The method was repeated 20 times for every parameter combination. Both the spherical and the axis-aligned model were used in the clustering method with all the data sets. However, the influence of the model was relatively small in these tests. The quality of the discovered clustering was mainly estimated using an information based distance [6]. The weakness of using this distance in our context is that the uniform cluster is handled symmetrically with the other clusters. When the plots of the two-dimensional clusterings were inspected visually, clusterings with a distance at most 0.5 bits from the original clustering looked intuitively nearly optimal.

The differences between the clusterings found by the three methods were typically quite small, and a direct comparison is difficult because of the different parameters. All the methods performed very well with 5 and 20-dimensional data from category 1, and with 2-dimensional data from the same category at least when $k < 12$. The 2-dimensional data with a large k had a very high density of uniform noise, and often some of the original Gaussian clusters were classified to the uniform component.

There were some interesting general observations. GEM-plus and GUEM-plus seemed to be quite unsensitive to the choice of m. Fairly good results could have been achieved just by using always the value $m = 30$. In contrast, GUEM needed a larger number of different parameter combinations in order to find good quality clusterings (Fig. 2). But also with GUEM m war mostly larger than the final number of Gaussian clusters, because the clusters could become empty in the greedy EM algorithm. Figure 1 illustrates which values of m led to the best clusterings that GEM-plus found in the two-dimensional case, and how the number of clusters in the final clustering corresponded to the "true" number of clusters.

GEM-plus suited poorly for the clustering of two-dimensional data from the category 2. The original uniform cluster was typically covered with many Gaussian clusters. The reason for this seems to be clear. The addition of a single point to the noise cluster can make the volume of the corresponding box much larger, leading to a greatly increased code length for the points of the noise cluster. Because the cluster cannot shrink thereafter, the method is extremely sensitive with this kind of data to the order in which the points are removed from the Gaussian clusters.

A minor but an apparent weakness of GUEM was that it often found clusterings in 5 and 20-dimensional data in which there were several excess clusters each containing only a very small number of points. As one would expect, GUEM-plus could prune these tiny clusters quite well. The best method for higher dimensional data was however in this test GEM-plus, for which the 20-dimensional test setting appeared to be trivial.

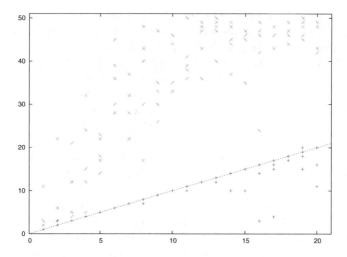

Fig. 1. The number of Gaussian clusters at the beginning (green crosses) and at the end (red pluses) of the GEM-plus runs resulting to smallest MDLs for different data sets. The "true" number of Gaussian clusters is on the x-axis. The 2-dimensional data sets belonged to the category *1A* and spherical Gaussian models were used in the search method.

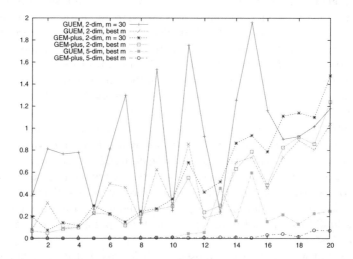

Fig. 2. The influence of m to the final clustering with GEM-plus and GUEM. Value $m = 30$ is compared with the best m. The "true" number of clusters is on the x-axis, the distance between the original and the discovered clustering in bits on the y-axis (averages over 5 data sets). The four uppermost lines refer to 2-dimensional data of the type *1S*, search methods used axis-aligned Gaussian models. The two lowest lines indicate that clustering of the corresponding 5-dimensional data was considerably easier.

8   Conclusion

We introduced a practical MDL based quality measure for a clustering consisting of Gaussian clusters and a uniform component. Experiments with synthetic data having a clear cluster structure hinted that the measure can be used succesfully with different clustering methods. Our simple heuristics GEM-plus and GUEM-plus had the remarkable quality in the experiments that they seemed to be quite unsensitive to the initial number of Gaussian clusters, given with the parameter m. Future plans include improving the clustering methods so that they would work well with a wider range of different data.

References

1. Arthur, D., Vassilvitskii, S.: k-means++: the advantages of careful seeding. In: SODA 2007: Proceedings of the Eighteenth Annual ACM-SIAM Symposium on Discrete algorithms, Philadelphia, PA, USA, pp. 1027–1035. Society for Industrial and Applied Mathematics (2007)
2. Dempster, A.P., Laird, N.M., Rubin, D.B.: Maximum likelihood from incomplete data via the em algorithm. Journal of the Royal Statistical Society. Series B (Methodological) 39(1), 1–38 (1977)
3. Grünwald, P.D.: The Minimum Description Length Principle. The MIT Press, Cambridge (2007)
4. Grünwald, P.D., Myung, I.J., Pitt, M.A. (eds.): Advances in Minimum Description Length Theory and Applications. The MIT Press, Cambridge (2005)
5. Kontkanen, P.: Computationally Efficient Methods for MDL-Optimal Density Estimation and Data Clustering. PhD thesis, University of Helsinki, Department of Computer Science (2009)
6. Meilă, M.: Comparing clusterings–an information based distance. Journal of Multivariate Analysis 98(5), 873–895 (2007)
7. Rissanen, J.: Modeling by shortest data description. Automatica 14(5), 465–471 (1978)
8. Rissanen, J.: A universal prior for integers and estimation by minimum description length. The Annals of Statistics 11(2), 416–431 (1983)
9. Rissanen, J.: Stochastic complexity. Journal of the Royal Statistical Society. Series B (Methodological) 49(3), 223–239 (1987)
10. Rissanen, J.: Fisher information and stochastic complexity. IEEE Transactions on Information Theory 42(1), 40–47 (1996)
11. Rissanen, J.: Information and Complexity in Statistical Modeling. Springer, New York (2007)
12. Szpankowski, W.: Average case analysis of algorithms on sequences. John Wiley & Sons, Chichester (2001)

Exploiting Code Redundancies in ECOC

Sang-Hyeun Park, Lorenz Weizsäcker, and Johannes Fürnkranz

Knowledge Engineering Group, TU Darmstadt, Germany
{park,lorenz,juffi}@ke.tu-darmstadt.de

Abstract. We study an approach for speeding up the training of error-correcting output codes (ECOC) classifiers. The key idea is to avoid unnecessary computations by exploiting the overlap of the different training sets in the ECOC ensemble. Instead of re-training each classifier from scratch, classifiers that have been trained for one task can be adapted to related tasks in the ensemble. The crucial issue is the identification of a schedule for training the classifiers which maximizes the exploitation of the overlap. For solving this problem, we construct a classifier graph in which the nodes correspond to the classifiers, and the edges represent the training complexity for moving from one classifier to the next in terms of the number of added training examples. The solution of the Steiner Tree problem is an arborescence in this graph which describes the learning scheme with the minimal total training complexity. We experimentally evaluate the algorithm with Hoeffding trees, as an example for incremental learners where the classifier adaptation is trivial, and with SVMs, where we employ an adaptation strategy based on adapted caching and weight reuse, which guarantees that the learned model is the same as per batch learning.

1 Introduction

Error-correcting output codes (ECOC) [5] are a well-known technique for handling multiclass classification problems, i.e., for problems where the target attribute is a categorical variable with $k > 2$ values. Their key idea is to reduce the k-class classification problem to a series of n binary problems, which can be handled by a 2-class classification algorithm, such as a SVM or a rule learner. Conventional ECOC always use the entire dataset for training each of the binary classifiers. Ternary ECOC [1] are a generalization of the basic idea which allows to train the binary classifiers on subsets of the training examples. For example, pairwise classification [8,9], which trains a classifier for each pair of classes, is a special case of this framework.

For many common general encoding techniques, the number of binary classifiers may exceed the number of classes by several orders of magnitude. This allows for greater distances between the code words, so that the mapping to the closest code word is not compromised by individual mistakes of a few classifiers. For example, for pairwise classification, the number of binary classifiers is quadratic in the number of classes. Thus, the increase in predictive accuracy comes with a corresponding increase in computational demands at classification time. In previous work [11], we focused on fast ECOC decoding methods, which tackled this problem. For example, for the special case of pairwise classification, the quadratic complexity can be reduced to $O(k \log k)$ in practice.

B. Pfahringer, G. Holmes, and A. Hoffmann (Eds.): DS 2010, LNAI 6332, pp. 266–280, 2010.

In this paper, we focus on the training phase, where overlaps of training instances in highly redundant codes are reduced without altering the models. This is done by identifying shared subproblems in the ensemble, which need to be learned only once, and by rescheduling the binary classification problems so that these subproblems can be reused as often as possible. This approach is obviously feasible in conjunction with incremental base learners, but its main idea is still applicable for the more interesting case when SVMs are used as base learners, by reusing computed weights of support vectors from related subproblems and applying an adapted ensemble caching strategy.

At first, we will briefly recapitulate ECOC with an overview of typical code designs and decoding methods (section 2) before we discuss their redundancies and an algorithm to exploit them in Section 3. The performance of this algorithm is then evaluated for Hoeffding trees and for SVMs as base classifiers (Section 4). Finally, we will conclude and elaborate on possible future directions.

2   Error-Correcting Output Codes

Error-correcting output codes [5] are a well-known technique for converting multi-class problems into a set of binary problems. Each of the k original classes receives a code word in $\{-1, 1\}^n$, thus resulting in a $k \times n$ coding matrix M. Each of the n columns of the matrix corresponds to a binary classifier where all examples of a class with $+1$ are positive, and all examples of a class with -1 are negative. Ternary ECOC [1] are an elegant generalization of this technique which allows 0-values in the codes, which correspond to ignoring examples of this class.

As previously mentioned, the well known *one-against-one* and *one-against-all* decomposition schemes for multiclass classification are particular codes within the framework of ECOC. Other well-known general codes include:

Exhaustive Ternary Codes cover all possible classifiers involving a given number of classes l. More formally, a (k, l)-exhaustive ternary code defines a ternary coding matrix M, for which every column j contains exactly l values, i.e., $\sum_{i \in K} |m_{i,j}| = l$. Obviously, in the context of multiclass classification, only columns with at least one positive $(+1)$ and one negative (-1) class are meaningful. These codes are a straightforward generalization of the exhaustive binary codes, which were considered in the first works on ECOC [5], to the ternary case. Note that $(k, 2)$-exhaustive codes correspond to pairwise classification.

In addition, we define a cumulative version of exhaustive ternary codes, which subsumes all (k, i)-codes with $i = 2 \dots l$, so up to a specific level l. In this case, we speak of (k, l)-*cumulative exhaustive codes*. For a dataset with k classes, (k, k)-cumulative exhaustive codes represent the set of all possible binary classifiers.

Random Codes are randomly generated codes, where the probability distribution of the set of possible symbols $\{-1, 0, 1\}$ can be specified. The zero probability parameter $r_{zp} \in [0, 1]$, specifies the probability for the zero symbol, $p(\{0\}) = r$, whereas the remainder is equally subdivided to the other symbols: $p(\{1\}) = p(\{-1\}) = \frac{1-r}{2}$. This type of code allows to control the degree of *sparsity* of the ECOC matrix. In accordance with the usual definition, we speak of *random dense codes* if $r_{zp} = 0$, which relates to binary ECOC.

3 Redundancies within ECOC

3.1 Code Redundancy

Many code types specify classifiers which share a common code configuration. For instance, in the case of exhaustive cumulative k-level codes, we can construct a *sub-classifier* by setting some $+1$ bits and some -1 bits of a specified classifier to zero. Clearly, the resulting classifier is itself a valid classifier that occurs in the ECOC matrix of this cumulative code. Furthermore, every classifier f of length $l < k$, is subclassifier of exactly $2 \cdot (n-l)$ classifiers with length $l+1$, since there are $n-l$ remaining classes and each class can be specified as positive or negative. Such redundancies also occur frequently in random codes with a probability of the zero-symbol smaller than 0.5, and therefore also in the special case of random dense codes, where the codes consists only of $+1$ and -1 symbols. On the other hand, the widely used one-against-one code has no code redundancy, and the redundancy of the one-against-all code is very low.

In general, the learning of a binary classifier is independent of the explicit specification, which class of instances is regarded as positive and which one as negative. So, from a learning point of view, the classifier specified by a column $\boldsymbol{m}_i = (m_{1i}, \ldots, m_{ki})$ is equivalent to $-\boldsymbol{m}_i$.

Formally, code redundancy can be defined as follows:

Definition 1 (Code Redundancy). *Let f_i and f_j be two classifiers and $(m_{1i}, \ldots, m_{ki})$ and $(m_{1j}, \ldots, m_{kj})$ their corresponding (ternary) ECOC columns. We say f_i and f_j are $p-$redundant, if for $a \in \{1 \ldots k\}$,*

$$p = \max(\#\{a \mid m_{ai} = m_{aj}, m_{ai} \neq 0\}, \#\{a \mid m_{ai} = -m_{aj}, m_{ai} \neq 0\})$$

Let $d = \max(d_H(\boldsymbol{m}_i, \boldsymbol{m}_j), d_H(-\boldsymbol{m}_i, \boldsymbol{m}_j))$, where d_H is the Hamming distance. Two classifiers f_i and f_j are p-redundant, if and only if $k-p = d-\#\{a \in \{1 \ldots k\} \mid m_{ai} = 0 \land m_{aj} = 0\}$. Thus, in essence, classifier redundancy is the opposite of Hamming distance except that bit positions with equal zero values are ignored. For convenience, similarly to the symmetric difference of sets, we denote for two classifiers f_i and f_j the set of classes which are only involved in one of their code configurations m_i and m_j as $f_i \triangle f_j$. More precisely, $f_i \triangle f_j = \{c_a \mid a \in \{1 \ldots k\} \land |m_{ai}| + |m_{aj}| = 1\}$. In addition, we speak of a *specified* classifier, if there exists a corresponding code-column in the given ECOC matrix.

3.2 Exploitation of Code Redundancies

Code redundancies can be trivially exploited by incremental base learners, which are capable of extending an already learned model on additional training instances. Then, repeated iterations over the same instances can be avoided, since shared subclassifiers only have to be learned once. The key issue is to find a training protocol that maximizes the use of such shared subclassifiers, and therefore minimizes the redundant computations. Note that the subclassifiers do not need to be specified classifiers, i.e., they do not need to correspond to a class code in the coding matrix.

This task may be viewed as a graph-theoretic problem. Let $G = (V, E)$ be a weighted directed graph with $V = \{n_r\} \cup \{f_i\} \cup \{f_s\}$, i.e., each classifier f_i and each possible

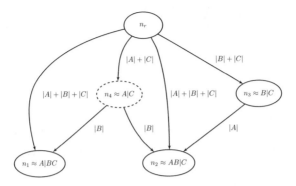

Fig. 1. A sample training graph. Three classifiers $f_1 = A|BC$, $f_2 = AB|C$ and $f_3 = B|C$ are specified. The non-specified classifier $f_4 = A|C$ is added because it is the maximal common subclassifier of f_1 and f_2. For each edge $e_{ij} = (n_i, n_j)$ the weights depict the training effort for learning classifier f_j based on classifier f_i ($|A|$ is the number of training instances of class A).

subclassifier f_s are in the set of nodes V. Furthermore, the special *root* node n_r is connected to every other node $n_i \in V$ with the directed edge (n_r, n_i). Besides, for each two non-root nodes n_i and n_j, there exists a directed edge (n_i, n_j), if and only if n_i is subclassifier of n_j. The weight of these edges is $f_j \triangle f_i$. For all edges (n_r, n_i), which are incident to the root node, the weight is the number of training instances involved in classifier f_i.

To elaborate, incident edges to the root node depict classifiers which are learned by batch learning. All other edges (n_i, n_j), which are edges between two (sub)-classifiers, represent incremental learning steps. Based on the learned model of classifier f_i, the remaining training instances of $f_j \triangle f_i$ are used to learn classifier f_j. The multiple possible paths to one particular classifier represents the possible ways to learn it. Each of these paths describe a different partitioning of training costs, represented by the number of edges (number of partitions) and edge weights (size of the partitions). Considering only one classifier, the cost for all paths are identical. But, by considering that paths of different redundant classifiers can overlap, and that shared subpaths are trained only once, the total training cost can be reduced. Another view at this graph is the following: every subgraph of G which is an arborescence consisting of all specified classifiers is a *valid* scheme for learning the ensemble, in the sense that it produces exactly the specified set of classifiers.

In this context, our optimization problem is to find a minimum-weight subgraph of G including all classifier nodes f_i, which relates to minimizing the processed training instances for the set of specified classifiers and therefore total training complexity of the ECOC ensemble. Note, this problem is known in graph theory as *Steiner problem in a directed graph*, which is NP-hard [14].

Figure 1 shows an example of such a *training graph* for a 3-class problem, where three classifiers $f_1 = A|BC$, $f_2 = AB|C$ and $f_3 = B|C$ are specified by a given ECOC matrix. A, B, C are symbol representatives for classes and $A|B$ describes the binary classifier which discriminates instances of class A against B. The standard

---

Algorithm 1. Training Graph Generation

Require: ECOC Matrix $M = (m_{i,j}) \in \{-1, 0, 1\}^{k \times n}$, binary classifiers $f_1, \ldots, f_n$
1: $V = \{n_r\}, E = \varnothing$
2: **for each** f_i **do**
3:      $V = V \cup \{n_i\}$ # Integration of all classifiers
4:      $e_{ri} = (n_r, n_i)$
5:      $w(e_{ri}) = I(f_i)$
6:      $E = E \cup \{e_{ri}\}$
7: **end for**
8: **for** $l = k$ **downto** 2 **do**
9:      $F = \{n \in V \setminus \{n_r\} \mid \text{length}(n) \geq l, \text{ seen}(n) = 0\}$ # level-wise subclassifier generation
10:     **for each** pair $(n_i, n_j) \in F \times F$ with $i \neq j$ **do**
11:         $n_s = \text{intersection}(n_i, n_j)$ # generate shared subclassifier of f_i and f_j
12:         **if** n_s is valid **then**
13:             **if** $n_s \notin V$ **then**
14:                 $V = V \cup \{n_s\}$ # classifier is new
15:                 $e_{rs} = (n_r, n_s)$
16:                 $w(e_{rs}) = I(f_s)$
17:                 $E = E \cup \{e_{rs}\}$
18:             **end if**
19:             $e_{si} = (n_s, n_i), e_{sj} = (n_s, n_j)$
20:             $w(e_{si}) = I(f_s \triangle f_i)$
21:             $w(e_{sj}) = I(f_s \triangle f_j)$
22:             $E = E \cup \{e_{si}, e_{sj}\}$
23:         **end if**
24:     **end for**
25:     $\forall n \in F.\text{seen}(n) = 1$ # mark as processed, see also note in text
26: **end for**
27: **return** $G = (V, E, w)$

---

training scheme, which learns each classifier separately, can be represented as a sub-graph $G_1 \subseteq G$ consisting of $V_1 = \{n_r, n_1, n_2, n_3\}$ and $E_1 = \{e_{r1}, e_{r2}, e_{r3}\}$. This scheme uses $2|A| + 3|B| + 3|C|$ training instances in total. An example where fewer training instances are needed is $G_2 = (V_1, E_2)$ with $E_2 = \{e_{r1}, e_{r3}, e_{32}\}$, which exploits that classifier f_2 can be incrementally trained from f_3, resulting in training costs $2|A| + 2|B| + 2|C|$. Another alternative is to add a non-specified classifier $f_4 = A|C$ to the graph, resulting in $G_3 = (V, E_3)$ with $E_3 = \{e_{r4}, e_{41}, e_{42}, e_{r3}\}$ with training costs $|A| + 3|B| + 2|C|$. It is easy to see that either G_2 or G_3 is the optimal Steiner Tree in this example and that both process fewer training examples than the standard scheme. Whether G_2 or G_3 is optimal, depends on whether $|A| > |B|$.

Since the optimal solution is in general hard to compute, we use a greedy approach. We first have to generate the training graph. Then, we iteratively remove local non-optimal edges, starting from the leaf nodes (specified classifiers) up to the root. Both methods are described in detail in the following subsections.

Generation of Training Graph. We consider an algorithm which is particularly tailored for exhaustive and exhaustive cumulative codes. Let F be the set of all classifiers f of a specific length l, which is successively decreased from k down to 2. For each pair $(f_i, f_j) \in F \times F$ the maximal common subclassifier f_s is determined and eventually integrated into the graph. Then, these classifiers are marked as processed (**seen**$(f) = 1$) and are not considered in the following steps of the generation algorithm. Level l is decreased and the algorithm repeats. The processed classifiers can be ignored, because for the systematic codes (exh. and exh. cumulative) all potential subclassifiers can be

---

Algorithm 2. Greedy Steiner Tree Computation

Require: Training Graph $G = (V, E, w)$, binary classifiers $f_1, \ldots, f_n$
 1: let Q be an empty FIFO-queue
 2: $\hat{V} = \varnothing, \hat{E} = \varnothing$
 3: **for each** f_i **do**
 4:     $Q.\text{push}(n_i), \hat{V} = \hat{V} \cup \{n_i\}$
 5: **end for**
 6: **while** $!Q.\text{isEmpty}()$ **do**
 7:     $n_i = Q.\text{pop}()$
 8:     $(n_x, n_i) = \text{argmin}_{(n_a, n_i) \in E} \, w((n_a, n_i))$
 9:     $\hat{E} = \hat{E} \cup (n_x, n_i), \hat{V} = \hat{V} \cup \{n_x\}$
10:     **if** $n_x \neq n_r$ **then**
11:         $Q.\text{push}(n_x)$
12:     **end if**
13: **end while**
14: **return** $\tilde{G} = (\hat{V}, \hat{E}, w)$

---

constructed using its immediate subclassifiers. This algorithm does not find all edges for random codes or general codes, but only for their inherent systematic code structures. For the sake of efficiency and also considering that we employ a greedy Steiner Tree Solving procedure afterwards, we neglect this fact.

A pseudo code is given in Algorithm 1. Note that there, the set F is populated with classifiers of length greater equal than l instead of exactly l, considering the special case that there can be multiple levels with zero classifiers or only one classifier. Also, classifiers should only be flagged as *processed* if they were actually checked at least once. The complexity of this version is exponential, but it will be later reduced to quadratic in combination with the greedy Steiner Tree algorithm.

In the beginning, for each specified classifier f_i a corresponding node n_i is generated in the graph and connected with the root node by the directed edge (n_r, n_i). In the main loop, which iterates over $l = k$ down to 2, for each pair (f_i, f_j) of classifiers of length l the maximum common subclassifier f_s is determined. If it is valid (i.e., it is non-zero and contains at least one positive and one negative class), two cases are possible:

- a corresponding node to f_s already exists in the tree: f_i and f_j are included to the set of childs of f_s, that means, two directed edges e_{si} and e_{js} with weights $I(f_i \triangle f_s)$, $I(f_j \triangle f_s)$ respectively are created, where $I(.)$ denotes the total number of training instances for a given code configuration.
- There exists no corresponding node to the subclassifier f_s: f_s is integrated into the tree by creating a corresponding node and by linking it to the root node with edge e_{rs} of weight $I(f_s)$. In addition, the same steps as in the first case are applied.

Greedy Computing of Steiner Trees. A Steiner Tree is, essentially, a minimum spanning tree of a graph, but it may contain additional nodes (which, in our case, correspond to unspecified classifiers). Minimizing the costs is equivalent to minimizing the total number of training examples that are needed to train all classifiers at the leaf of the tree from its root. As mentioned previously, we tackle this problem in a greedy way.

Let f_i be a specified classifier and E_i the set of incident incoming edges. We compute the minimum-weight edge and remove all other incoming edges. The outgoing node of this minimum edge is stored to repeat the process on this node afterwards, e.g. by adding it into a FIFO-queue. This is done until all classifiers and connected subclassifiers have been processed. Note, some subclassifiers are never processed, since all outgoing edges may have been removed. A pseudocode of this simple greedy approach is depicted in algorithm 2. In the following, we will refer to it as the *min-redundant training scheme*, and to the calculated approximate Steiner Tree as $\hat{G}$.

This greedy approach can be combined with the generation method of the training graph, such that the resulting Steiner Graph is identical and such that the overall complexity is reduced to polynomial time. Recall the first step of the generation method: all pairs of classifiers of length k are checked for common subclassifiers and eventually integrated into G. After generating $O(n^2)$ subclassifiers, for each classifier f_i (of length k) the minimal incoming edge[1] is marked. All unmarked edges and also the corresponding outgoing nodes, if they have no other child, are removed. In the next step of the iteration, $l = k - 1$, the number of nodes with length $l - 1$ are now at most n, since only maximally n new subclassifiers were included into the graph G. This means, for each level, $O(n^2)$ subclassifiers are generated, where the generation/checking of a subclassifier has cost of $\Theta(k)$, since we have to check k bits. So, in total, each level costs $O(n^2 \cdot k)$ operations. And, since we have k levels, the total complexity is $O(k^2 \cdot n^2)$. The implementation of the combined greedy method is straight-forward, so we omit a pseudocode and we will refer to it as GSTEINER.

3.3   Incremental Learning with Training Graph

Given a Steiner Tree of the training graph, learning with an incremental base learner is straight-forward. The specific training scheme is traversed in preorder depth-first-manner, i.e. at each node, the node is first evaluated and then its subtrees are traversed in left-to-right order. Starting from the root node, the first classifier f_1 is learned in batch mode. In the next step, if f_1 has a child, i.e. f_1 is subclassifier of another classifier f_2, f_1 is copied and incrementally learned with instances of $f_2 \triangle f_1$, yielding classifier f_2 and so on.

After the learning process, all temporary learned classifiers, which served as subclassifiers and are not specified in the ECOC matrix, are removed, and the prediction phase of the ECOC ensemble remains the same.

In this paper, we use *Hoeffding Trees* [6] as an example for an incremental learner. It is a very fast incremental decision tree learner with anytime properties. One of its main features is that its prediction is guaranteed to be asymptotically nearly identical to that of a corresponding batch learned tree. We used the implementation provided in the Massive Online Analysis Framework [2].

3.4   SVM Learning with Training Graph

While incremental learners are obvious candidates for our approach to save training time, the problem actually does not demand full incrementality because we always add

---

[1] The weights of the edges are identical to the corresponding ones in the fully generated training graph, since it only depends on the total number of training instances, computable by the code configuration of the subclassifier, and not on the actual partitioning.

batches of examples corresponding to different classes to the training set. Thus, the incremental design of a training algorithm might retard the training compared to an algorithm that can naturally incorporate larger groups of additional instances. Therefore, we decided to study the applicability of this approach to a genuine batch learner, and selected the Java-implementation of LIBSVM [4]. The adaption of this base learner consists of two parts: First, the previous model (subclassifier) is used as a starting point for the successor model in the training graph, and second, the caching strategy is adapted to this scenario.

Reuse of Weights. A binary SVM model consists of a weight vector α containing the weights α_i for each training instance (x_i, y_i) and a real-valued threshold b. The latter is derived from α and the instances without significant costs. The weights α are obtained as the solution of a quadratic optimization problem with a quadratic form $\alpha^T(y^T K y)\alpha$ that incorporates the inputs through pairwise evaluations $K_{ij} = k(x_i, x_j)$ of the kernel function k. The first component to speed up the training is to use the weights α of the parent model as start values for optimizing the child weights $\bar{\alpha}$. That is, we set $\bar{\alpha}_i = \alpha_i$, if instance i belongs to the parent model and $\bar{\alpha}_i = 0$ otherwise.

The mutual influence of different instances on their respective weights is twofold. There is a local mutual influence due to the fact that an instance can stand in the shadow of another instance closer to the decision boundary. And there is a weaker, global mutual influence that also takes effect on more unrelated instances communicating through the error versus regularization trade-off in the objective.

If we add additional instances to the training set we might expect that there is only a modest alternation of the old weights, because many of the new instances will have little direct effect on the local influence among previous instances. On the other hand, if the new instances do interfere with some subsets of the previous instances, the global influence can strongly increase as well. In any case, we are more interested in the question whether the parent initialization of the weights does speed up the optimization step.

Cache Strategy. It is well-known that caching of kernel evaluations provides significant speed-up for the learning with SVMs [10]. LIBSVM uses a *Least-Recently-Used* (LRU) Cache, which stores columns of the matrix K respective its signed variant $Q = y^T K y$. Since we use an ensemble of classifiers which potentially overlap in terms of their training instances and therefore also in their matrices K, it is beneficial to replace their local caches, which only keep information for each individual classifier, with an ensemble cache, which allows to transfer information from one classifier to the next.

Typically, each classifier receives a different subset of training instances $T_l \subset T$, specified by its code configuration. In order to transfer common kernel evaluations K_{ab} from classifier f_i to another classifier f_j, the cached columns have to be transformed, since they can contain evaluations of irrelevant instances. Each K_{ab} has to be removed, if instances a or b are not contained in the new training set and also the possible change in the ordering of instances has to be considered in the columns. The main difficulty is the implementation of an efficient mapping of locally used instance ids to the entire training set and its related transformation steps, otherwise, the expected speed-up of an ensemble caching strategy is undone.

Two ensemble cache strategies were evaluated, which are based on the local cache implementation of LIBSVM. The first one reuses *nearly all* reusable cached kernel evaluations from one classifier to another. For each classifier, two mapping tables $m_a(.)$ and $m_o(.)$ are maintained, where m_a associates each local instance number with its corresponding global instance number in order to have a unique addressing used in the transformation step. The table m_o is the mapping table from the previous learning phase. Before using the old cache for the learning of a new classifier, all cached entries are marked (as to be converted). During querying of the cache two cases can occur:

- **a cached column is queried:** If the entry is marked, the conversion procedure is applied. Using the previous and actual mapping table m_o, m_a, the column is transformed to contain only kernel evaluations for relevant instances, which can be done in $O(|m_o| \cdot \log |m_o|)$. Missing kernel evaluations are marked with a special symbol, which are computed afterwards. In addition, the mark is removed.
- **an uncached column is queried:** If the free size of the cache is sufficient, the column is computed and normally stored. Otherwise, beforehand, the least recently used entry is repeatedly removed until the cache has sufficient free space.

Since the columns are converted only on demand, unnecessary conversions are avoided and their corresponding entries are naturally replaced by new incoming kernel evaluations due to the LRU strategy. But, this tradeoff has the disadvantage that kernel evaluations that have been computed and cached at some point earlier may have to be computed again if they are requested later. The marked entries are carried maximally only over two iterations, otherwise it would be necessary for each additional iteration to carry another mapping table. We denote this ensemble caching method as *Short-Term Memory* (STM). One beneficial feature is the compatibility to any training scheme, in particular to the standard and the min-redundant training scheme.

The second ensemble caching method is particularly tailored to the use with a min-redundant training scheme. It differs from the previous one only in its transformation step. Recall that the learning phase traverses the subgraph in preorder depth-first manner. That means that during the learning procedure only the following two cases can occur: either the current classifier f_i is the child of a subclassifier f_j, or the current classifier is directly connected with the root node.

This information can be used for a more efficient caching scheme. For the first case, the set of training instances of f_i is superset of f_j, i.e. $T_j \subseteq T_i$. That means, $|T_j|$ rows and columns can be reused and also importantly without any costly transforming method. The columns and rows have to be simply trimmed to size $|T_j|$ for the reuse in the current classifier. Trimming is sometimes necessary, since they can contain further kernel evaluations from previously learned sibling nodes, i.e. nodes which share the same subclassifier f_j. So, the cache for the current classifier is prepared by removing Q_{ab} with $a > |T_j| \lor b > |T_j|$. In the second case, we know beforehand that no single kernel evaluation can be reused in the actual classifier. So, the cache is simply cleared. We denote this ensemble cache method as *Semi-Local* (SL) cache.

4 Experimental Evaluation

4.1 Experimental Setup

As we are primarily concerned with computational costs and not with predictive accuracy, we applied pre-processing based on all available instances instead of building a pre-processing model on the training data only. First, missing values were replaced by the average or majority value for numeric or ordinal attributes respectively. Second, all numeric values were normalized, such that the values lie in the unit interval.

Our experiment consisted of following parameters and parameter ranges:

- **6+2 multiclass classification datasets** from the UCI repository [7], where 6 relatively small datasets in terms of instances (up to ca. 4000) were used in conjunction with LIBSVM and two large-scale datasets, *pokerhand* and *covtype* consisting of $581,012$ and $1,025,010$ instances, were used with Hoeffding Trees. The number of classes lie in the range between 4 and 11.
- **3 code types**: exhaustive k-level codes, exhaustive cumulative k-level codes, random codes of up to length 500 with $k = 3, 4$ and $r_{zp} = 0.2, 0.4$
- **2 learn methods**: min-redundant and standard training scheme
- **2 base learners**: incremental learner Hoeffding Trees and batch learner LIBSVM (no parameter tuning, RBF-kernel) for which following parameters were evaluated:
  - **3 cache methods**: two ensemble cache methods, namely STM and SL, and the standard local cache of LIBSVM
  - **4 cache sizes**: $25\%, 50\%, 75\%, 100\%$ of the number of total kernel evaluations

All experiments with LIBSVM were conducted with 5-fold cross-validation and for Hoeffding Trees a training-test split of 66% to 33% was used. The parameters of the base learners were not tuned, because we were primarily interested in their computational complexity.[2]

4.2 Hoeffding Trees

Table 1 shows a comparison between the standard training scheme and the greedy computed min-redundant scheme with respect to the total amount of training instances. It shows that even with the suboptimal greedy procedure a significant amount of training instances can be saved. In this evaluation, the worst case can be observed for dataset *covtype* with 3-level exhaustive codes, for which the ratio to the standard training scheme is 22%. In absolute numbers, this relates to processing 3.8 million training instances instead of 17.2 million. In summary, the improvements range from 78% to 98% or in other words, 4 to 45 times less training instances are processed.

Table 2 shows the corresponding total training time. It shows that the previous savings w.r.t. the number of training instances do not transfer directly to the training time. One reason is that the constant factor in the linear complexity of Hoeffding Trees regarding the number of training instances decreases for increasing number of training

---

[2] Tuning of the SVM parameters of the base learners can be relevant here because it may affect the effectiveness of reusing and caching of models. However, this would add additional complexity to the analysis of total cost and was therefore omitted to keep the analysis simple.

Table 1. Total number of processed training instances of standard and min-redundant training scheme. The italic values show the ratio of both. The datasets *pokerhand* and *covtype* consist of $581,012$ and $1,025,010$ instances respectively, from which 66% was used as training instances.

dataset	standard	min-redundant	standard	min-redundant
	EXHAUSTIVE CUMULATIVE CODES			
		$l = 3$		$l = 4$
pokerhand	79,151,319	9,429,611 (*0.119*)	476,937,435	10,479,451 (*0.022*)
covtype	19,556,868	3,807,748 (*0.195*)	73,242,388	5,354,720 (*0.073*)
	EXHAUSTIVE CODES			
		$l = 3$		$l = 4$
pokerhand	73,062,756	9,429,591 (*0.129*)	397,786,116	10,478,523 (*0.026*)
covtype	17,256,060	3,796,818 (*0.220*)	53,685,520	5,191,055 (*0.097*)
	RANDOM CODES			
		$r_{zp} = 0.4$		$r_{zp} = 0.2$
pokerhand	258,035,711	10,205,330 (*0.040*)	311,051,271	8,990,547 (*0.029*)
covtype	153,519,616	6,744,692 (*0.044*)	95,483,532	5,300,005 (*0.056*)

Table 2. Training time in seconds. This table shows training performances for the standard and the min-redundant learning scheme. The italic values shows the ratio of both.

dataset	standard	min-redundant	standard	min-redundant
	EXHAUSTIVE CUMULATIVE CODES			
		$l = 3$		$l = 4$
pokerhand	261.27	127.33 (*0.487*)	1530.06	542.57 (*0.355*)
covtype	118.70	40.89 (*0.344*)	463.09	93.71 (*0.202*)
	EXHAUSTIVE CODES			
		$l = 3$		$l = 4$
pokerhand	236.52	131.12 (*0.554*)	1337.00	522.18 (*0.391*)
covtype	101.50	34.65 (*0.341*)	330.97	83.58 (*0.253*)
	RANDOM CODES			
		$r_{zp} = 0.4$		$r_{zp} = 0.2$
pokerhand	896.41	356.43 (*0.398*)	1089.99	537.11 (*0.493*)
covtype	1106.48	157.61 (*0.142*)	695.84	107.12 (*0.154*)

instances. Furthermore, some overhead is incurred for copying the subclassifiers before each incremental learning step. In total, exploiting the redundancies yields a run-time reduction of about $44.6\% - 85.8\%$.

The running-time for GSTEINER (constructing the graph and greedily finding the Steiner tree, without evaluation of the classifiers) is depicted in Table 3. For the *systematic* code types, exhaustive and its cumulative version, the used time is in general negligible compared to the total training time. The only exception is for dataset *pokerhand* with random codes and $r_{zp} = 0.2$: About 106 seconds were used and contributes therefore one-fifth to the total training time in this case.

4.3   LibSVM

Table 4 shows a comparison of training times between LIBSVM and its adaptions with weight reusing and ensemble caching strategies. M1 and M2 use the standard training scheme, where M1 is standard LIBSVM with local cache and M2 uses the ensemble caching strategy STM. M3 and M4 utilize a min-redundant training scheme with STM and SL respectively. The underlined values depict the best value for each dataset and code-type combination. The results confirm that the weight reuse and ensemble caching

Table 3. GSTEINER running time in seconds

	EXH. CUMULATIVE		EXHAUSTIVE		RANDOM	
	$l = 3$	$l = 4$	$l = 3$	$l = 4$	$r_{zp} = 0.4$	$r_{zp} = 0.2$
pokerhand	0.82	4.63	4.56	3.57	22.09	105.97
covtype	0.24	3.01	0.14	0.17	0.67	0.52

Table 4. Training time in seconds using a cache size of 25%

	optdigits	page-blocks	segment	solar-flare-c	vowel	yeast
	EXHAUSTIVE CUMULATIVE CODES					
			$l = 3$			
M1	92.28 ± 0.36	8.73 ± 0.19	6.56 ± 0.05	3.47 ± 0.07	5.80 ± 0.02	5.43 ± 0.03
M2	80.70 ± 0.37	8.32 ± 0.37	6.00 ± 0.03	4.30 ± 0.08	4.90 ± 0.02	5.62 ± 0.02
M3	76.93 ± 0.60	6.90 ± 0.18	6.94 ± 0.05	3.13 ± 0.16	6.28 ± 0.04	5.77 ± 0.03
M4	53.37 ± 0.40	2.93 ± 0.27	4.19 ± 0.05	1.70 ± 0.25	3.51 ± 0.01	2.98 ± 0.02
			$l = 4$			
M1	833.12 ± 14.98	24.66 ± 0.43	33.98 ± 0.21	18.61 ± 0.35	47.61 ± 0.08	40.42 ± 0.09
M2	666.02 ± 1.54	21.19 ± 0.80	28.69 ± 0.14	22.94 ± 0.52	36.72 ± 0.08	41.19 ± 0.11
M3	680.75 ± 8.23	18.30 ± 0.51	36.91 ± 0.39	15.08 ± 1.71	51.61 ± 0.15	41.79 ± 0.10
M4	410.44 ± 6.08	5.32 ± 0.53	17.18 ± 0.13	8.59 ± 1.27	25.26 ± 0.06	22.01 ± 0.10
	EXHAUSTIVE CODES					
			$l = 3$			
M1	87.42 ± 0.35	7.63 ± 0.39	6.02 ± 0.03	3.17 ± 0.05	5.51 ± 0.03	5.11 ± 0.02
M2	75.28 ± 0.29	6.76 ± 0.12	5.48 ± 0.03	3.95 ± 0.07	4.58 ± 0.03	5.28 ± 0.01
M3	75.61 ± 1.04	7.09 ± 0.27	6.91 ± 0.04	3.13 ± 0.14	6.25 ± 0.03	5.83 ± 0.05
M4	53.13 ± 0.39	2.90 ± 0.21	4.13 ± 0.03	1.71 ± 0.25	3.48 ± 0.02	3.00 ± 0.02
			$l = 4$			
M1	735.76 ± 9.63	15.31 ± 0.49	27.13 ± 0.31	15.14 ± 0.28	41.78 ± 0.09	34.99 ± 0.08
M2	570.69 ± 1.93	12.72 ± 0.45	22.76 ± 0.13	18.72 ± 0.42	31.92 ± 0.06	35.73 ± 0.06
M3	646.6 ± 11.98	16.39 ± 0.44	34.24 ± 0.36	14.69 ± 1.59	49.75 ± 0.10	41.09 ± 0.10
M4	397.79 ± 5.07	4.76 ± 0.46	15.88 ± 0.09	8.45 ± 1.17	24.55 ± 0.10	21.71 ± 0.06
	RANDOM CODES					
			$r_{zp} = 0.4$			
M1	1654.0 ± 22.6	25.7 ± 1.1	156.5 ± 1.7	34.7 ± 1.5	37.5 ± 0.6	46.9 ± 1.2
M2	1424.4 ± 32.8	24.3 ± 0.5	162.9 ± 0.8	46.1 ± 1.9	39.7 ± 0.7	52.1 ± 1.3
M3	1609.2 ± 44.3	22.6 ± 0.3	190.6 ± 3.8	39.9 ± 5.4	65.8 ± 2.3	79.1 ± 2.0
M4	1378.8 ± 34.4	5.7 ± 0.3	140.6 ± 3.0	25.9 ± 3.7	57.1 ± 2.5	64.5 ± 2.4
			$r_{zp} = 0.2$			
M1	2634.6 ± 59.5	10.2 ± 0.3	123.0 ± 0.9	48.2 ± 2.0	49.6 ± 0.4	67.2 ± 1.2
M2	2281.7 ± 29.6	8.6 ± 0.5	129.7 ± 1.4	63.2 ± 3.1	53.0 ± 0.4	74.1 ± 1.3
M3	3049.0 ± 48.3	12.7 ± 0.2	157.9 ± 1.4	57.6 ± 13.3	153.0 ± 2.0	157.5 ± 2.1
M4	2594.0 ± 64.8	3.6 ± 0.2	128.5 ± 2.4	39.1 ± 9.4	144.6 ± 1.7	144.0 ± 2.2

techniques can be used to exploit code redundancies for LIBSVM. For exhaustive codes and its cumulative variant, M4 dominates all other approaches and achieves an improvement of $31.4\% - 78.4\%$ of the training time. However, the results for random codes are not so clear.

For the datasets *vowel* and *yeast* both methods employing the min-redundant training schemes (M3 and M4) use significantly more time. This can be explained with the relative expensive cost for generating and solving the Steiner Tree in these cases, as depicted in Table 5 (89 and 52 sec for vowel and yeast). Contrary to the the results on *optdigits*, for these datasets the tree generation and solving has a big impact on the total training time. Nevertheless, this factor is decreasing for increasing number of instances, since the complexity of GSTEINER only depends on k and n. Besides, based on the results with various cache sizes, which we omit due to space restrictions (we refer to [12] for all results), the cache size has a greater impact on the training time for random

Table 5. GSTEINER running time for random codes in seconds

	optdigits	page-blocks	segment	solar-flare-c	vowel	yeast
$r_{zp} = 0.4$	8.93	< 0.01	0.12	0.50	15.10	8.56
$r_{zp} = 0.2$	53.60	< 0.01	0.12	1.26	89.34	52.80

Table 6. Training time in seconds of random codes using a cache size of 75%

	optdigits	page-blocks	segment	solar-flare-c	vowel	yeast
			RANDOM CODES, CACHE=75%			
			$r_{zp} = 0.4$			
M1	1603.4 ± 22.2	25.8 ± 0.5	153.7 ± 1.5	34.3 ± 1.5	36.0 ± 0.6	45.6 ± 1.1
M2	1317.4 ± 16.3	23.0 ± 0.4	136.9 ± 1.0	45.1 ± 1.9	35.9 ± 0.6	51.2 ± 1.3
M3	1364.6 ± 53.6	22.4 ± 0.2	148.7 ± 1.3	35.8 ± 4.5	60.9 ± 2.0	82.6 ± 2.2
M4	1162.6 ± 27.6	5.5 ± 0.3	70.3 ± 0.4	7.9 ± 0.8	42.8 ± 2.0	27.4 ± 1.2
			$r_{zp} = 0.2$			
M1	2507.0 ± 33.8	10.3 ± 0.3	119.8 ± 1.2	47.6 ± 2.0	47.6 ± 0.4	65.3 ± 1.2
M2	1826.2 ± 21.3	8.5 ± 0.6	98.7 ± 0.6	61.3 ± 3.1	44.3 ± 0.4	70.6 ± 1.2
M3	2093.7 ± 38.6	12.4 ± 0.2	116.9 ± 0.8	51.0 ± 11.0	139.9 ± 1.8	163.7 ± 2.6
M4	1632.5 ± 40.2	3.9 ± 0.1	56.8 ± 0.3	10.0 ± 1.6	118.6 ± 1.6	87.7 ± 2.2

codes than for the systematic ones. Table 6 shows as an example the performance for random codes with a cache size of 75%. Notice the reduction of the training time for the different methods in comparison to Table 4, where a cache size of 25% was used. M4 achieves the best efficiency increase and by subtracting the time for generating and solving the tree, M4 dominates again all other methods.

Table 7 shows the number of optimization iterations of LIBSVM, which can be seen as an indicator of training complexity. The ratio values are averaged over all datasets and show that the reuse of weights in the pseudo-incremental learning steps lead to a reduction of optimization iterations. Once again, the effect on the ensemble caching strategy can be seen in Table 8, showing a selection of the results, here for cache sizes 25% and 75%. The first column of each block describes the number of kernel evaluation calls. The consistent reduction for min-redundant schemes M3 and M4 is accredited to the weight-reusing strategy. Except for random codes with $r_{zp} = 0.2$ and cache size=25% all methods using an ensemble cache strategy (M2, M3 and M4) outperform the baseline of LIBSVM with a local cache. Among these three methods, M3 and M4 both outperform M2 in absolute terms, but not relative to the number of calls. For the special case (random codes, $r_{zp} = 0.2$, M3, M4), one can again see the increased gain of a bigger cache size for the min-redundant training schemes.

Even though all ensemble caching strategies almost always outperform the baseline in terms of hit-miss measures, the corresponding time complexities of Table 4 show that only M4, which uses a min-redundant training scheme and the SL caching strategy, is reliably reducing the total training time. The rather costly transformation cost of STM is the cause for the poor performance of M2 and M3.

5   Related Work

In [3], an efficient algorithm for cross-validation with decision trees is proposed, which also exploits training set overlaps, but focuses on a different effect, namely that in this

Table 7. Comparison of LIBSVM Optimization iterations. The values show the ratio of optimization iterations of a min-redundant training scheme with weight reusing to standard learning.

	EXH. CUMULATIVE		EXHAUSTIVE		RANDOM	
	$l = 3$	$l = 4$	$l = 3$	$l = 4$	$r_{zp} = 0.4$	$r_{zp} = 0.2$
	0.673	0.576	0.768	0.745	0.701	0.773

Table 8. Cache efficiency and min-redundant training scheme impact: averaged mean ratio values of kernel evaluation calls (first column) and actual computed kernel evaluations (second column) to the baseline: standard LIBSVM (M1). The values of M1 are set to 1 and the following values describe the ratio of corresponding values of M2, M3 and M4 to M1.

	EXH.CUMULATIVE				EXHAUSTIVE				RANDOM			
	$l = 3$		$l = 4$		$l = 3$		$l = 4$		$r_{zp} = 0.4$		$r_{zp} = 0.2$	
					CACHE $= 25\%$							
M2	1.00	0.68	1.00	0.61	1.00	0.66	1.00	0.60	1.00	0.83	1.00	0.84
M3	0.78	0.56	0.71	0.52	0.87	0.63	0.88	0.67	0.84	0.83	0.95	1.01
M4	0.78	0.56	0.71	0.51	0.87	0.63	0.88	0.65	0.84	0.83	0.95	1.00
					CACHE $= 75\%$							
M2	1.00	0.59	1.00	0.48	1.00	0.56	1.00	0.44	1.00	0.64	1.00	0.56
M3	0.78	0.43	0.71	0.34	0.87	0.47	0.88	0.42	0.84	0.41	0.95	0.47
M4	0.78	0.44	0.71	0.32	0.87	0.48	0.88	0.41	0.84	0.42	0.95	0.49

case the generated models tend to be similar, such that often identical test nodes are generated in the decision tree during the learning process. This approach is not applicable here, since during the incremental learning steps, the inclusion of new classes may lead to significant model changes. Here, a genuine incremental learner or in the case of LIBSVM different approaches are necessary. However, the main idea, to reduce redundant computations is followed also here.

Pimenta et al. [13] consider the task of optimizing the size of the coding matrix so that it balances effectivity and efficiency. Our approach is meant to optimize efficiency for a given coding matrix. Thus, it can also be combined with their approach if the resulting *balanced* coding matrix is code-redundant.

6   Conclusion

We studied the possibility of reducing the training complexity of ECOC ensembles with highly redundant codes such as exhaustive cumulative, exhaustive and random codes. We proposed an algorithm for generating a so-called training graph, in which edges are labeled with training cost and nodes represent (sub-)classifiers. By finding an approximate Steiner Tree of this graph in a greedy manner, the training complexity can be reduced without changing the prediction quality. An initial evaluation with Hoeffding Trees, as an example for an incremental learner, yielded time savings in the range of 44.6% to 85.8%. Subsequently, we also demonstrated how SVMs can be adapted for this scenario by reusing weights and by employing an ensemble caching strategy. With this approach, the time savings for LIBSVM ranged from 31.4% to 78.4%. In general, we can expect higher gains for incremental base learners whose complexity grows more steeply with the number of training instances. The presented approach is useful for all considered high-redundant code types, and also for random codes, for which the impact of the GSTEINER algorithm decreases with increasing training instances. In addition, the generation of a min-redundant training scheme could be seen as a pre-processing step, such that it is not counted or only counted once for the total training time of an ECOC ensemble, because it is reusable and independent of the base learner.

However, this approach has its limitations. GSTEINER can be a bottleneck for problems with a high class count, since its complexity is $O(n^2 \cdot k^2)$ and the length n for common code types such as exhaustive codes grow exponentially in the number of classes k. And, this work considers only highly redundant code types, which are not unproblematic. First, usually in conjunction with ECOC ensembles, one prefers diverse classifiers, which are contrasting the redundant codes in our sense. The more shared code configurations exist in an ensemble, the less independent are its classifiers. Secondly, these codes are not as commonly used as the low-redundant decompositions schemes one-against-all and one-against-one.

Acknowledgments. This work was supported by the *German Science Foundation (DFG)*.

References

1. Allwein, E.L., Schapire, R.E., Singer, Y.: Reducing multiclass to binary: A unifying approach for margin classifiers. J. Mach. Learn. Res (JMLR) 1, 113–141 (2000)
2. Bifet, A., Holmes, G., Kirkby, R., Pfahringer, B.: MOA: Massive Online Analysis. J. Mach. Learn. Res., JMLR (2010), http://sourceforge.net/projects/moa-datastream/
3. Blockeel, H., Struyf, J.: Efficient algorithms for decision tree cross-validation. J. Mach. Learn. Res. (JMLR) 3, 621–650 (2003)
4. Chang, C.-C., Lin, C.-J.: LIBSVM: a library for support vector machines (2001), Software, available at http://www.csie.ntu.edu.tw/~cjlin/libsvm
5. Dietterich, T.G., Bakiri, G.: Solving multiclass learning problems via error-correcting output codes. J. Artif. Intell. Res. (JAIR) 2, 263–286 (1995)
6. Domingos, P., Hulten, G.: Mining high-speed data streams. In: KDD, Boston, MA, USA, pp. 71–80. ACM, New York (2000)
7. Frank, A., Asuncion, A.: UCI machine learning repository (2010)
8. Friedman, J.H.: Another approach to polychotomous classification. Technical report, Department of Statistics, Stanford University, Stanford, CA (1996)
9. Fürnkranz, J.: Round robin classification. J. Mach. Learn. Res. (JMLR) 2, 721–747 (2002)
10. Joachims, T.: Making large-scale SVM learning practical. In: Schölkopf, B., Burges, C., Smola, A. (eds.) Advances in Kernel Methods - Support Vector Learning, pp. 169–184. MIT Press, Cambridge (1999)
11. Park, S.-H., Fürnkranz, J.: Efficient decoding of ternary error-correcting output codes for multiclass classification. In: Buntine, W.L., Grobelnik, M., Mladenić, D., Shawe-Taylor, J. (eds.) ECML/PKDD-09, Part II, Bled, Slovenia, pp. 189–204. Springer, Heidelberg (2009)
12. Park, S.-H., Weizsäcker, L., Fürnkranz, J.: Exploiting code-redundancies in ECOC for reducing its training complexity using incremental and SVM learners. Technical Report TUD-KE-2010-06, TU Darmstadt (July 2010)
13. Pimenta, E., Gama, J., Carvalho, A.: Pursuing the best ecoc dimension for multiclass problems. In: Wilson, D., Sutcliffe, G. (eds.) FLAIRS Conference, pp. 622–627. AAAI Press, Menlo Park (2007)
14. Wong, R.: A dual ascent approach for steiner tree problems on a directed graph. Mathematical Programming 28(3), 271–287 (1984)

Concept Convergence in Empirical Domains

Santiago Ontañón and Enric Plaza

IIIA, Artificial Intelligence Research Institute
CSIC, Spanish Council for Scientific Research
Campus UAB, 08193 Bellaterra, Catalonia, Spain
{santi,enric}@iiia.csic.es

Abstract. How to achieve shared meaning is a significant issue when more than one intelligent agent is involved in the same domain. We define the task of *concept convergence*, by which intelligent agents can achieve a shared, agreed-upon meaning of a concept (restricted to empirical domains). For this purpose we present a framework that, integrating computational argumentation and inductive concept learning, allows a pair of agents to (1) learn a concept in an empirical domain, (2) argue about the concept's meaning, and (3) reach a shared agreed-upon concept definition. We apply this framework to marine sponges, a biological domain where the actual definitions of concepts such as orders, families and species are currently open to discussion. An experimental evaluation on marine sponges shows that concept convergence is achieved, within a reasonable number of interchanged arguments, and reaching short and accurate definitions (with respect to precision and recall).

1   Introduction

How to achieve shared meaning is a significant issue when more than one intelligent agent is involved in the same domain. In this paper we focus on empirical domains, where intelligent agents are able to learn, in an individual way, the concepts that are relevant to describe that domain from examples. In this scenario, two or more agents will require some process for sharing, comparing, critiquing and (eventually) agreeing on the meaning of the concepts of a domain. Our proposal is that an agent communication process based on argumentation supports the required aspects to find a shared, agreed-upon meaning of concepts.

For instance, in zoology, the definition of "manta ray" (the largest species of ray) has been a subject of debate; another example is in the domain of astronomy, where the definition of "planet" has been subject of recent debate. If more than one expert is to collaborate in these domains, they need to reach a shared definition of these concepts. Notice that these examples do not deal with the issue of *ontology alignment* (where different names or terms for the same concept are aligned); rather, the debate is about the meaning and scope (with respect to an empirical domain) of a particular concept. In this article we propose a framework intended to model a particular kind of process to reach this shared meaning we call concept convergence.

B. Pfahringer, G. Holmes, and A. Hoffmann (Eds.): DS 2010, LNAI 6332, pp. 281–295, 2010.

We will define the task of *concept convergence* as follows: Given two or more individuals which have individually learned non-equivalent meanings of a concept C from their individual experience, find a shared, equivalent, agreed-upon meaning of C. Two agents achieve concept convergence when (a) they share a concept C within some shared terminology, (b) their individual meanings for C are equivalent in a field of application, and (c) each agent individually accepts this agreed-upon meaning. Notice that concept convergence is less general than the complex discussion on how many species of manta ray should be recognized or how should be defined the concept of planet; however, it is more clearly specified and we will show it can be automated for empirical domains[1].

The task of concept convergence can be performed by the integration of computational argumentation and inductive concept learning. We have developed A-MAIL, a framework allows the agents to argue about the concept they learn using induction [7]. A-MAIL is a unified framework where autonomous agents learn from experience, solve problems with their learnt hypotheses, autonomously generate arguments from experience, communicate their inductive inferences, and argue about them in order to reach agreements with other agents.

The remainder of this paper is organized as follows. First we formally define concept convergence. Then our empirical argumentation framework A-MAIL is described. Then we motivate the usefulness of concept convergence in the biological domain of marine sponges, including an experimental evaluation of two inductive agents arguing about definitions of several concepts. The paper closes with related work, conclusions and future work.

2   Concept Convergence

Our approach integrates notions and techniques from two distinct fields of study — namely inductive learning and computational argumentation— to develop a new approach to achieve concept convergence. We will define the meaning and definition of concepts in the framework of inductive concept learning, which is the process by which given an *extensional definition* of a concept C then an *intensional* definition of a concept C expressed in an ontology $\mathcal{O}$ is found.

Let $\mathcal{E} = \{e_1...e_M\}$ be a field of application composed of M individuals described in an ontology $\mathcal{O}$ and let $C \in \mathcal{O}$ be a concept: an *extensional description* of C is a subset of individuals $E^+ \subset \mathcal{E}$ that are instances of C. E^+ are called (positive) examples of C, while the rest of the examples $E^- = \mathcal{E} - E^+$ are called counterexamples (or negative examples).

---

[1] Notice that ontology alignment (or matching) is a related topic but it focuses on determining correspondences between concepts [3]. As such, alignment's main goal is to establish a "concept name correspondence" relationship such that a semantic interoperability is achieved by being capable of substituting a concept name by a corresponding name. Concept convergence is different, we assume that the individual members of a multiagent system have a common concept vocabulary, but they still do not share a precise shared *definition* of some concept(s).

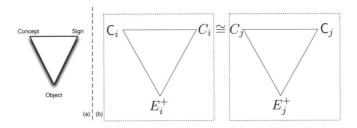

Fig. 1. (a) Semiotic triangle; (b) schema for two agents where a concept sign (C) is shared $(C_i \cong C_j)$ while concept descriptions may be divergent $(\mathsf{C}_i \not\cong \mathsf{C}_i)$

Definition 1. *An* intensional definition C *of a concept* C *is a well formed formula built using the concepts in* O *such that it subsumes* $(\sqsubseteq)$ *all positive examples of* C *and no counterexample of* C:

$$\forall e_i \in E^+ : \mathsf{C} \sqsubseteq e_i \ \wedge \ \forall e_j \in E^- : \mathsf{C} \not\sqsubseteq e_j$$

For simplicity, we will shorten the previous expression as follows: $\mathsf{C} \sqsubseteq E^+ \wedge \mathsf{C} \not\sqsubseteq E^-$. In this framework, we will define the task of concept convergence between 2 agents based on the notion of *semiotic triangle*. The well-known semiotic triangle in Fig. 1(a) expresses meaning as the relationship between sign, concept, and object. Specifically:

1. A *sign* is a designation of the concept in some ontology (in our framework the name of the concept $C \in \mathcal{O}$);
2. A *concept* is "A unit of thought constituted through abstraction on the basis of properties common to a set of objects" [ISO 5963:1985] (in our framework the intensional description C)
3. An *object* is a material or immaterial part of the perceived world (in our framework, the objects in $\mathcal{E}$)

Now, concept convergence between 2 agents means that each one has its own semiotic triangle concerning a particular concept, as shown in Fig. 1(b). We assume that both agents share the designation of the concept C in an ontology, which in Fig. 1(b) is expressed by the equivalence $C_i \cong C_j$. The agents do not share their intensional definitions of the concept —which we'll assume are consistent with their extensional representations of concepts E_i^+ and E_j^+. Moreover, the agents do not share their individual collections of examples E_i and E_j.

Definition 2. Concept Convergence *(between 2 agents) is defined as follows:*

Given *two agents* $(A_i$ *and* $A_j)$ *that agree on the sign* C *denoting a concept* $(C_i \cong C_j)$ *and with individually different intensional* $(\mathsf{C}_i \not\cong \mathsf{C}_i)$ *and extensional* $(E_i^+ \neq E_j^+)$ *definitions of that concept,*

Find *a convergent, shared and agreed-upon intensional description* $(\mathsf{C}_i' \cong \mathsf{C}_j')$ *for* C *that is consistent for each individual with their extensional descriptions.*

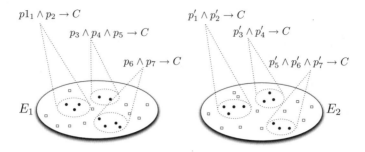

Fig. 2. The relationship of concept definitions for two inductive agents

For example, in this paper we used the domain of marine sponge identification. The two agents need to agree on the definition of the target concept $C =$ *Hadromerida*, among others. While in ontology alignment the focus is on establishing a mapping between the ontologies of the two agents, here we assume that the ontology is shared, i.e. both agents share the concept name *Hadromerida*. Each agent has experience in a different area (one in the Atlantic, and the other in the Mediterranean), so they have collected different samples of Hadromerida sponges, those samples constitute their extensional definitions (which are different, since each agent has collected sponges on their own). Now they want to agree on an intensional definition C, which describes such sponges. In our experiments, one such intensional definition reached by one of the agents is: $C =$ "all those sponges which do not have gemmules in their external features, whose megascleres had a tylostyle smooth form and that do not have a uniform length in their spikulate skeleton".

2.1   Empirical Argumentation for Concept Convergence

Concept convergence in empirical domains is modeled by agents that perform induction to achieve intensional definition of one or more concepts. Figure 2 shows the relationship of concept definitions for two inductive agents concerning a concept C. Each agent has a sample of examples of C and examples that are not C. The task of concept convergence is to find a shared and mutually acceptable definition for C that is consistent with the examples each agent has. The information exchanged during argumentation about how C should be defined is the information that will enact a process of belief revision in each individual agent until an agreed-upon definition is achieved. This paper focuses on 2-agent argumentation, leaving concept convergence among more agents for future work.

In the A-MAIL framework, an intensional definition of a concept C is represented as a *disjunctive description* $C = r_1 \vee ... \vee r_n$, where each of the conjuncts r_i will be called a *generalization*, such that each positive example of C is subsumed by at least one of the generalizations, and no generalizations subsume any counterexample of C. When an example is subsumed by a generalization in C, we will say that the example is *covered*. Each one of these generalizations is a well formed formula representing a *generalization* of a set of examples. We assume that a *more-general-than* relation (*subsumption*) exists among generalizations, and when a generalization r_1 is more general than another generalization r_2 we write $r_1 \sqsubseteq r_2$. Additionally, if a generalization r is a generalization of an example e, we will also say that r is more general than e, or that r subsumes or covers e, noting it as $r \sqsubseteq e$. Moreover, for practical purposes the intensional definitions are allowed to subsume less than 100% of positive examples.

Concept convergence is assessed individually by an agent A_i by computing the *individual degree of convergence* among two definitions C_i and C_j as:

Definition 3. *The* individual degree of convergence *among two intensional definitions C_i and C_j for an agent A_i is:*

$$K_i(C_i, C_j) = \frac{|\{e \in E_i | C_i \sqsubseteq e \wedge C_j \sqsubseteq e\}|}{|\{e \in E_i | C_i \sqsubseteq e \vee C_j \sqsubseteq e\}|}$$

where K_i is 0 if the two definitions are totally divergent, and 1 when the two definitions are totally convergent. The degree of convergence corresponds to the ratio between the number examples covered by both definitions (intersection) and the number of examples covered by at least one definition (union). The closer the intersection is to the union, the more similar the definitions are.

Definition 4. *The* joint degree of convergence *of two intensional definitions C_i and C_j is:*

$$K(C_i, C_j) = min(K_i(C_i, C_j), K_j(C_j, C_i))$$

Concept convergence is defined as follows:

Definition 5. *Two intensional definitions are* convergent ($C_i \cong C_j$) *if $K(C_i, C_j)$ $\geq 1 - \epsilon$, where $0 \leq \epsilon \leq 1$ is a the degree of divergence allowed.*

3   Empirical Argumentation

An argumentation framework $AF = \langle A, R \rangle$ is composed by a set of arguments A and an attack relation R among the arguments. In our approach we will adopt the semantics based on dialogical trees [1]. For a wider explanation the formal model underlying our framework see [5].

There are two kinds of arguments in A-MAIL:

- A *rule argument* $\alpha = \langle r, \overline{C} \rangle$ is a pair where r is a generalization and $\overline{C} \in \{C, \neg C\}$. An argument $\langle r, C \rangle$ states that induction has found a rule such that $r \rightarrow C$ (i.e. that examples covered by r belong to C), while $\langle r, \neg C \rangle$ states that induction has found a rule such that $r \rightarrow \neg C$ (i.e. that examples covered by r do not belong to C).
- An *example argument* $\alpha = \langle e, \overline{C} \rangle$ consists of an example $e \in \mathcal{E}$, which can be a positive or a negative example of C, i.e. $\overline{C} \in \{C, \neg C\}$.

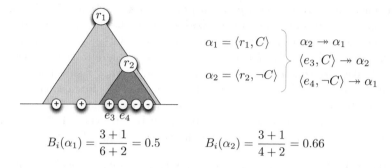

$$\alpha_1 = \langle r_1, C \rangle \left.\begin{array}{c} \\ \\ \end{array}\right\} \quad \begin{array}{c} \alpha_2 \twoheadrightarrow \alpha_1 \\ \langle e_3, C \rangle \twoheadrightarrow \alpha_2 \\ \langle e_4, \neg C \rangle \twoheadrightarrow \alpha_1 \end{array}$$

$$\alpha_2 = \langle r_2, \neg C \rangle$$

$$B_i(\alpha_1) = \frac{3+1}{6+2} = 0.5 \qquad B_i(\alpha_2) = \frac{3+1}{4+2} = 0.66$$

Fig. 3. Exemplification of several arguments, their confidences, and attack relations

Moreover, we allow rules to cover some negative examples, while defining a confidence measure as follows:

Definition 6. *The confidence $B_i(\alpha)$ of a rule argument α for an agent A_i is:*

$$B_i(\alpha) = \begin{cases} \frac{|\{e \in E_i^+ | \alpha.r \sqsubseteq e\}|+1}{|\{e \in E_i | \alpha.r \sqsubseteq e\}|+2} & \textit{if } \alpha.\overline{C} = C \\[2ex] \frac{|\{e \in E_i^- | \alpha.r \sqsubseteq e\}|+1}{|\{e \in E_i | \alpha.r \sqsubseteq e\}|+2} & \textit{if } \alpha.\overline{C} = \neg C \end{cases}$$

$B_i(\alpha)$ is the ratio of examples correctly covered by α over the total number examples covered by α. Moreover, we add 1 to the numerator and 2 to the denominator following the Laplace probability estimation procedure. Other confidence measures could be used, our framework only requires some confidence measure that reflects how much a set of examples endorses the argument.

Definition 7. *A rule argument α is τ-acceptable for an agent A_i if $B_i(\alpha) \geq \tau$, where $0 \leq \tau \leq 1$.*

In our framework, only τ-acceptable generalizations are allowed for a predetermined threshold τ. To ensure only highly quality rules are considered. Next, we will define attacks between arguments.

Definition 8. *An attack relation ($\alpha \twoheadrightarrow \beta$) between arguments α, β holds when:*
1. *$\langle r_1, \widehat{C} \rangle \twoheadrightarrow \langle r_2, \overline{C} \rangle \Longleftrightarrow \widehat{C} = \neg \overline{C} \wedge r_2 \sqsubset r_1$, or*
2. *$\langle e, \widehat{C} \rangle \twoheadrightarrow \langle r, \overline{C} \rangle \Longleftrightarrow \widehat{C} = \neg \overline{C} \wedge r \sqsubseteq e$*
(where $\overline{C}, \widehat{C} \in \{C, \neg C\}$)

Notice that a rule argument α only attacks another argument β if $\beta.r \sqsubset \alpha.r$, i.e. when β is a strictly more general argument than α. This is required since it implies that all the examples covered by α are also covered by β, and thus if they support opposing concepts, they must be in conflict.

Figure 3 exemplifies some arguments and with their corresponding attacks. Positive examples of the concept C are marked with a positive sign, whereas negative examples are marked with a negative sign. Rule arguments are represented as triangles covering examples; when an argument α_1 subsumes another argument α_2, we draw α_2 inside of the triangle representing α_1. Argument α_1 has a generalization r_1 supporting C, which covers 3 positive examples and 3 negative examples, and thus has confidence 0.5, while argument α_2 has a generalization r_2 supporting $\neg C$ with confidence 0.66, since it covers 3 negative examples and only one positive example. Two example arguments are shown: $\langle e_3, C \rangle$ and $\langle e_4, \neg C \rangle$. Now, $\alpha_2 \twoheadrightarrow \alpha_1$ because α_2 supports $\neg C$, α_1 supports C and $r_1 \sqsubset r_2$. Additionally $\langle e_3, C \rangle \twoheadrightarrow \alpha_2$, since e_3 is a positive example of C, α_2 supports $\neg C$ and $r_2 \sqsubseteq e_3$.

Next we will summarily define when arguments *defeat* other arguments, based on the idea of argumentation lines [1]. An *Argumentation Line* $\alpha_n \twoheadrightarrow \alpha_{n-1} \twoheadrightarrow \ldots \twoheadrightarrow \alpha_1$ is a sequence of arguments where α_i attacks α_{i-1} and α_1 is called the *root*. Notice that odd arguments are generated by the agent whose generalization is under attack (the *proponent*) and the even arguments are generated by the agent attacking that generalization (the *opponent*).

Moreover, an α-*rooted argumentation tree* T is a tree where each path from the root node α to one of the leaves constitutes an argumentation line rooted on α. Therefore, a set of argumentation lines rooted in the same argument α_1 can be represented as an argumentation tree, and vice versa. Notice that example arguments may appear only in the leaves of an argumentation tree. The example-free argumentation tree T^f corresponding to T is a tree rooted in α that contains the same rule arguments of T but no example arguments.

In order to determine whether the root argument α is warranted (undefeated) or defeated the nodes of the α-rooted tree are marked U (undefeated) or D (defeated) according to the following (cautious) rules: (1) every leaf node is marked U; (2) each inner node is marked U iff all of its children are marked D, otherwise it is marked D.

Finally we will define the status of the argumentation among two agents A_i and A_j at an instant t as the tuple $\langle R_i^t, R_j^t, G^t \rangle$, consisting of:

- $R_i^t = \{\langle r, C \rangle | r \in \{r_1, ..., r_n\}\}$, the set of rule arguments representing the current intensional definition $C_i^t = r_1 \vee ... \vee r_n$ for agent A_i.
- G^t contains the collection of arguments generated before t by either agent, and belonging to a tree rooted in an argument in $R_i^{t'}$, where $t' \leq t$.

R_j^t is the same for agent A_j. Now we can turn to integrate inductive learning with computational argumentation.

3.1   Argument Generation through Induction

Agents need two kinds of argument generation capabilities: generating an initial intensional definition from examples, and generating attacks to arguments.

When an agent A_i that wants to generate an argument β that attacks another argument α, β has to satisfy four conditions: a) support the opposite concept than α, b) have a high confidence $B_i(\beta)$ (at least being τ-acceptable), c) satisfy

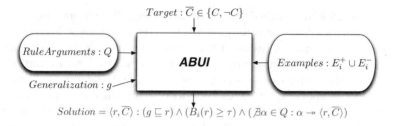

$$Target : \overline{C} \in \{C, \neg C\}$$

$$RuleArguments : Q$$

ABUI

$$Examples : E_i^+ \cup E_i^-$$

$$Generalization : g$$

$$Solution = \langle r, \overline{C} \rangle : (g \sqsubseteq r) \wedge (B_i(r) \geq \tau) \wedge (\nexists \alpha \in Q : \alpha \twoheadrightarrow \langle r, \overline{C} \rangle)$$

Fig. 4. ABUI is an inductive concept learning algorithm which can take additional background knowledge, in the form of arguments, into account

$\beta \twoheadrightarrow \alpha$, and d) β should not be defeated by any argument previously generated by any of the agents. Existing inductive learning techniques cannot be applied out of the box for this process, because of the additional restrictions imposed. For this purpose, we developed the Argumentation-based Bottom-up Induction (ABUI) algorithm, capable of performing such task [7]. However, any algorithm which can search the space of rules, looking for one which satisfies the four conditions stated before would work in our framework.

ABUI is an inductive method for concept learning which, in addition to training examples, can take into account additional background knowledge in the form of arguments (see Fig. 4). ABUI is a bottom-up inductive learning method, which tries to generate rules that cover positive examples by starting from a positive example and generalizing it as much as possible in order to cover the maximum number of positive examples and while covering the minimum number of negative examples possible. During this generalization process, ABUI only considers those generalization which will lead to arguments not being defeated by any rule in the background knowledge. Specifically, ABUI takes 4 input parameters: a target concept $\overline{C} \in \{C, \neg C\}$, a set of examples $E_i^+ \cup E_i^-$, a generalization g, and a set of arguments Q which both agents have agreed to be true. ABUI finds (if it exists) an argument $\beta = \langle r, \overline{C} \rangle$ such that: $(g \sqsubseteq r) \wedge (B_i(r) \geq \tau) \wedge (\nexists \alpha \in Q : \alpha \twoheadrightarrow \langle r, \overline{C} \rangle)$.

To generate a β such that $\beta \twoheadrightarrow \alpha$, the agent calls ABUI with $g = \alpha.r$ and with the set of agreed upon arguments Q (the subset of arguments in G^t which are undefeated).

- If ABUI returns an individually τ-acceptable β, then β is the attacking argument to be used.
- If ABUI fails to find an argument, then A_i looks for examples attacking α in E_i. If any exist, then one such example is randomly chosen to be used as an attacking argument.

Otherwise, A_i is unable to generate any argument attacking α.

3.2   Belief Revision

During argumentation, agents exchange arguments which contain new rules and examples. The Belief Revision process of an agent A_i triggered at an instant t, with an argumentation state $\langle R_i^t, R_j^t, G^t \rangle$ works as follows:

1. Each example argument in G_i^t is added to E_i, i.e. A_i expands its extensional definition of C.
2. Since E_i might have changed, the confidence in any argument in R_i^t or G^t might have changed. If any of these arguments becomes not individually τ-acceptable they removed from R_i^{t+1} and G^{t+1}.
3. If any argument α in R_i^t became defeated, and A_i is not able to expand the argumentation tree rooted in α to defend it, then the rule $\alpha.r$ will be removed from C_i. As a consequence, some positive examples in E_i will not be covered by C_i any longer. Then ABUI is called with the now uncovered examples to find new rules that cover them and that will be added to C_i.

3.3   Concept Convergence Argumentation Protocol

The concept convergence argumentation process follows an iterative protocol composed of a series of rounds, during which two agents argue about the individual rules that compose their intensional definitions of a concept C. At every round t of the protocol, each agent A_i holds a particular intensional definition C_i^t, and only one agent will hold a *token*. The holder of the token can assert new arguments in the current round. At the end of each round the token is passed on to the other agent. This cycle continues until $C_i \cong C_j$.

The protocol starts at round $t = 0$ and works as follows:

1. Each agent A_i communicates their current intensional definition by sharing R_i^0. The token goes to one agent at random, and the protocol moves to 2.
2. The agents share $K_i(C_i, C_j)$ and $K_j(C_j, C_i)$, their individual convergence degrees. If $C_i \cong C_j$ the protocol ends with success; if no agent has produced a new attack in the last two rounds then the protocol ends with failure; otherwise it moves to 3.
3. If modified by belief revision, the agent with the token, A_i, communicates its current intensional definition R_i^t. Then, the protocol moves to 4.
4. If any argument $\alpha \in R_i^t$ is defeated, and A_i can generate an argument α' to defend α, α' is sent to A_j. Also, if any of the undefeated arguments $\beta \in R_j^t$ is not individually τ-acceptable for A_i, and A_i can find an argument β' to extend any β-rooted argumentation line, in order to attack β, then β' is sent to A_j. If any of these arguments was sent, a new round $t+1$ starts; the token is given to the other agent, and the protocol moves back to 2. Otherwise the protocol moves to 5.
5. If there is any example $e \in E_i^+$ such that $C_j^t \not\sqsubseteq e$, A_i sends e to A_j (since the intentional definition of A_j does not cover e). A new round $t+1$ starts, the token is given to the other agent, and the protocol moves to 2.

Moreover, in order to ensure termination, no agent is allowed to send twice the same argument. A-MAIL ensures that the convergence of the resulting concepts is at least τ if (1) the number of examples is finite, (2) the number of rules that can be generated is finite. Convergence higher than τ cannot be ensured, since $100 \times (1 - \tau)\%$ of the examples covered by a τ-acceptable rule might be negative.

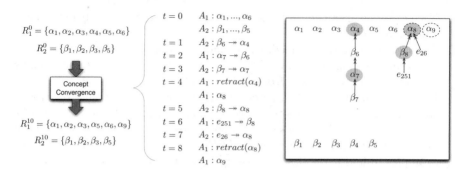

Fig. 5. An example concept convergence argumentation. Left hand side shows starting point and result. The middle shows the list of messages exchanged during the protocol. Right hand side shows the resulting argumentation trees.

Even when both agents use different inductive algorithms, convergence is assured since by assumption they are using the same finite generalization space, and there is no rule τ-acceptable to one agent that could not be τ-acceptable to the other agent when both know the same collection of examples.

An example process of concept convergence is shown in Fig. 5. On the left hand side are the arguments (concept definition) of each agent before and after. In the middle, Fig. 5 shows the messages exchanged during the protocol, and on the right hand side the argumentation trees used. We can see that in round $t = 0$ the agents just exchange the arguments that compose their concept definitions. Then, in rounds 1, 2 and 3, the agents are arguing about the argument α_4, when ends up being defeated (shaded node). As a consequence, agent A_1 retracts α_4 and proposes a new one, α_8 (dashed node). The agents argue about α_8 in rounds 5 to 7, and eventually α_8 is defeated. Finally, agent A_1 retracts α_8, and proposes a new argument α_9, which is accepted (not attacked) by A_2. In this example, A_1 does not attack any argument in the definition of agent A_2.

4   Concept Convergence for Marine Sponges

Marine sponge classification poses a challenge to benthologists because of the incomplete knowledge of many of their biological and cytological features, and due to the morphological plasticity of the species. Moreover, benthology specialists are distributed around the world and they have experience in different benthos that spawn species with different characteristics due to the local habitat conditions. Due to these problems, the classification or sponges into different classes is a challenging problem which is still under discussion among specialists.

The problem that we use as our test bed is that of learning which are the features that distinguish the different orders of sponges among each other, i.e. finding their intensional definition. We will focus on the scenario where two different experts have collected sponges in different locations and that these sponges are properly classified into their respective orders. Now, the two experts

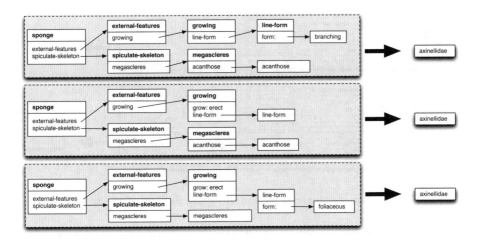

Fig. 6. Original concept definition learnt by an agent for the Axinellidae class, and composed of three rules

are interested in having a specific agreed definition of each of the different order of sponges, so that their classification is clear in the future.

We have designed an experimental suite with a collection of 280 marine sponges pertaining to three different orders of the Demospongiae class (Astrophorida, Hadromerida and Axinellidae), taken from the Demospongiae dataset from the UCI repository. For our evaluation, we divide this collection of sponges in two disjoint sets, and give each set to one agent, which corresponds to an expert. Given a target order, say Axinellidae, each agent learns by induction a definition which characterizes all the sponges belonging to that order, and does not cover any sponge from any other order. After that, both agents argue about those definitions to reach an agreement using A-MAIL. The expected result is that the definition they reach after argumentation is better than the definitions they found individually (it is in agreement with the data known to both agents), and that it is achieved without exchanging large amounts of information.

Figure 6 shows an example definition of Axinellidae found by one agent in our experiments. The definition is composed of three rules. The first one, for instance states that "all the sponges which have a branching line-form growing and acanthose in the megascleres in the spikulate-skeleton" are Axinellidae.

Figure 7 shows two arguments (α_3 and β_4) as generated in one of our experiments by 2 agents while arguing about the definition of the Axinellidae order. An agent A_1 had proposed α_3, stating that "all the sponges which have a branching line-form growing and megascleres in the spikulate skeleton" are Axinellidae. This was so, since this rule was consistent with A_1 knowledge, i.e. with the set of sponges A_1 knew. However, this rule turned out to be too general, since it covered some sponges known to the other agent, A_2, which were not Axinellidae. In order to attack this rule, agent A_2 generated the argument β_4, which states that "all the sponges which have a branching line-form growing, a hand, and a

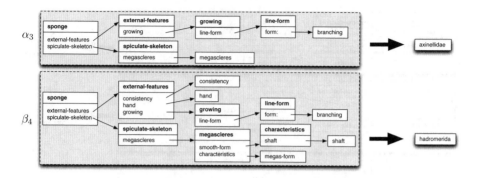

Fig. 7. Examples of arguments α_3 and β_4, where β_4 is attacking argument α_3

Table 1. Precision (P), Recall (R) and degree of convergence (K) for the intensional definitions obtained using different methods

Concept	Centralized		Individual			A-MAIL		
	P	R	P	R	K	P	R	K
Axinellidae	0.98	1.00	0.97	0.95	0.80	0.97	0.95	0.89
Hadromerida	0.85	0.98	0.89	0.91	0.78	0.92	0.96	0.97
Astrophorida	0.98	1.00	0.97	0.97	0.93	0.98	0.99	0.97

shaft in the smooth form of the megascleres" are actually Hadromeridae. Since A_1 could not attack β_4, α_3 is defeated.

4.1   Experimental Evaluation

We perform concept convergence on each of the 3 orders in the marine sponges data set: Astrophorida, Hadromerida and Axinellidae. In an experimental run, we randomly split the data among the two agents and, given a target concept, the goal of the agents was to reach a convergent definition of such concept. We compare the results of A-MAIL with respect to agents which do not perform argumentation (*Individual*), and to the result of centralizing all the examples and performing centralized concept learning (*Centralized*). Comparing the results of *Individual* agents and agents using A-MAIL provides a measure of the benefits of A-MAIL, whereas comparing with *Centralized* gives a measure of the quality of the outcome. All the results are the average of 10 executions, $\epsilon = 0.05$ and $\tau = 0.75$. We used the same induction algorithm, ABUI, for all the experiments.

Table 1 shows one row for each of the 3 concepts we used in our evaluation; for each one we show three values: precision, (P, how many examples covered that are actually positive examples); recall, (R, how many positive examples in the data set are covered by the definition); and convergence degree (K, as defined in Definition 4). The first thing we see is that indeed A-MAIL is able to increase convergence from the Individual setting. Moreover, for all concepts except for Axinellidae the convergence degree is higher than 0.95 (i.e. $1 - \epsilon$).

Table 2. Comparison of the cost and quality of obtaining intensional definition from examples using different settings. Cost is measured in time (in seconds), and for A-MAIL, also the average number of example arguments (NE) and rule arguments (NR) exchanged. Quality is measured by the average number of rules (R) in intensional definitions.

Concept	Centralized		Individual		A-MAIL			
	time	R	time	R	time	R	NE	NR
Axinellidae	82.3s	7	40.8s	4.10	65.2s	6.65	10.7	15.6
Hadromerida	173.3s	11	75.6s	6.15	164.8s	9.2	18.5	32.6
Astrophorida	96.7s	6	47.7s	7.00	50.6s	4.1	4.1	9.7

100% convergence is not reached because $\tau = 0.75$ in our experiments. This means that acceptable rules can cover some negative examples, which allows for the appearance of some divergence. Increasing τ could improve convergence but makes finding rules by induction more difficult, and thus recall might suffer. Finally, notice that argumentation also improves precision and recall that reach values close to the ones achieved by Centralized.

Table 2 shows the average cost of each of the three settings. Column *time* shows the average CPU time used in each execution; when there are 2 agents (in the Individual and A-MAIL settings) individual time is obtained dividing 2. The Centralized setting uses more time on average than either Individual or A-MAIL settings. Table 2 also shows the average number of examples and of rule arguments exchanged among the agents, showing that A-MAIL only requires the exchange of a small amount of examples and arguments in order to converge.

Quality of solution is estimated by compactness of concept descriptions. The definitions found by A-MAIL are more compact (have less rules) than the definitions found by a Centralized approach. For instance, for the concept Astrophorida, the Centralized setting obtains a definition consisting of 6 rules, whereas A-MAIL generates only 4.1 rules on average.

In summary, we can conclude that A-MAIL successfully achieves concept convergence. In addition to improve the quality of the intensional definition (precision and recall), this is achieved by exchanging only a small percentage of the examples the agents know (as opposed to the centralized strategy where all the examples are given to a single agent, which might not be feasible in some applications). Moreover, the execution time of A-MAIL is on average lower than that of a centralized strategy. An interesting implication of this is that A-MAIL could be used for distributed induction, since it achieves similar results than a centralized approach, but at a lower cost, and in a distributed fashion.

5   Related Work

In our approach to concept convergence, we used our A-MAIL framewok [7]. A-MAIL is a framework which integrates inductive learning techniques with computational argumentation. In previous work, we applied A-MAIL to the task of distributed inductive learning, where agents are interested in benefitting from data known to other agents in order to improve performance. In this paper, we have used A-MAIL for a different task: concept convergence, where the goal is for two agents to *coordinate* their definitions of specific concepts. This process can be used, as we have shown, to model the process of argumentation between biology specialists about the definition of specific species. However, A-MAIL can be used for other tasks such as joint deliberation (when agents what to reach an agreement on a specific decision to a particular problem).

The integration of arguments into a machine learning framework is a recent idea, receiving increasing attention, as illustrated by the argument-based machine learning framework [4]. The main difference between this framework and A-MAIL is that in argument-based machine learning, arguments are given as the *input* of the learning process, while A-MAIL *generates* arguments by induction and uses them to reach agreements among agents.

Our work is also related to multiagent inductive learning. One of the earliest in this area was MALE [9], in which a collection of agents tightly cooperated during learning, effectively operating as if there was a single algorithm working on all data. Similar to MALE, DRL [8] is a distributed rule learning algorithm based on finding rules locally and then sending them to the other agents for evaluation. The idea of merging theories for concept learning has been also studied in the framework of Version Spaces [2].

6   Conclusions

This paper has presented the task of concept convergence. Concept convergence is different from ontology alignment in that we are not trying to find correspondence between ontologies, but reach shared definitions to known concepts. Since concept convergence is a broad subject we have focused on empirical domains. We have proposed to use inductive learning techniques to represent concepts and computational argumentation to regulate the communication process. For this purpose we have summarized A-MAIL, a framework that integrates inductive learning and computational argumentation; this integration is achieved by (1) considering rules learned by inductive learning as arguments, and (2) developing inductive learning techniques that are able to find new generalizations that are consistent with or attack a given set of arguments.

We have motivated the approach in the biological domain of marine sponges, where definitions of taxonomic concepts are still under debate. Experiments in this domain show that computational argumentation integrated with induction is capable of solving the concept convergence task, and the process is efficient (in the sense of the number of arguments that need to be exchanged).

As part of our future work, we intend to investigate more complex settings of concept convergence, and other tasks than can be performed by integrating induction with argumentation. Concerning concept convergence, we have started by focusing on the 2-agent scenario, but we intend to investigate concept convergence for n agents. Since computational argumentation is traditionally modeled as a dialogue between 2 agents, moving to a n-agents scenario requires more complex interaction models, such as those of committees (following argumentation-based deliberation in committees as in [6]). Another avenue of research is convergence on more than one concept; when these concepts are interdependent we surmise our current approach would work when dependencies are not circular; circular dependencies would require a more sophisticated approach.

Moreover, integrating induction with argumentation allows other kinds of tasks, such are using argumentation among agents to improve the individual inductive model [7]; another task is deliberative agreement, where 2 or more agents disagree on whether a situation or object is an instance of a concept C and user argumentation to reach an agreement on that issue.

Acknowledgments. This research was partially supported by projects Next-CBR (TIN2009-13692-C03-01), Aneris (PIF08-015-02) and Agreement Technologies (CONSOLIDER CSD2007-0022).

References

[1] Chesñévar, C.I., Simari, G.R., Godo, L.: Computing dialectical trees efficiently in possibilistic defeasible logic programming. In: Baral, C., Greco, G., Leone, N., Terracina, G. (eds.) LPNMR 2005. LNCS (LNAI), vol. 3662, pp. 158–171. Springer, Heidelberg (2005)

[2] Hirsh, H.: Incremental version-space merging: a general framework for concept learning. PhD thesis, Stanford University, Stanford, CA, USA (1989)

[3] Kalfoglou, Y., Schorlemmer, M.: Ontology mapping: The state of the art. In: Kalfoglou, Y., Schorlemmer, M., Sheth, A., Staab, S., Uschold, M. (eds.) Semantic Interoperability and Integration, Dagstuhl Seminar Proceedings, Dagstuhl, Germany, vol. 04391 (2005)

[4] Mozina, M., Zabkar, J., Bratko, I.: Argument based machine learning. Artificial Intelligence 171(10-15), 922–937 (2007)

[5] Ontañón, S., Dellunde, P., Godo, L., Plaza, E.: Towards a logical model of induction from examples and communication. In: Proceedings of the 13th International Conference of the Catalan Association for Artificial Intelligence. Frontiers in Artificial Intelligence. IOS Press, Amsterdam (in press, 2010)

[6] Ontañón, S., Plaza, E.: An argumentation-based framework for deliberation in multi-agent systems. In: Rahwan, I., Parsons, S., Reed, C. (eds.) ArgMAS 2007. LNCS (LNAI), vol. 4946, pp. 178–196. Springer, Heidelberg (2008)

[7] Ontañón, S., Plaza, E.: Multiagent inductive learning: an argumentation-based approach. In: ICML 2010. Omnipress (2010), http://www.icml2010.org/papers/284.pdf

[8] Provost, F.J., Hennessy, D.: Scaling up: Distributed machine learning with cooperation. In: Proc. 13th AAAI Conference, pp. 74–79. AAAI Press, Menlo Park (1996)

[9] Sian, S.S.: Extending learning to multiple agents: Issues and a model for multi-agent machine learning (MA-ML). In: Kodratoff, Y. (ed.) EWSL 1991. LNCS, vol. 482, pp. 440–456. Springer, Heidelberg (1991)

Equation Discovery for Model Identification in Respiratory Mechanics of the Mechanically Ventilated Human Lung[*]

Steven Ganzert[1],[**], Josef Guttmann[2], Daniel Steinmann[2], and Stefan Kramer[1]

[1] Institut für Informatik I12, Technische Universität München, D-85748 Garching b. München, Germany
[2] Department of Anesthesiology and Critical Care Medicine, University Medical Center, Freiburg, Hugstetter Str. 55, D-79106 Freiburg, Germany
steven.ganzert@in.tum.de

Abstract. Lung protective ventilation strategies reduce the risk of ventilator associated lung injury. To develop such strategies, knowledge about mechanical properties of the mechanically ventilated human lung is essential. This study was designed to develop an equation discovery system to identify mathematical models of the respiratory system in time-series data obtained from mechanically ventilated patients. Two techniques were combined: (i) the usage of declarative bias to reduce search space complexity and inherently providing the processing of background knowledge. (ii) A newly developed heuristic for traversing the hypothesis space with a greedy, randomized strategy analogical to the GSAT algorithm. In 96.8% of all runs the applied equation discovery system was capable to detect the well-established equation of motion model of the respiratory system in the provided data. We see the potential of this semi-automatic approach to detect more complex mathematical descriptions of the respiratory system from respiratory data.

Keywords: Equation discovery, declarative bias, GSAT algorithm, respiratory mechanics, mechanical ventilation, lung protective ventilation.

1 Introduction

Mechanical ventilation is the live-saving therapy in intensive care by all means. However, inadequate ventilator settings can induce or aggravate lung injury during mechanical ventilation. To prevent such ventilator associated lung injury (VALI), lung protective ventilation strategies are essential. Such strategies have been shown to considerably improve the outcome of critically ill patients [25]. A prerequisite for such strategies is the analysis of the respiratory mechanics under mechanical ventilation. Such analyses are generally based on mathematical models to describe and interpret the associated mechanisms [11,2,3,15,1]. The equation of motion (EOM) [14] is a commonly accepted mathematical model of

---

[*] http://wwwkramer.in.tum.de/Members/ganzert/publications/
ganzert_etal_ds2010_suppl_material.pdf
[**] Corresponding author.

B. Pfahringer, G. Holmes, and A. Hoffmann (Eds.): DS 2010, LNAI 6332, pp. 296–310, 2010.
© Springer-Verlag Berlin Heidelberg 2010

the respiratory system and provides the basis for most clinically applied methods of respiratory mechanics analysis [30,22]. However, this model has restrictions since it does not include aspects of non-linearity or inhomogeneity. Consequently, refined methods and models are required to analyze respiratory mechanics. Especially monitoring techniques under the dynamic condition of continuous mechanical ventilation at bedside are of interest [23,13].

Equation discovery has been introduced [20] to identify functional relations from parameters observed over the course of time. Equation discovery systems have to face a variety of problems. Fitting time series is eminently affected by the initial parameter settings within the fitting process and also by noise in the data. Various attempts have been made to address this issue [26,27,18,33]. Another problem is the representation of the results. LAGRANGE [9], for example is capable to generate models consisting of higher order differential equations, however the task of solving them remains part of post-processing the results. Another general problem in data mining and machine learning also affecting equation discovery systems is the handling of vast search spaces. Several approaches have been introduced such as coping with irrelevant variables [10] and detecting inconsistencies of units [10,17]. In the same context the use of declarative bias, which is well known from the field of inductive logic programming [24], was introduced to equation discovery [31].

The purpose of this study was to implement an equation discovery system applicable to model respiratory mechanics. An important system requirement was the capability to process background knowledge given by previously developed mechanical lung models and to present such background knowledge in a flexible manner. This would be the methodological prerequisite to identify even more elaborated models within respiratory data. Respiratory data from 13 mechanically ventilated patients with normal lungs was analyzed. A previously developed system [31] was modified. The original system traverses the hypothesis space in a specific sequential order which showed to have a biased impact on the performance of the system. The benchmark test for the modified system consisted of the re-identification of a well known model of respiratory mechanics.

2   Medical Background: The EOM Model

The EOM for the respiratory system describes the relation between airway pressure P_{aw}, applied volume V and respiratory airflow $\dot{V}$ by

$$P_{aw} = V/C + \dot{V} \times R + PEEP \tag{1}$$

where C denotes the so-called compliance (volume distensibility) of the lung and is defined by the quotient of volume change divided by pressure change, i.e. $C = \Delta V / \Delta P$. R denotes the resistance of the respiratory system and $PEEP$ (positive end-expiratory pressure) the offset of the airway pressure at the end of expiration. In medical practice, a commonly accepted strategy for lung protective ventilation is to ventilate the lung in a pressure-volume range where lung compliance is maximal. In this way, less pressure per milliliter of inflated gas volume is needed.

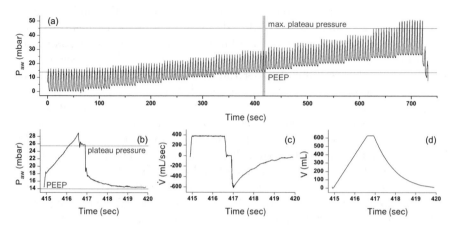

Fig. 1. Sample data of one patient. (a) PEEP-wave maneuver with successively increasing PEEP level in steps of 2 mbar up to a maximum plateau pressure of 45 mbar. (b) Sample pressure curve of one breath at PEEP level of 14 mbar. As it is the last breath before the next change of PEEP, this breath was used for analysis. The plateau pressure level is approximated at the end of the zero-flow phase after inflation. (c), (d) Corresponding flow and volume curve, respectively.

This can be obtained by adequately adjusting the pressure offset (i.e. $PEEP$) and the breath-by-breath applied (tidal) volume at the ventilator.

3   Methodological Background: LAGRAMGE and GSAT

3.1   LAGRAMGE

Declarative bias was introduced to the field of equation discovery by Todorovski and Džeroski with the LAGRAMGE system [31]. The system uses context free grammars to reduce the hypothesis space. The grammars restrict the choice of mathematical models by defining applicable model fragments. Parse trees — representing mathematical equations — are derived from the given grammar and span the hypothesis space. Briefly summarizing, the system operates as follows on an input consisting of two parts:

- The data $D = (M, Var, v_d)$ with $M = \{(t_{1..m}, v_1(t_{1..m}), \ldots, v_k(t_{1..m}))_{1..l}\}$ as set of one ore more time series data sets (Fig. 1), $Var = \{v_1, v_2, \ldots, v_k\}$ as set of domain variables and $v_d \in Var$ as dependent variable.
- A context free grammar $G = (N, T, P, S)$ with N as set of non-terminal symbols, T as set of terminal symbols, P as set of production rules $\{p_1, \ldots, p_l\}$ and $S \in N$ as non-terminal start symbol (Fig. 2).

Model Evaluation. Each production rule $p \in P$ is of the form $A \rightarrow \alpha$. $A \in N$ is called the left side and $\alpha \in (N \cup T)^*$ the right side of the production rule. In the following, P_A denotes the set of production rules $\{p_{A,1}, \ldots, p_{A,l}\}$ having

A as left side. Each parse tree $\mathcal{T}_S$ derivable from the grammar G with the non-terminal start symbol S as root node represents a model, which is fit to the given data by a fitting method $\mathcal{F}$. The quality of the model is estimated by a heuristic function $\mathcal{F}_h$.

Model Refinement. $\mathcal{T}_S$ is refined to $\mathcal{T}_S'$ by choosing a non-terminal node A in $\mathcal{T}_S$, applying the succeeding production rule $p_{A,i+1}$ of the previously applied rule $p_{A,i}$ and terminating the resulting subtree $\mathcal{T}_A$ by iteratively applying the first production rule $p_{B,1} \in P_B$ for all non-terminal leaf nodes B in the expanded successor trees of $\mathcal{T}_A$ until no more non-terminals are left in the leaves.

Beam Search. Applying a beam search [4] of width n, in each iteration successively for all n beam elements, i.e. parse trees $\mathcal{T}_S$, each possible refinement $\mathcal{T}_S'$ is generated by separately refining each non-terminal node A in $\mathcal{T}_S$. From the set of all refinements $\{\mathcal{T}_{S,1}', \cdots, \mathcal{T}_{S,m}'\}$ of all beam elements unified with the set of the beam elements $\{\mathcal{T}_{S,1}, \cdots, \mathcal{T}_{S,n}\}$ of the previous iteration, the n elements with the lowest error in terms of $\mathcal{F}_h$ are stored in the beam for the next iteration. The algorithm stops, if within an iteration none of the beam elements was substituted.

Search Heuristic. Besides the reduction of the search space complexity, the declarative bias applied in this system in terms of a context free grammar implies a specific guidance through the hypothesis space. Preferring short hypothesis, the production rules (and therefore the productions as well) of the grammar are ordered by the minimal height of a parse tree $\mathcal{T}_A$ derived with the production rule $p_{A,i} \in P_A$ at its root node A and thus the productions P_A are rather ordered m-tuples $P_A = (p_1, \ldots, p_m)$ than sets of m production rules $\{p_1, \ldots, p_m\}$. Furthermore, production rules having the same height are ordered by their input sequence given by the grammar just like the domain variables, whose input sequence is given by their listing in the time series data sets.

$$
\begin{aligned}
E &\rightarrow E + F \quad | \quad E - F \quad | \quad F \\
F &\rightarrow F \times T \quad | \quad F \div T \quad | \quad T \\
T &\rightarrow const \quad | \quad Var \quad | \quad (E)
\end{aligned}
$$

Fig. 2. Universal grammar $G = (N, T, P, S)$ allowing to apply the four basic mathematical operands $+, -, \times$ and $\div$ with $N = \{E, F, T, Var\}$, $T = \{+, -, \times, \div, const, (,)\}$, $S = E$ and $P = \{p_{E,1}, \ldots, p_{T,3}\}$ (e.g., $p_{E,1} = E \rightarrow E + F$). Note the system specific interpretation of the '|' symbol. In the LAGRAMGE system, production rules having the same height are ordered by their listing sequence in the grammar. Therefore, '|' can be interpreted as 'before'-relation. In the modified system, '|' is used in its original meaning of 'OR'.

3.2 GSAT

The GSAT algorithm [29] was developed to solve hard satisfiability problems (SAT) of propositional formulae in conjunctive normal form (CNF). In brief, it iteratively processes two steps:

1. A truth assignment for the clause is randomly generated. The number of tries, i.e. iterations, is limited by a maximum number of repetitions.
2. Find the variable with the largest increase in the total number of satisfied clauses if its assignment is reversed and flip its assignment. Within one iteration, this is repeated for a maximum number of flips.

In summary, GSAT performs as a local greedy search and the method has been found to perform effectively in applications such as graph coloring, N-queens encoding and Boolean induction problems.

4 Materials and Methods

4.1 System Modifications

The refinement of the parse trees as well as the processing of the beam search has been modified. (The implementation of the modified system is provided for download in the supplementary material.) Following GSAT, the improvement of a model is processed in two steps:

1. *Model refinement*: Instead of choosing the production rules for the refinement of a parse tree T_S and the termination of the subtrees T_A in the predefined sequential order, these are selected randomly. Instead of terminating the subtrees T_A for all non-terminals A in the parse tree T_S by application of the first successor rule $p_{A,i+1}$ of the previously applied rule $p_{A,i}$, the succeeding rule $p_{A,j} \in P_A$ is randomly chosen. The in this way derived subtree T_A is terminated by iteratively applying a randomly chosen production rule $p_{B,i} \in P_B$ to all non-terminal leaf nodes B in T_A until no more non-terminal is left in the leaves of T_A and thus deriving T_S'. Referring to GSAT, this mimics the initial randomly generated truth assignment of the variables of a Boolean formula. For detailed information, see algorithm 1 in the supplementary material.
2. *Search heuristic*: For the model T_S' refined in this way, the system tries to minimize $\mathcal{F}_h$. Successively for all non-terminals B in T_S', all $p_{B,i} \in P_B$ are applied and the in this manner derived subtrees T_B are randomly refined again as described above. After each of such a random refinement, $\mathcal{F}_h$ is evaluated. T_S' with the lowest error is determined as the refined model. Again referring to GSAT, this mimics the step of flipping the variable assignment with the objective of finding the variable with the largest increase in the total number of satisfied clauses. For detailed information, see algorithm 2 in the supplementary material.

Processing a beam search of width n, each single beam element $\mathcal{T}_S$ is evaluated separately. That is, not the maximum number of n improved models of $\mathcal{T}_S$ compared to all n beam elements are replaced within the beam. Instead, only the actually evaluated beam element is replaced if applicable. The algorithm stops if none of the beam elements $\mathcal{T}_{S,1..n}$ can be improved or none of the elements can be refined anymore. For detailed information, see algorithm 3 in the supplementary material.

To avoid redundant model evaluations, the sequence $seq(\mathcal{T}_S) = (i,\ldots,q)$ of applied production rules is stored in a lookup table during the derivation of a parse tree $\mathcal{T}_S$. As during the iterative termination process the non-terminals within the right side $\alpha \in (N \cup T)^*$ of a production rule $A \to \alpha$ are identified from left to right, each parse tree derivation can be identified by a unique sequence of the applied production rule indices. Before evaluation of a refined parse tree, the system checks for redundancy in the lookup table.

4.2   Patients and Data Sets

The study included data of 13 mechanically ventilated patients under preoperative anesthesia. Data was obtained by automated respiratory maneuvers. During such maneuvers, continuous mechanical ventilation was applied at successively increasing PEEP-levels. Starting from 0 mbar, PEEP was increased in steps of 2 mbar up to a maximum plateau pressure of 45 mbar (Fig. 1, a, b). For detailed information about the subjects and medication, see [12].

For analysis, the last breathing cycle before increasing the PEEP to the next level was extracted from the data. Thus, each patient data set consisted of 11 to 14 (13.7 $\pm$ 0.95 [mean $\pm$ SD]) breathing cycles at different pressure levels. Additionally to the airway pressure P_{aw}, the flow $\dot{V}$ and the volume V were recorded as time series (Fig. 1c and d).

4.3   Experiments

The benchmark test to validate the modified system was to identify the EOM model in the respiratory data. The performance of the modified system was tested against the original algorithm. Special attention was paid to the effect of different input sequences of the domain variables and production rules. The quality of a model was determined by two parameters. (i) A model equation should represent the EOM model, i.e re-identify the EOM from the data. A model was assumed to represent the EOM model, if after simplification the resulting equation consisted of the three additive terms $const \times V/C$, $const \times \dot{V} \times R$ and $const \times PEEP$ (eqn.1) with optional constant multipliers $const$. Additional constant additive terms were allowed to be included and thus a positively identified model was of the form

$$P_{aw} = [const \times] V/C + [const \times] \dot{V} \times R + [const \times] PEEP [+ const]. \quad (2)$$

Note that $const$, C and R could take both positive or negative values, but had to lead to pure positive additive terms aside from the additional constant additive terms [+ const], which could also take negative values after simplification of the equation. (ii) Secondly, the root mean squared error (RMSE) of the model fit to the raw data should be minimized.

The input data sets consisted of time series data for the four domain variables P_{aw}, $\dot{V}$, V and $PEEP$ with P_{aw} as dependent variable. The $PEEP$ level was given as constant value during a single breathing cycle. The input sequence of the independent variables, i.e. their listing in the data files, was permuted and thus, for each patient the six sequences [$(\dot{V}, V, Peep)$, $(\dot{V}, Peep, V)$, $(V, \dot{V}, Peep)$, $(V, Peep, \dot{V})$, $(Peep, \dot{V}, V)$, $(Peep, V, \dot{V})$] were provided. The input grammar consisted of a universal grammar, allowing to apply the four basic mathematical operands $+$, $-$, $\times$ and $\div$ (Fig. 2). While a parse tree depth of five would have been sufficient to derive the EOM model, resulting in a search space size of 7300, we allowed a maximum parse tree depth of six, increasing the search space complexity to 14,674,005 different parse trees. The production rules were provided in an initial sequence and its reverse [$initial, reverse$]. Consequently, twelve different input combinations of variable sequences in the data files and production rule sequences in the grammar were analyzed for each of the 13 patient data sets.

To identify the statistical correlation between the input sequence and the quality of the analysis results, a two-factorial analysis of variance (ANOVA) was performed. For both systems (original and modified) it was tested if (i) the identification of the EOM and (ii) if the RMSE of the provided model solutions depended on the input sequence of the variables and production rules. The median values of the RMSE calculated over all experimental runs were compared by a Wilcoxon rank-sum test. The significance level was set to $p < 0.05$ for 'significant' and $p < 0.003$ for 'highly significant'.

Data preprocessing and statistical analysis were performed by application of the Matlab® software package version R2006b (The MathWorks, Natick, MA). The two versions of the applied equation discovery systems were implemented in the C programming language. Experiments were run on a standard laptop computer (1001.7 MiB memory, 1.83 GHz processor) running under the operating system Ubuntu (version 8.04).

5   Results

5.1   Benchmark Test (i): Identification of the EOM

The original system identified the EOM in 37.2%, the modified system in 96.8% of all runs. For the original system, the two factorial ANOVA revealed a highly significant dependence of the identification rate from the input sequence of the domain variables as well as from the input sequence of the production rules. The interaction between these two factors was found to be statistically significant. In contrast, the modified system did not show such dependencies. The original system showed a higher variance of the results (see Fig. 3 and Table 1 in the supplementary material).

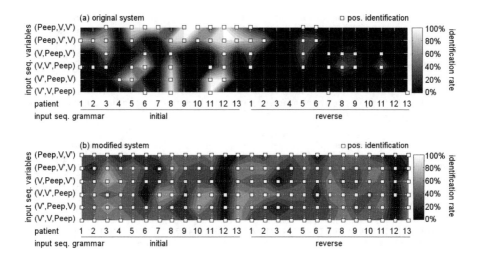

Fig. 3. Experimental results for (a) original and (b) modified system with respect to identification of the EOM in the data. The *identification rate* indicates the percentage of beam elements (i.e. models) representing the EOM upon termination of the algorithm. The contour plot (interpolation for visualization purposes) represents the *identification rate* for all thirteen patients with respect to the input sequences of the domain variables and the production rules. The square markers indicate that at least one beam element provides a *positive identification* of the EOM. The dependence of the original system on the input sequences is indicated by the far more inhomogeneous gray scale coloration. Some peaks can be clearly identified within the range of the initial input sequence of the productions rules. Note that a darker color indicates a lower *identification rate*.

5.2   Benchmark Test (ii): RMSE of Model Fits

A statistically highly significant dependence of the RMSE on the input sequence of the production rules was observed for the original system. The dependence of the RMSE on the input sequence of the domain variables as well as the interaction between the two factors was found to be statistically significant. The modified system did not show such dependencies. The original system showed a higher variance of the RMSE (see Fig. 4 and Table 2 in the supplementary material). The modified system generated more precise models with respect to the RMSE in the early as well as in the final state of the iterative model derivation process. The maximum number of iteration steps amounts to 27 for the original and to 23 for the modified system (Fig. 5). The medians of the RMSE are statistically significantly lower in the modified system (see insert of Fig. 5).

5.3   General Performance

The original algorithm performed about 20% more iterations than the modified system before termination. The modified system evaluated 9 times more model

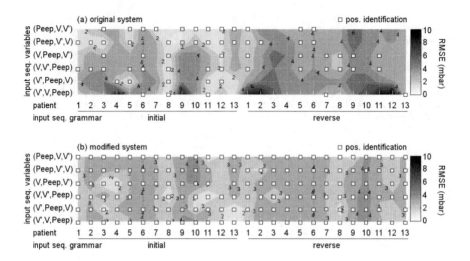

Fig. 4. Experimental results for (a) original system and (b) modified system with respect to the error variance of the generated models. The error variance is represented by the median *RMSE* of all beam elements (i.e. models) upon termination of the algorithm. The contour plot (interpolation for visualization purposes) represents this median *RMSE* for all thirteen patients with respect to the input sequences of the domain variables and the production rules. The square markers indicate that at least one beam element provides a *positive identification* of the EOM. The independence of the modified system on the input sequences is indicated by the more homogeneous gray scale coloration. Note that (a) a brighter color indicates a lower *RMSE*, (b) the interpolation is exclusively for better visualization, (c) adjacency of entries on the x- and y-axis is arbitrary as they merely represent the datasets and sequential orders respectively.

equations than the original system. About 24% of the model equations generated by the modified system were redundant derivations and thus not re-evaluated (see Table 3 in the supplementary material).

6   Discussion

The main results of this study are: (i) the modified equation discovery system was able to identify the well-known EOM-model in real-world data obtained from mechanically ventilated patients and (ii) the system's performance was independent of the input sequences of the model variables and the production rules of the context free grammar representing domain-dependent background knowledge.

Originally, equation discovery approaches were data-driven [26,21,5,16]. Depending on the size of the dataset and the hypothesis space, the trial-and-error strategy of this domain-independent [27] policy to fit functional formulae is generally time-consuming. Besides this computational problem – inherent in most

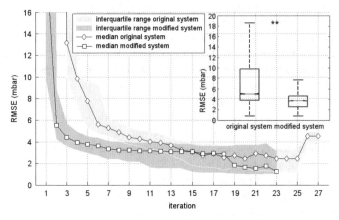

Fig. 5. *RMSE* (root mean squared error) during the course of iterations: the *RMSE* is calculated as the median *RMSE* over all beam elements after each iteration. It is based on the results for all patients and all twelve input sequence combinations (grammar and domain variables). Note that the increase of the median *RMSE* for the original system at the end of the presented curve is caused by the modality of calculating the median over the results for all patients and input sequences, only separated by the iteration step. (During a single run, the algorithm would stop before an increase of the *RMSE*, i.e. a worsening of the model fit.) *insert*: Median of the *RMSE* for the original and the modified system calculated over all results for all patients and independent of the input sequences. As a Wilcoxon rank-sum test showed, the medians differ with high statistical significance (** $p < 0.003$).

machine learning and data mining tasks – Schaffer [28] highlighted that it is not the mere identification of relationships in data which equation discovery has to deal with. In fact, equation discovery systems should exclusively identify relationships of real scientific relevance, a problem which has not been considered for a long time. He also raised concerns about systems being exclusively demonstrated on few hand-selected cases often involving artificial data [27]. Consequently, the E* algorithm he introduced was tested on a vast set of measured data. In spite of searching an infinite space of possible functional formulae, the search space was restricted to a fixed, finite set of potential relationships. In a way, this was a first step to a knowledge-driven, domain-dependent approach in equation discovery. Focussing on the reduction of search space complexity, declarative bias was introduced to equation discovery [31]. Besides the benefit concerning computational complexity, this approach provides the capability to process domain-dependent background knowledge. Moreover, declarative bias represents such background knowledge in a very flexible way and therefore provides the feasibility to describe a problem task as a trade-off between exploration and exploitation: if the hypothesis space is bounded by tight constraints in terms of detailed background knowledge, the presented results might be scientifically more relevant while part of the exploratory discovery aspect of finding new systematic relations in the data might be lost. On the other hand, if the constraints are not tight enough, one has not only to face time constraints caused by a vast hypothesis space, but potentially also a loss of relevance in the provided results.

For the domain-specific task to identify a physiological lung model in measured respiratory time series data, an equation discovery system had to: (i) be able to handle background knowledge, (ii) derive (correct) results in the least possible time and (iii) be able to handle noisy data in a robust way. According to (i), we found the LAGRAMGE system being well applicable to our task. However, first experiments on simulated data revealed that the ordering of the production rules according to the height of derivable parse trees implied a particular bias when traversing the hypothesis space which consequently affected the identification of suitable models. We found two technical reasons responsible for these effects. Firstly, in addition to the requested reduction of the hypothesis space by the declarative bias, the hypothesis space was kind of 'cut' in a second way. The sequential application of the production rules implies the uniqueness of the derivation paths. The refinement of a parse tree T_S requires the termination of its subtree T_A, to which the production rule $p_{A,1}$ has been applied, by applying the first production rule $p_{B,1}$ to all nonterminals B in T_A. Therefore, if T_S is not kept as one of the beam elements for the next refinement iteration, none of the successor trees of T_A applying rule $p_{B,i \neq 1}$ instead of $p_{B,1}$ to any of the nonterminals B can be generated and thus are lost for evaluation. This effect is even aggravated by the implementation of the beam search. Within an iteration, the element with the highest error with respect to $\mathcal{F}_h$ of the actual beam is immediately substituted with a refinement having a lower error. Therefore, a refinement T_S' is not exclusively compared to its origin T_S, but always to the set of all beam elements. This implies the possibility, that even if a refinement could improve its origin in terms of $\mathcal{F}_h$, it might not be stored in the beam for further refinements. This property could be compensated to a certain degree by a larger beam width, however resulting in an increasing memory demand. On the other hand, if a beam element was not improved but also not substituted by the refinement of another element, the same production rules are applied in the next iteration step again and thus the same refinements are re-evaluated. Secondly, the sequential transition of the hypothesis space might occasionally cause the algorithm to meet the stopping criterion in a local minimum. Refining subtree T_A by applying production rule $p_{A,i+1}$ might impair the model in terms of $\mathcal{F}_h$ compared to the application of $p_{A,i}$ and thus T_A might be excluded from further refinements — depending on the beam width. However, refining T_A by $p_{A,i+2}$ in turn might improve the model, but in this case is not processed anymore.

Thus, the task was to adequately modify the search heuristic of the algorithm. Although being simple, the local greedy search performed by the GSAT algorithm [29] has been shown to perform effectively. Yet, the performance of GSAT was hard to explain. It has been speculated that the crucial factor is to have an approximate solution which can be refined iteratively. We found this approach perfectly matching our task: starting with a randomly generated model, i.e parse tree, this was attempted to be improved by systematically applied random refinements of the iteratively derived parse trees. According to our results, this strategy to traverse the hypothesis space was found to be effective. The previously mentioned 'cutting' of the hypothesis space – biased by the sequential application of production rules and thus ordered parse tree refinements – could obviously be avoided by the strategy of randomized parse tree refinements. Applied in context of the modified beam search, the system includes aspects of a momentum and a lookahead: each single beam element is evaluated and replaced as necessary. On the other hand, the stopping criterion is not met before none of all beam elements can be improved with respect to the heuristic function. In this way, individual elements are refined again (lookahead) although possibly not having been improved in a preceding refinement iteration (momentum).

Testing our system on real-world data obtained from mechanically ventilated patients without any preprocessing of the data in terms of noise reduction, the results indicate a robust performance. The additional bias affecting the search strategy was eliminated. The data noise as usually found in measured data of physiological systems did not show any effect on the performance within our experimental settings. However, caused by the momentum and the lookahaed, the modified system evaluated a much higher number of models compared to the original system. Resulting in an increased evaluation time, this is a hard constraint concerning the potential application at bedside under the aspect of individualized model inference for the patient as diagnostic support. Nevertheless, an average of about 8550 evaluated models is still a small fraction of the overall number of more than 1.4 million models spanning the hypothesis space.

7 Related Work

To the best of our knowledge, this is the first application of an equation discovery system to measured respiratory data from intensive care medicine. However, efforts have been made to improve equation discovery systems which are capable to process domain-dependent background knowledge since they have been introduced. Promoting the view of such 'domain-dependent' equation discovery, Langley *et al.* [19,7] introduced the approach of inductive process modeling (IPM). IPM combines the objectives of model inference from time series data, knowledge representation in established scientific formalisms and incorporation of domain knowledge embedded in a simulation environment. The system incorporates LAGRAMGE to take advantage of declarative bias though not using the full scope of LAGRAMGE's abilities concerning the representation of background knowledge in terms of context free grammars. IPM combines processes into a model, and background knowledge consists of the provision of generic processes. Models are induced under the assumption that the combination of any set of generic processes produces valid model structures. Consequently, the model space might contain candidates non-compliant to the expectations of a domain expert. Thus, Todorovski *et al.* [32] further developed the IPM approach to hierarchical inductive process modeling (HIPM), where processes are represented in a hierarchical order. A general problem in equation discovery – as well as in the entire field of machine learning – is that of overfitting the data. An approach addressing this point explicitly is the FUSE (forming unified scientific explanations) system [6]. A main aspect of FUSE is a data preprocessing step related to bagging which enables the system to infer several different models by application of HIPM. Amongst those, the subprocesses are ranked by their frequency and combined into a final model. In another approach discussing overfitting in equation discovery, De Pauw *et al.* [8] introduced a model identifiability measure to generate models with an optimized complexity with respect to the given data. Background knowledge is represented in form of an initial model being either general or overly complex.

A main requirement on an equation discovery system applied to the specific domain of modeling lung mechanics was a flexible representation of background knowledge. The system should be capable to infer models from rather low-level prior knowledge as well as from more detailed knowledge. This is important to the domain expert, who might start modeling from scratch (no background knowledge) or who might want to improve existing models (background knowledge given). Although having proven to perform well in multiple applications, we assume the aforementioned systems demand rather detailed prior insight into the processes to be modeled. Again, the representation of background knowledge by context free grammars is highly flexible and furthermore can be used for reducing the search space. Depending on the amount of background knowledge provided, the analysis of the data can be designed to have more exploratory or more exploitative characteristics. It is arguable to which degree model complexity could be influenced by an appropriate structure of the applied grammar which in turn could help to reduce overfitting but potentially would require again specific prior knowledge. It also arguable, if for some applications the order dependent approach of the original LAGRAMGE system might even be helpful, as it could be used as a specific feature to bring in background knowledge. As our study was designed in the context of a specific real-world problem domain, our approach was of a rather pragmatic character. We focused on improving the robustness of the applied system concerning reproducibility of model inference in the context of sparse prior knowledge, accepting the previously mentioned drawbacks of a comparatively basic system. However, the optimization of model complexity and the integration of a model identifiability measure is an important aspect to be considered in future work, and further experiments could be conducted with such a system incorporating the search strategy presented in this study.

8   Conclusion

We presented an equation discovery approach based on the LAGRAMGE system, being able to handle domain-dependent background knowledge. To decouple the search heuristic for traversing the hypothesis space from the presentation of the background knowledge, we implemented a randomized hill-climbing search heuristic resembling the GSAT algorithm. This novel system was shown to be well applicable in the domain of real-world time series data, as it was robust concerning the mode of the data input, the representation of background knowledge and noise inherent in the measured data. This is a prerequisite for further applications in respiratory physiology.

References

1. Bates, J.H.T.: A recruitment model of quasi-linear power-law stress adaptation in lung tissue. Ann. Biomed. Eng. 35, 1165–1174 (2007)
2. Bates, J.H.T., Brown, K.A., Kochi, T.: Identifying a model of respiratory mechanics using the interrupter technique. In: Proceedings of the Ninth American Conference I.E.E.E. Engineering Medical Biology Society, pp. 1802–1803 (1987)
3. Beydon, L., Svantesson, C., Brauer, K., Lemaire, F., Jonson, B.: Respiratory mechanics in patients ventilated for critical lung disease. Eur. Respir. J. 9(2), 262–273 (1996)
4. Bisiani, R.: Beam search. In: Shapiro, S. (ed.) Encyclopedia of Artificial Intelligence, pp. 56–58. Wiley & Sons, Chichester (1987)
5. Bradshaw, G.L., Langley, P., Simon, H.A.: Bacon.4: The discovery of intrinsic properties. In: Proceedings of the Third Biennial Conference of the Canadian Society for Computational Studies of Intelligence, pp. 19–25 (1980)
6. Bridewell, W., Asadi, N.B., Langley, P., Todorovski, L.: Reducing overfitting in process model induction. In: Proceedings of the 22nd International Conference on Machine Learning, pp. 81–88 (2005)
7. Bridewell, W., Langley, P., Todorovski, L., Džeroski, S.: Inductive process modeling. Mach. Learn. 71, 1–32 (2008)
8. DePauw, D.J.W., DeBaets, B.: Incorporating model identifiability into equation discovery of ode systems. In: Proceedings of the 2008 GECCO Conference Companion on Genetic and Evolutionary Computation, pp. 2135–2140 (2008)
9. Džeroski, S., Todorovski, L.: Discovering dynamics: From inductive logic programming to machine discovery. J. Intell. Inf. Syst. 4, 89–108 (1994)
10. Falkenhainer, B.C., Michalski, R.S.: Integrating quantitative and qualitative discovery in the ABACUS system. In: Machine Learning: An Artificial Intelligence Approach, pp. 153–190. Morgan Kaufman, San Mateo (1990)
11. Fung, Y.C.: Biomechanics. Mechanical Properties of Living Tissues. Springer, New York (1981)
12. Ganzert, S., Möller, K., Steinmann, D., Schumann, S., Guttmann, J.: Pressure-dependent stress relaxation in acute respiratory distress syndrome and healthy lungs: an investigation based on a viscoelastic model. Crit. Care 13(6) (2009)
13. Grasso, S., Terragni, P., Mascia, L., Fanelli, V., Quintel, M., Herrmann, P., Hedenstierna, G., Slutsky, A.S., Ranieri, V.M.: Airway pressure-time curve profile (stress index) detects tidal recruitment/hyperinflation in experimental acute lung injury. Crit. Care Med. 32(4), 1018–1027 (2004)
14. Haberthür, C., Guttmann, J., Osswald, P.M., Schweitzer, M.: Beatmungskurven - Kursbuch und Atlas. Springer, Heidelberg (2001)
15. Hickling, K.G.: The pressure-volume curve is greatly modified by recruitment. a mathematical model of ards lungs. Am. J. Respir. Crit. Care Med. 158(1), 194–202 (1998)
16. Koehn, B.W., Zytkow, J.M.: Experimenting and theorizing in theory formation. In: Proceedings ACM SIGART International Symposium on Methodologies for Intelligent Systems, pp. 296–307 (1986)

17. Kokar, M.M.: Determining arguments of invariant functional descriptions. Mach. Learn. 1(4), 403–422 (1986)
18. Križman, V., Džeroski, S., Kompare, B.: Discovering dynamics from measured data. In: Working Notes of the MLnet Workshop on Statistics, Machine Learning and Knowledge Discovery in Databases, pp. 191–198 (1995)
19. Langley, P., Sanchez, J., Todorovski, L., Džeroski, S.: Inducing process models from continuous data. In: Proceedings the Nineteenth International Conference on Machine Learning, pp. 347–354 (2002)
20. Langley, P.W.: Bacon: A production system that discovers empirical laws. In: Proceedings of the Fifth International Joint Conference on Artificial Intelligence, p. 344 (1977)
21. Langley, P., Zytkow, J.M.: Data-driven approaches to empirical discovery. Artif. Intell. 40, 283–310 (1989)
22. Macintyre, N.R.: Basic principles and new modes of mechanical ventilation. In: Crit Care Med: Perioperative Management, pp. 447–459. Lippincott Williams & Wilkins, Philadelphia (2002)
23. Mols, G., Brandes, I., Kessler, V., Lichtwarck-Aschoff, M., Loop, T., Geiger, K., Guttmann, J.: Volume-dependent compliance in ARDS: proposal of a new diagnostic concept. Intens. Care Med. 25(10), 1084–1091 (1999)
24. Nédellec, C., Rouveirol, C., Adé, H., Bergadano, F., Tausend, B.: Declarative bias in ILP. In: DeRaedt, L. (ed.) Advances in Inductive Logic Programming, pp. 82–103. IOS Press, Amsterdam (1996)
25. Network, T.A.R.D.S.: Ventilation with lower tidal volumes as compared with traditional tidal volumes for acute lung injury and the acute respiratory distress syndrome. the acute respiratory distress syndrome network. N. Engl. J. Med. 342(18), 1301–1308 (2000)
26. Nordhausen, B., Langley, P.: A robust approach to numeric discovery. In: Proceedings of the Seventh International Conference on Machine Learning, pp. 411–418 (1990)
27. Schaffer, C.: A proven domain-independent scientific function-finding algorithm. In: Proceedings of the 8th National Conference on Artificial Intelligence, pp. 828–833 (1990)
28. Schaffer, C.: Bivariate scientific function finding in a sampled, real-data testbed. Mach. Learn. 12, 167–183 (1991)
29. Selman, B., Levesque, H.J., Mitchell, D.: A new method for solving hard satisfiability problems. In: Proceedings of the Tenth National Conference on Artificial Intelligence, pp. 440–446 (1992)
30. Tobin, M.J.: Ventilator monitoring, and sharing the data with patients. Am. J. Respir. Crit. Care Med. 163(4), 810–811 (2001)
31. Todorovski, L., Džeroski, S.: Declarative bias in equation discovery. In: Proceedings of Fourteenth Internationl Conference on Machine Learning, pp. 376–384 (1997)
32. Todorovski, L., Bridewell, W., Shiran, O., Langley, P.: Inducing hierarchical process models in dynamic domains. In: Proceedings of the Twentieth National Conference on Artificial Intelligence, AAAI 2005, pp. 892–897 (2005)
33. Zembowicz, R., Zytkow, J.M.: Automated discovery of empirical equations from data. In: Raś, Z.W., Zemankova, M. (eds.) ISMIS 1991. LNCS, vol. 542, pp. 429–440. Springer, Heidelberg (1991)

Mining Class-Correlated Patterns for Sequence Labeling

Thomas Hopf and Stefan Kramer

Institut für Informatik/I12, Technische Universität München, Boltzmannstr. 3,
D-85748 Garching bei München, Germany
mail@thomas-hopf.de, kramer@in.tum.de

Abstract. Sequence labeling is the task of assigning a label sequence to an observation sequence. Since many methods to solve this problem depend on the specification of predictive features, automated methods for their derivation are desirable. Unlike in other areas of pattern-based classification, however, no algorithm to directly mine class-correlated patterns for sequence labeling has been proposed so far. We introduce the novel task of mining class-correlated sequence patterns for sequence labeling and present a supervised pattern growth algorithm to find all patterns in a set of observation sequences, which correlate with the assignment of a fixed sequence label no less than a user-specified minimum correlation constraint. From the resulting set of patterns, features for a variety of classifiers can be obtained in a straightforward manner. The efficiency of the approach and the influence of important parameters are shown in experiments on several biological datasets.

Keywords: Sequence mining, correlated pattern mining, label problem, sequence labeling, pattern-based classification, pattern growth.

1 Introduction

The task of assigning a label sequence to an observation sequence is a machine learning problem which occurs in various areas such as predicting properties of biological sequences or natural language processing [1]. For solving the problem, a variety of different methods can be employed including Hidden Markov Models (HMMs, [2]), Conditional Random Fields (CRFs, [3]) or Support Vector Machines (SVMs, [4,5]). While differing in their underlying approach, many of these methods are similar regarding the fact that the relevant characteristics of the primary input data have to be described by features (e.g. SVMs) or are extracted by the model itself via feature functions (e.g. CRFs).

The fundamental idea of pattern-based classification is that patterns in the data can be used to generate these features while yielding more accurate and comprehensible models [6]. Ideally, the distribution of the used patterns differs between classes, thus allowing for discrimination by the model. Considerable research effort has been spent in this area towards the supervised mining of itemsets and structured patterns discriminating between classes, where in both

B. Pfahringer, G. Holmes, and A. Hoffmann (Eds.): DS 2010, LNAI 6332, pp. 311–325, 2010.

cases one class label is associated with the itemset or structured object as a whole (for an overview, see [6]). However, comparatively little attention has been paid to the identification of sequential patterns for sequence labeling purposes, where each position in the observation sequence is assigned its own label. Birzele and Kramer devised a method for the prediction of protein secondary structure based on the mining of frequent sequential amino acid patterns and subsequent post-processing using a χ^2 and a precision-recall filter [7]. But to the best of our knowledge, no method to directly mine sequential patterns indicative of certain sequence labels has been published so far.

In this paper, we introduce an algorithm to mine sequential patterns in observation sequences which correlate with the occurrence of certain sequence labels according to a user-defined minimum correlation threshold. The algorithm is the result of integrating approaches for correlated itemset mining [8,9] and pattern growth based sequence mining [10].

The paper is organized as follows. Section 2 gives a brief overview of related work, before Section 3 shortly recapitulates existing concepts for correlated pattern mining on which our algorithm is based. Subsequently, we formalize the problem of mining label-correlated sequence patterns and devise an algorithm to solve the problem (Section 4). After giving an experimental validation on biological sequences in Section 5, we conclude in Section 6.

2   Related Work

Our work is related to research in the areas of frequent sequential pattern mining and correlation mining. The first algorithms for sequential pattern mining have been introduced by Agrawal and Srikant [11,12] employing ideas from their earlier Apriori algorithm [13]. Later work increased efficiency by employing different approaches such as the vertical data format (SPADE [14]), pattern growth in projected databases (PrefixSpan [10]) or depth-first search using bitmap representations (SPAM [15]). Considerable research has also been targeted towards the incorporation of different types of constraints [16].

The problem of correlation mining has been introduced first by Bay and Pazzani [17] and Morishita and Sese, who showed it is possible to compute a tight upper bound on the value of convex correlation measures when mining the pattern lattice [8]. Later contributions to the field have been made by Nijssen and Kok [9] and Nijssen et al. [18] by transforming minimum correlation constraints in ROC space. Note that correlation mining is also closely related to other problems such as subgroup discovery and emerging pattern mining [6].

At the interface of both fields, several methods for mining discriminative subsequences exist [19,20]. However, a common property of all these methods is that sequences in the database are assigned one class label as a whole, but not a sequence of labels corresponding to each element in the observation sequence.

3   Existing Approaches for Correlated Pattern Mining

Since our method for mining label-correlated sequence patterns is based on the work of Morishita and Sese [8] and Nijssen and Kok [9], we shortly review the central ideas of these approaches.

The task of correlated pattern mining is to find rules $p \rightarrow t$ with $\mu(p \rightarrow t)$ satisfying a certain threshold θ, where μ is a correlation measure such as accuracy, information gain or χ^2, p is a pattern over the examples according to some pattern language $\mathcal{L}$, and t is a class label. While we restrict this paper to the χ^2 measure, other convex correlation measures could be used with both approaches [8,9].

For a binary classification problem, consider a database D consisting of examples which belong either to the *target* class t, for which correlated patterns are mined, or to the *non-target* class $\bar{t}$. Selecting the subsets of examples belonging to class t and $\bar{t}$ yields the *target database* (D_t) and *non-target database* ($D_{\bar{t}}$). The degree of correlation between the occurrence of a pattern p and a class label t can be calculated from a contingency table [9]:

$\alpha_1(p)n_1$	$(1 - \alpha_1(p))n_1$	n_1
$\alpha_2(p)n_2$	$(1 - \alpha_2(p))n_2$	n_2
$\alpha_1(p)n_1 + \alpha_2(p)n_2$	$n_1 + n_2 - \alpha_1(p)n_1 - \alpha_2(p)n_2$	$n_1 + n_2$

Here, n_1 and n_2 denote the number of examples in class t and $\bar{t}$. $\alpha_1(p)$ is the relative frequency of examples having class t covered by a rule $p \rightarrow t$, whereas $\alpha_2(p)$ is the relative frequency of examples in class $\bar{t}$ covered by pattern p and thus wrongly predicted to be in class t by the rule. $N = n_1 + n_2$ is the total number of examples in the database. Using the entries of the table, the χ^2 value of a pattern p is given by

$$\chi^2(p) := \chi^2(\alpha_1(p)n_1, \alpha_2(p)n_2, n_1, n_2) := \sum_{i,j \in \{1,2\}} \frac{(O_{ij} - E_{ij})^2}{E_{ij}} \quad (1)$$

where $E_{i1} = (\alpha_1(p)n_1 + \alpha_2(p)n_2)n_i/N$, $E_{i2} = ((1 - \alpha_1(p))n_1 + (1 - \alpha_2(p))n_2)n_i/N$, $O_{i1} = \alpha_i(p)n_i$ and $O_{i2} = (1 - \alpha_i(p))n_i$. Unfortunately, the χ^2 measure and other correlation measures are neither monotonic nor anti-monotonic [8], which prevents the use of the Apriori pruning strategy. However, Morishita and Sese showed that the χ^2 value of any superset q of an itemset p ($q \supseteq p$) is bounded above by $u(p) = \max \{\chi^2(\alpha_1(p)n_1, 0, n_1, n_2), \chi^2(0, \alpha_2(p)n_2, n_1, n_2)\}$. This finding allows to mine correlated itemsets in an anti-monotonic fashion by searching for itemsets with $u(p) \geq \theta$ (*promising itemsets*). Since any correlated itemset is also promising, all correlated itemsets can be derived from the set of promising itemsets. All supersets of promising itemsets have to be evaluated, because there exists the possibility that some or all of them are correlated to a degree of no less than θ. On the other hand, if an itemset p is not promising ($u(p) < \theta$), all supersets containing p can be pruned away, because $\chi^2(q) \leq u(p) < \theta$ follows for all $\{q \,|\, q \supseteq q\}$.

Nijssen and Kok showed that it is possible to transform a minimum correlation constraint into a disjunction of minimum frequency constraints [9]. In a binary setting with target class t, this allows to mine patterns fulfilling a single induced minimum frequency constraint in the target database D_t using any technique for frequent pattern mining. For the χ^2 measure, this constraint is $\theta_t = \frac{\theta N}{N^2 - n_1 N + \theta n_1}$. The resulting set of frequent patterns has to be post-processed by computing the missing entries in the contingency table in the non-target database $D_{\bar{t}}$ for the calculation of the correlation value. All patterns having a correlation of at least θ form the final result.

The central idea of our algorithm is to find all frequent (respectively promising) patterns according to the following

Lemma 1. *A pattern can only be frequent (promising), if all of its prefixes are frequent (promising).*

This follows immediately from the anti-monotonicity of the minimum frequency constraint and from the proof by Morishita and Sese that a pattern can only be promising, if all of its subpatterns are promising [8]. The precise definition of a prefix is given in the following section.

4   Mining Label-Correlated Sequence Patterns

In this section, we show how to mine label-correlated sequence patterns based on the presented approaches for correlated itemset mining.

First, we need to formalize the problem of mining correlated sequence patterns (Fig. 1) with the following definitions.

Definition 1. *A region $r = (a, b)$ is a set $\{a, \ldots, b - 1\}$ of relative offsets, $a, b \in \mathbb{Z}, a < b$. Together with an absolute sequence index j, r refers to the absolute sequence indices $\{j + a, \ldots, j + b - 1\}$.*

A region describes a set of continuous positions relative to a central position (Fig. 1) and can thus be used to generalize the concept of the *sliding window*, where an offset indexes only one position. In our approach, a set of relevant regions is defined in advance by the user.

Definition 2. *Given a minimum correlation constraint (μ, θ), a set of labeled sequences $T = \{(\boldsymbol{x}^{(i)}, \boldsymbol{y}^{(i)})\}$, a pattern language $\mathcal{L}$, a label alphabet $\mathcal{Y}$ and a set of regions R, $\forall r = (a, b) \in R$, the task of mining label-correlated patterns in the sequences $(\boldsymbol{x}, \boldsymbol{y}) \in T$ is to find all patterns $p \in \mathcal{L}$ which induce a rule "p matches $\boldsymbol{x}$ starting in $\{j + a, \ldots, j + b - 1\} \longrightarrow y_j = t$" for a fixed target label $t \in \mathcal{Y}$. "$\longrightarrow$" means that the occurrence of $y_j = t$ is correlated with the occurrence of p in r no less than θ according to the correlation measure μ.*

We assume that all $x_j \in \boldsymbol{x}$ are contained in the observation alphabet $\mathcal{X}$ to choose an appropriate pattern language $\mathcal{L}$. We define the pattern alphabet $\Sigma \subseteq \{\sigma \mid \sigma \in \mathcal{X}^+ \wedge \forall i, j, i \neq j : \sigma_i \neq \sigma_j\}$ as subset of the set of all single literals from the observation alphabet $\mathcal{X}$ and groups of at least two different literals, effectively matching groups of regular expressions (e.g. DE means D or E). Patterns are expected to consist of at least one element from the pattern alphabet, yielding the pattern language $\mathcal{L} = \{p \mid p \in \Sigma^+\}$. We say a pattern $p = \sigma_0...\sigma_n$ matches at position j of a sequence, iff $\forall o = 0, \ldots, n : x_{j+o} \in \sigma_o$. A pattern q is a *prefix* of pattern $p = \sigma_0...\sigma_n$, iff $q = \sigma_0...\sigma_l$ with $l \in \{0, \ldots, n - 1\}$.

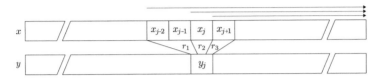

Fig. 1. Mining label-correlated patterns in sequences. In this example we consider sequence position j and regions $r_1 = (-2, 0)$, $r_2 = (0, 1)$, $r_3 = (1, 2)$, referring to observations $\{x_{j-2}, x_{j-1}\}$, $\{x_j\}$ and $\{x_{j+1}\}$, respectively. For each region r_i, the task is to find patterns starting in the positions referred to by region r_i that correlate with the assignment of class label $y_j = t$. Patterns of arbitrary length may only start in the positions referred to by the region and can extend down to the end of the sequence.

To solve the problem of finding label-correlated sequence patterns with the existing approaches for correlated pattern mining, the database of labeled sequences must be transformed into a database with one class label per entry (Fig. 2). This transformation has to meet three requirements: First, the same sequence pattern might have different correlation values in different regions. Therefore, each given region $r \in R$ has to be viewed as an independent mining problem on a separate database $D^{(r)}$. Second, each transaction may only have exactly one class label. Since we want to mine patterns which correlate with the label at position j, the label of a database entry is y_j. Third, when evaluating if a pattern is present in the transaction for the absolute position j in some labeled sequence (x, y) and a given region $r = (a, b)$, the pattern may only start from indices $\{j + a, \ldots, j + b - 1\}$ in x.

Based on these requirements, we define the size of the target and non-target database as well as the frequency of a pattern in either database for each region. For some region $r = (a, b)$, the size of the target database is

$$|D_t^{(r)}| = \sum_{\substack{(x,y)\in T}} \sum_{\substack{j=0 \\ y_j=t}}^{|y|-1} \begin{cases} 1 & \text{if } (0 \leq j + a) \wedge (j + b \leq |y|) \\ 0 & \text{otherwise} \end{cases} . \tag{2}$$

Analogously, using $y_j \neq t$ gives the size of the non-target database. We decide to count only transactions where the positions indexed by the region fit completely into the observation sequence. For statistical soundness, the decrease in possible start positions due to longer patterns is not considered. Assuming $M_p(x)$ is the list of all absolute start positions of matches of p in an observation sequence x,

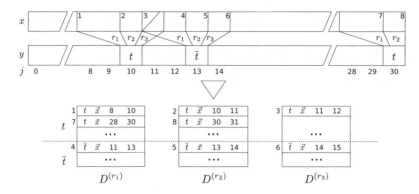

Fig. 2. Explicit sequence database transformation. The transformation builds one separate database $D^{(r_i)}$ for each given region r_i with tuples containing the sequence label, the observation and valid starting positions for matches. For all positions j in the labeled sequences, one transaction is created per region database as long as the region fits completely into the observation. Depending on the label y_j, each transaction belongs to the target database (t) or to the non-target database ($\bar{t}$). The regions in this example are the same as given in Fig. 1.

the frequency of a pattern p in the target database for a region $r = (a, b)$ is given by

$$f_t^{(r)}(p) = \sum_{\substack{(\boldsymbol{x}, \boldsymbol{y}) \in T}} \sum_{\substack{j=0 \\ y_j = t}}^{|\boldsymbol{y}|-1} \begin{cases} 1 & \text{if } \exists m \in M_p(\boldsymbol{x}) : (j + a \le m < j + b) \\ & \wedge \, (0 \le j + a) \wedge (j + b \le |\boldsymbol{y}|) \\ 0 & \text{otherwise} \end{cases} \tag{3}$$

Besides evaluating if there is at least one match starting in the positions referred to by j and r, we ensure that the region fits completely into the observation sequence to obtain counts which are consistent with the database sizes defined by Equation (2). Accordingly, $y_j \ne t$ yields the non-target database frequency of pattern p. Based on Equations (2) and (3), it is possible to calculate the χ^2 value for any pattern p in some region r.

The central idea of our algorithm (Alg. 1) is to exploit the prefix anti-monotonic property (Lemma 1) to mine all correlated patterns recursively by pattern growth in projected databases, employing the two major ideas of the **PrefixSpan** algorithm [10]. It works as follows.

The algorithm is started by **pg-mine**$(\epsilon, R, \emptyset, T, t, \theta, \Sigma)$ with the empty prefix ϵ and the set of all specified regions R as initial parameters. First, the target and non-target database sizes as well as the minimum frequency threshold for each region $r \in R$ are determined using Equation (2), whereas the else-case is skipped in the initial call. Then, for each element σ of the pattern alphabet Σ, all matches of σ to the observation sequences $x^{(i)}$ in T are determined and stored in the match list $\mathcal{M}_\sigma$ as 4-tuples $(i, j, \Gamma_{ij}, \Delta_{ij})$ containing the index of the matched sequence i, the index of the start position j as well as the count and correction terms Γ_{ij} and Δ_{ij}. The terms Γ_{ij} and Δ_{ij} are immediately calculated

---

Algorithm 1: pg-mine(prefix, R', $\mathcal{M}$, T, t, θ, Σ)

Data: Current *prefix*, frequent regions R' of prefix, match list $\mathcal{M}$ of prefix occurrences, dataset $T = \{(\boldsymbol{x}^{(i)}, \boldsymbol{y}^{(i)})\}$, target class $t \in \mathcal{Y}$, minimum correlation threshold θ, pattern alphabet Σ

1 **begin**
2   **if** *prefix* $= \epsilon$ **then**   // check if first function call (empty prefix)
3      $\lfloor \; \forall r \in R'$: calculate $|D_t^{(r)}|$, $|D_{\bar{t}}^{(r)}|$ and $\theta_t^{(r)}$ // DB sizes, thresholds
4   **else**   // non-empty prefix: check prefix extension of each match,
5   **foreach** $m = (i, j, \Gamma_{ij}, \Delta_{ij}) \in \mathcal{M}$ **do**   // append match to projection
6   **if** $j + |prefix| < |\boldsymbol{x}^{(i)}|$ **then**   // of symbols matching extension
7          $\lfloor \; \forall \sigma \in \Sigma : \boldsymbol{x}_{j+|\text{prefix}|}^{(i)} \in \sigma$: append m to $\mathcal{M}_\sigma$

8   **foreach** $\sigma \in \Sigma$ **do**   // consider all possible prefix extensions
9   **if** *prefix* $= \epsilon$ **then**   // empty prefix: create new match list for σ
10        $\lfloor \; \mathcal{M}_\sigma = \left[(i, j, \Gamma_{ij}, \Delta_{ij}) \middle| \boldsymbol{x}_j^{(i)} \in \sigma \right]$ // incl count/correction terms
11   **foreach** $(i, j, \Gamma_{ij}, \Delta_{ij}) \in \mathcal{M}_\sigma$ **do**   // all occurrences of prefix·σ
12   **foreach** $r \in R'$ **do**   // update counts for each region
13          $f_t^{(r)}(\sigma) = f_t^{(r)}(\sigma) + \Gamma_{ij}(r, t) - \Delta_{ij}(r, t, \mathcal{M}_\sigma)$ // target DB
14          $\lfloor \; f_{\bar{t}}^{(r)}(\sigma) = f_{\bar{t}}^{(r)}(\sigma) + \Gamma_{ij}(r, \bar{t}) - \Delta_{ij}(r, \bar{t}, \mathcal{M}_\sigma)$ // non-target DB

15      $\tilde{R}' = \left\{ r \in R' \middle| f_t^{(r)}(\sigma) \geq \theta_t^{(r)} \right\}$ // determine frequent regions
       // output pattern sequence and correlated regions (if any)
16   print$\left(\text{prefix} \cdot \sigma, \left\{ r \in \tilde{R}' \middle| \chi^2 \left(f_t^{(r)}(\sigma), f_{\bar{t}}^{(r)}(\sigma), |D_t^{(r)}|, |D_{\bar{t}}^{(r)}| \right) \geq \theta \right\} \right)$
17   **if** $\tilde{R}' \neq \emptyset$ **then**   // check if at least one region is frequent
18        $\lfloor$ pg-mine(prefix $\cdot \sigma, \tilde{R}', \mathcal{M}_\sigma, T, t, \theta, \Sigma)$ // grow recursively

---

after obtaining *all* matches to an observation sequence $x^{(i)}$, as explained below. Since the match list $\mathcal{M}_\sigma$ is created by matching sequence after sequence, it is automatically sorted in ascending order according to i and j. The match list contains *all* possible start positions of patterns having the prefix σ. When growing the initial prefix σ by any $\sigma' \in \Sigma$, usually only a subset of all matches will also be matches for $\sigma\sigma'$ and thus relevant for frequency calculation. This allows to create projections of the initial match list $\mathcal{M}_\sigma$ which become smaller and smaller after extending σ with additional symbols.

For a match $m \in \mathcal{M}_\sigma$ starting at position j of sequence i and some region $r = (a, b) \in R$, $\Gamma_{ij}(r, t)$ and $\Gamma_{ij}(r, \bar{t})$ are the numbers of covered target and non-target database transactions, i.e. how much the occurrence of match m increases the target and non-target database frequencies of the current pattern for region r. For example, consider two matches of σ occurring at indices $j_1 = 12$ and $j_2 = 13$ in $\boldsymbol{x}^{(i)}$ and region $r = (-2, 0)$. Moreover, let $\boldsymbol{y}_{13}^{(i)} \neq t$, $\boldsymbol{y}_{14}^{(i)} = t$, $\boldsymbol{y}_{15}^{(i)} = t$. Then the first match covers label positions 13 and 14 ($\Gamma_{ij_1}(r, t) = 1$, $\Gamma_{ij_1}(r, \bar{t}) = 1$), the second match covers positions 14 and 15 ($\Gamma_{ij_2}(r, t) = 2$, $\Gamma_{ij_2}(r, \bar{t}) = 0$). Formally, $\Gamma_{ij}(r, t)$ is defined by

$$\left| \left\{ k \mid \boldsymbol{y}_k^{(i)} = t \wedge max(j - (b-1), -a, 0) \le k \le min(j - a, |\boldsymbol{y}^{(i)}| - b, |\boldsymbol{y}^{(i)}| - 1)) \right\} \right|. \tag{4}$$

Accordingly, $\Gamma_{ij}(r, \bar{t})$ can be obtained by considering only k with $\boldsymbol{y}_k^{(i)} \ne t$. The lowest valid index k of a covered label is given by maximization over the following conditions: the lowest index which can be covered by a match $(j - (b-1))$, the complete fit of the region within the observation $(-a)$ and that the index lies within the label sequence (0). Analogously, the highest valid covered label index follows from minimization over the highest index which can be covered by a match$(j - a)$, complete fit of the region within the observation $(|\boldsymbol{y}^{(i)}| - b)$ and that the index lies within the label sequence $(|\boldsymbol{y}^{(i)}| - 1)$.

For some label sequence positions, neighboring matches in an observation sequence might be counted multiple times, leading to statistical incorrectness. In the example above, σ is counted twice for the one transaction of label index 14. Thus, it may be necessary to subtract correction terms $\Delta_{ij}(r, t, \mathcal{M}_\sigma)$ for each match (note that this problem cannot occur for regions $r = (a, a+1)$ of size 1). It is not known beforehand which of the initial matches will be contained in later projections for prefix extensions. Due to this problem, we have to calculate the correction information according to the following procedure.

Let $\mathcal{M}_\sigma^{(i)}$ be the sublist of $\mathcal{M}_\sigma$ containing all matches of σ in sequence i in ascending order according to the match starting position j. Then, for each $m_e \in \mathcal{M}_\sigma^{(i)}$ with $e \in \{0, ..., |\mathcal{M}_\sigma^{(i)}| - 2\}$ we can calculate the *correction function* Δ_{ij} as follows. For each region $r = (a, b)$, we determine the set of other matches occurring *after* m_e in the i-th sequence which cover at least one position in the label sequence together with m_e, i.e. the set $\{m_f | m_f \in \mathcal{M}_\sigma^{(i)} \wedge e < f \le |\mathcal{M}_\sigma^{(i)}| - 1 \wedge start_f \le end_e\}$, where $start_f = max(m_f.j - (b-1), -a, 0)$ and $end_e = min(m_e.j - a, |\boldsymbol{y}^{(i)}| - b, |\boldsymbol{y}^{(i)}| - 1))$. The corresponding necessary correction in the target database frequency if m_e and m_f are occurring together can be computed as $|\{k \mid \boldsymbol{y}_k^{(i)} = t \wedge start_f \le k \le end_e\}|$ and accordingly for the non-target database by employing $\boldsymbol{y}_k^{(i)} \ne t$. These target and non-target frequency correction offsets for all co-occurring m_f are stored in a list (attached to m_e) which is sorted by ascending f. One such list $m_e.C^{(r)}$ is created independently per match m_e and region $r \in R$.

The target and non-target database frequencies of a pattern can be calculated by iterating over all matches and adding the affected target and non-target label positions given by $\Gamma_{ij}(r, t)$ and $\Gamma_{ij}(r, \bar{t})$ for the current match and each region. Multiple counts are removed subsequently by subtracting the correction function Δ_{ij}. Besides the region and whether to use the target or non-target label, the correction function takes into account the current database projection $\mathcal{M}_\sigma$ (in the beginning, the projection trivially is the full match list of σ). When correcting the added database counts of a match m_e in region r by $\Delta_{ij}(r, t, \mathcal{M}_\sigma)$ and $\Delta_{ij}(r, \bar{t}, \mathcal{M}_\sigma)$, we have to search for the *first* match m_f in $m_e.C^{(r)}$ which also occurs in the projection $\mathcal{M}_\sigma$ and subtract the corresponding target and non-target database correction terms. Any other entry $m_{f'}$ in the list after m_f which also would co-occur with m_e (only possible for regions having $b - a \ge 3$) must be

skipped since the corresponding correction takes place when the iteration proceeds to the next match m_f. The right correction term for each match m_e can be
obtained with only few comparisons because $m_e.C^{(r)}$ is sorted and the highest
co-occurring match index is known, which allows to leave out unnecessary comparisons. Using the target database frequencies, it is possible to determine the
regions $\tilde{R}' \subseteq R'$ where the pattern σ fulfills the minimum frequency constraint
on the target database. The set of regions where the pattern is correlated no less
than the minimum correlation constraint θ can be determined by computing the
χ^2 measure for all $r \in \tilde{R}'$. Together with the pattern sequence σ, the correlated
associated regions are immediately output as result (unless empty). The algo-
rithm is then applied recursively to the current prefix σ of length 1 by passing
the full match list $\mathcal{M}_\sigma$ as an argument.

In any recursive call (i.e., non-empty prefix), the *else* case (Line 4) first iterates over all entries m of the given projection $\mathcal{M}$ to create one new projected
match database $\mathcal{M}_\sigma$ per symbol σ in the pattern alphabet. Checking which extension of the given prefix occurs for each match m, m is appended to all $\mathcal{M}_\sigma$
where σ matches onto $x^{(i)}_{j+\text{offset}}$ (i.e. σ is the single literal or is a matching group
containing the literal). $\mathcal{M}_\sigma$ then contains all occurrences of the prefix extended
by σ (*prefix·σ*). Intuitively, the prefix is grown to longer patterns in projected
databases. Note that for the actual implementation, we use pseudo-projections
[10] containing pointers to the initial match list for efficiency reasons.

For each symbol σ, the frequencies of the prefix extended by σ can then be
determined using the new projections $\mathcal{M}_\sigma$. This yields the subset $\tilde{R}'$ of given
regions R' where the pattern is still frequent. In turn, calculating the correlation
measure for the pattern in all $r \in \tilde{R}'$ gives the regions where it is correlated. The
full pattern sequence (*prefix·σ*) together with the correlated regions is written
to the output unless the region set is empty. Then, the algorithm is applied
recursively to the grown prefix if there is at least one frequent region. Hence,
a prefix is extended recursively until the frequency in the target database falls
below the minimum frequency threshold in all regions $r \in R$.

Another version of the algorithm using the approach of Morishita and Sese
(in the following *smp*) can be obtained by two slight modifications. First, there
is no necessity to calculate minimum frequency constraints (Line 3). Second, the
pruning of regions by the minimum frequency criterion is replaced by checking
the upper bound on the correlation value of any extension of the current pattern
(Line 15). This set of *promising* regions of the pattern $c = \text{prefix} \cdot \sigma$ is given by

$$\tilde{R}' = \left\{ r \in R' \middle| max \left(\begin{matrix} \chi^2(f_t^{(r)}(c), 0, |D_t^{(r)}|, |D_{\bar{t}}^{(r)}|), \\ \chi^2(0, f_{\bar{t}}^{(r)}(c), |D_t^{(r)}|, |D_{\bar{t}}^{(r)}|) \end{matrix} \right) \geq \theta \right\} . \tag{5}$$

Instead of visiting all patterns fulfilling the minimum frequency constraint on the
target database, the algorithm now traverses all promising patterns. Note that
smp also finds patterns which are negatively correlated with the target label.
The approach by Nijssen and Kok (in the following *fpp*) finds such patterns in
some cases (patterns which are frequent in the target database, but have even
higher frequency in the non-target database). As we are interested in positively

correlated patterns and to give consistent results, our algorithm outputs only positively correlated regions of patterns, i.e. those where the pattern is over-represented in the target database ($O_{11} > E_{11}$).

The correctness of both versions of the algorithm can be shown by adopting the proof for the `PrefixSpan` algorithm [10]. Starting from the empty prefix ϵ and all regions $r \in R$, due to the prefix anti-monotonicity of the problem (Lemma 1) the algorithm visits all frequent (promising) patterns by extending frequent (promising) prefixes. This holds for all regions, since for any pattern sequence, the recursion proceeds until not a single region is frequent (promising) anymore. As the correlation measure is calculated for all frequent (promising) regions of a pattern, the algorithm finds all patterns in all regions with a minimum correlation of θ.

After mining, the identified correlated patterns and the induced rules can be transformed into features for many types of machine learning models for sequence labeling in a straightforward manner.

5   Experiments

For the assessment of the efficiency and practicability of our approach, we perform a number of experiments on different problems related to labeling biological sequences. The tasks involve protein secondary structure (*sec*), solvent accessibility (*acc*), β-turn (*beta*) and transmembrane helix (*tm*) prediction on a variety of datasets (Table 1). We define the alphabets $\Sigma_3 = \{$A, C , D, E, F, G, H, I, K, L, M, N, P, Q, R, S, T, V, W, Y$\}$ consisting of single amino acid literals, $\Sigma_2 = \Sigma_3 \cup \{$HKR, DE, FYW, VIL, STDNGA, DENQRS, VILMFA$\}$, which adds groups of amino acids with similar chemical properties, and $\Sigma_1 = \Sigma_2 \cup \{$EAL, VIYWF, PGND$\}$, which also contains groups of amino acids frequently found in helices, sheets and coils [7]. We evaluate several region sets which cover different amounts of sequence around the central position with increasing region sizes ($w15$-$s1 = \{(-7,6), \ldots, (7,8)\}$, $w15$-$s2 = \{(-7,-5), \ldots, (5,7)\}$, $w30$-$s1 = \{(-15,-15), \ldots, (15,16)\}$, $w30$-$s2 = \{(-15,-13), \ldots, (13,15)\}$, $w30$-$s3 = \{(-15,-12), \ldots, (12,15)\}$, $w30$-$s4 = \{(-15,-11), \ldots, (13,17)\}$, $w30$-$s5 = \{(-15,-10), \ldots, (10,15)\}$).

Both versions of our algorithm were implemented in C++ and compiled with gcc using the -O3 compiler flag. All experiments were performed on a Sun Fire X2200 M2 x64 dual-core node running Ubuntu Linux, using one single AMD Opteron 2.6 GHz CPU and 8 GB RAM. The implementation and all data are available for download from http://wwwkramer.in.tum.de/research/data_mining/pattern_mining/sequence_mining. We performed extensive experiments on combinations of the above parameters and datasets. Since the underlying trends are similar for the different problems and due to lack of space, we restrain this section to a discussion of the results for secondary structure prediction (*sec*). Note that all figures show the aggregated information from one run of the algorithm for each target label, i.e. the sum of runtimes and pattern numbers and the maximum memory usage.

Table 1. Overview of used datasets. Sequence labels are: (*sec*) helix H, sheet E, coil C; (*acc*) buried B, exposed E; (*beta*) turn T, non-turn N; (*tm*) inside I, membrane M, outside O.

problem	dataset	sequences	positions	label distribution	reference
sec	scopsfr	939	157813	H (36.8%), E (22.8%), C (40.4%)	[7]
	cb513	513	84119	H (34.6%), E (22.7%), C (42.8%)	[21]
	cb396	396	62189	H (35.4%), E (22.7%), C (41.8%)	[21]
	rs117	117	21930	H (32.0%), E (22.5%), C (45.5%)	[21]
acc	cb513	513	84119	B (55.8%), E (44.2%)	[21]
beta	bteval	426	96339	T (24.6%), N (75.4%)	[22]
tm	tm160	160	63114	I (26.4%), M (24.1%), O(49.5%)	[23]

As one would expect, we find that the runtime (Fig. 3(a)) increases with the number of matching groups in the alphabet and lower correlation thresholds (in the following, we will focus on experiments using the most complex alphabet Σ_1). The same holds for increasing region sizes (Fig. 3(b)) while keeping the amount of covered sequence constant (*w30-x*). When using the corresponding set of regions of the same size, but covering only 15 sequence positions (*w15-x*), the runtime is lower because the influence of matches has to be evaluated for a smaller number of regions (data not shown). In all performed experiments, the runtime closely follows the number of mined patterns (Fig. 3(c)), i.e. the algorithm works well for a wide range of minimum correlation thresholds. With regard to runtime, the *fpp*-based algorithm outperforms its *smp* counterpart. To investigate this difference, we log the number of recursive function calls to measure the amount of traversed search space (Fig. 3(d)). Surprisingly, the number of recursive calls is lower for *smp* which suggests that the imposed bound is stricter than the minimum frequency constraint of *fpp*. Profiling of both variants shows that the overhead for *smp* is generated by the higher number of χ^2 value calculations which are more expensive than checking a single frequency constraint.

A comparison of the results on datasets of increasing size (*rs117, cb396, cb513, scopsfr*) suggests that memory usage of the algorithm scales linearly with increasing dataset size (Fig. 4(a)). Memory usage is very similar for *fpp* and *smp*. Interestingly, it also differs only slightly when using different χ^2 thresholds, demonstrating the effectiveness of using projections of the initial match list $\mathcal{M}_\sigma$ in the recursive calls. As previously, the increase in runtime closely follows the increase in the number of identified patterns (Fig. 4(b), 4(c)). With increasing dataset size, we observe a stronger increase in the number of patterns and therefore runtimes. The increase is most pronounced for lower χ^2 thresholds and larger region sizes (Fig. 4(c), 4(d)), indicating additional information gain the larger the dataset is.

For a quantitative analysis of correlated patterns, we compute the coverage of positions in the sliding window (*w30-s1*) by increasing per pattern the count for each position a pattern overlaps, starting from the region(s) the pattern matches in (Fig. 5(a)). Highest coverage is found for the central position in the window, dropping rapidly from there. This corresponds well to the fact that

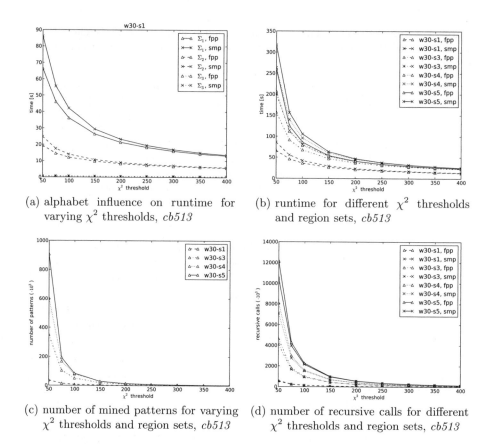

(a) alphabet influence on runtime for varying χ^2 thresholds, *cb513*

(b) runtime for different χ^2 thresholds and region sets, *cb513*

(c) number of mined patterns for varying χ^2 thresholds and region sets, *cb513*

(d) number of recursive calls for different χ^2 thresholds and region sets, *cb513*

Fig. 3. Influence of alphabet, region set and χ^2 threshold choice on runtime of the algorithm and the number of identified label-correlated patterns

many methods for secondary structure prediction use a sliding window of size 15. The length distribution of patterns (Fig. 5(a)) has its maximum in the range of 3 to 5 for different thresholds, which is concurrent with the locality of secondary structure elements. Comparing the results for *sec* to the other problems (*acc*, *beta*, *tm*), different distributions of pattern lengths and starting points can be observed (data not shown). For instance, *tm* patterns are considerably longer and numerous because of the distinct hydrophobic amino acid composition of transmembrane segments.

In summary, we find that our algorithm scales well (1) with increasing dataset sizes and (2) decreasing correlation thresholds. (3) Runtime is mostly determined by the number of identified sequential patterns, whereas (4) memory usage mainly depends on dataset size, but not on the chosen correlation threshold.

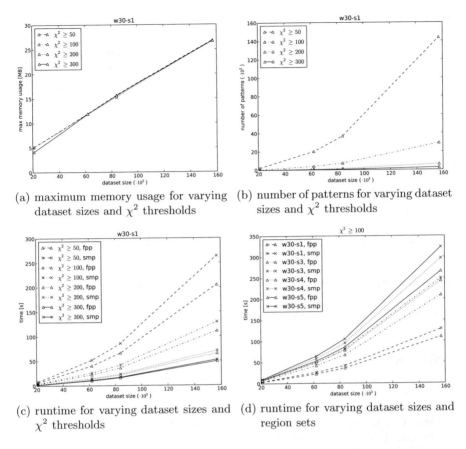

(a) maximum memory usage for varying dataset sizes and χ^2 thresholds

(b) number of patterns for varying dataset sizes and χ^2 thresholds

(c) runtime for varying dataset sizes and χ^2 thresholds

(d) runtime for varying dataset sizes and region sets

Fig. 4. Influence of dataset size on runtime, number of patterns and maximum memory usage

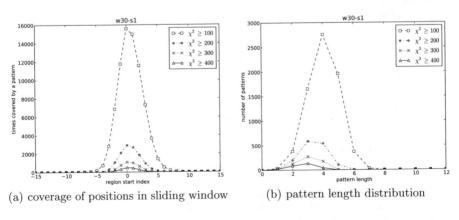

(a) coverage of positions in sliding window

(b) pattern length distribution

Fig. 5. Quantitative description of patterns mined on *cb513* with *w30-s1* region set

6   Conclusion

In this paper, we formally introduced the novel task of mining label-correlated sequence patterns. We presented an effective pattern growth based algorithm to solve the problem and showed its effectiveness in various experiments. While evaluated on biological sequences, the algorithm should be useful for many different variants of the sequence labeling problem. With only slight modifications, it would also be possible to adopt the algorithm to a top-k setting, other convex correlation measures or a multi-class setting with more than one target label [8,9]. Another promising option should be to use combinations of adjacent sequence labels as target labels when applying the method together with linear-chain CRFs, which can take into account previously assigned labels in their feature functions. Moreover, additional prefix anti-monotonic constraints and support for different thresholds on each region could be added to customize the identification of interesting patterns depending on the investigated problem.

Acknowledgments. We would like to thank Jun Sese for providing the improved AprioriSMP implementation and Anders Pedersen and Fabian Birzele for providing various datasets.

References

1. Dietterich, T.G.: Machine learning for sequential data: a review. In: Caelli, T.M., Amin, A., Duin, R.P.W., Kamel, M.S., de Ridder, D. (eds.) SPR 2002 and SSPR 2002. LNCS, vol. 2396, pp. 15–30. Springer, Heidelberg (2002)
2. Rabiner, L.R.: A tutorial on Hidden Markov Models and selected applications in speech recognition. Proc. of the IEEE 77(2), 257–286 (1989)
3. Lafferty, J., McCallum, A., Pereira, F.: Conditional Random Fields: probabilistic models for segmenting and labeling sequence data. In: Proc. of the 18th Int. Conf. on Machine Learning (ICML 2001), pp. 282–289. Morgan Kaufmann, San Francisco (2001)
4. Vapnik, V.N.: Statistical learning theory. Wiley, New York (1998)
5. Tsochantaridis, I., Joachims, T., Hofmann, T., Altun, Y.: Large margin methods for structured and interdependent output variables. Journal of Machine Learning Research 6, 1453–1484 (2005)
6. Bringmann, B., Nijssen, S., Zimmermann, A.: Pattern-based classification: a unifying perspective. In: ECML/PKDD-09 Workshop From Local Patterns to Global Models (2009)
7. Birzele, F., Kramer, S.: A new representation for protein secondary structure prediction based on frequent patterns. Bioinformatics 22, 2628–2634 (2006)
8. Morishita, S., Sese, J.: Traversing itemset lattices with statistical metric pruning. In: Proc. of the 19th ACM SIGMOD-SIGACT-SIGART Symposium on Principles of Database Systems (PODS 2000), pp. 226–236. ACM, New York (2000)
9. Nijssen, S., Kok, J.N.: Multi-class correlated pattern mining, extended version. In: Bonchi, F., Boulicaut, J.-F. (eds.) KDID 2005. LNCS, vol. 3933, pp. 165–187. Springer, Heidelberg (2006)

10. Pei, J., Han, J., Mortazavi-Asl, B., Pinto, H., Qiming, C., Dayal, U., Hsu, M.C.: PrefixSpan: mining sequential patterns efficiently by prefix-projected pattern growth. In: Proc. of the 17th Int. Conf. on Data Engineering (ICDE 2001), pp. 215–224. IEEE Computer Science, Washington (2001)
11. Agrawal, R., Srikant, R.: Mining sequential patterns. In: Proc. of the 11th Int. Conf. on Data Engineering (ICDE 1995), pp. 3–14. IEEE Computer Society, Washington (1995)
12. Srikant, R., Agrawal, R.: Mining sequential patterns: generalisations and performance improvements. In: Apers, P.M.G., Bouzeghoub, M., Gardarin, G. (eds.) EDBT 1996. LNCS, vol. 1057, pp. 3–17. Springer, Heidelberg (1996)
13. Agrawal, R., Srikant, R.: Fast algorithms for mining association rules. In: Proc. of the 20th Int. Conf. on Very Large Data Bases (VLDB 1994), pp. 487–499. Morgan Kaufmann, San Francisco (1994)
14. Zaki, M.J.: SPADE: an efficient algorithm for mining frequent sequences. Machine Learning 42, 31–60 (2001)
15. Ayres, J., Gehrke, J., Yiu, T., Flannik, J.: Sequential pattern mining using a bitmap representation. In: Proc. of the 8th ACM SIGKDD Int. Conf. on Knowledge Discovery and Data Mining (KDD 2002), pp. 429–435. ACM, New York (2002)
16. Han, J., Cheng, H., Xin, D., Yan, X.: Frequent pattern mining: current status and future directions. Data Mining and Knowledge Discovery 15(1), 55–86 (2007)
17. Bay, S.D., Pazzani, M.J.: Detecting change in categorical data: mining contrast sets. In: Proc. of the 5th ACM SIGKDD Int. Conf. on Knowledge Discovery and Data Mining (KDD 1999), pp. 302–306. ACM, New York (1999)
18. Nijssen, S., Guns, T., De Raedt, L.: Correlated itemset mining in ROC space: a constraint programming approach. In: Proc. of the 15th ACM SIGKDD Int. Conf. on Knowledge Discovery and Data Mining, pp. 647–656. ACM, New York (2009)
19. Hirao, M., Hoshino, H., Shinohara, A., Masayuki, T., Setsuo, A.: A practical algorithm to find the best subsequence patterns. In: Morishita, S., Arikawa, S. (eds.) DS 2000. LNCS (LNAI), vol. 1967, pp. 141–154. Springer, Heidelberg (2000)
20. Fischer, J., Mäkinen, V., Välimäki, N.: Space efficient string mining under frequency constraints. In: Proc. of the 8th Int. Conf. on Data Mining (ICDM 2008), pp. 193–202. IEEE Computer Society, Washington (2008)
21. Cuff, J.A., Barton, G.J.: Evaluation and improvement of multiple sequence methods for protein secondary structure prediction. Proteins 34(4), 508–519 (1999)
22. Kaur, H., Raghava, G.P.S.: An evaluation of beta-turn prediction methods. Bioinformatics 18, 1508–1514 (2002)
23. Sonnhammer, E.L.L., von Heijne, G., Krogh, A.: A Hidden Markov Model for predicting transmembrane helices in protein sequences. In: Proc. of the 6th Int. Conf. on Intelligent Systems for Molecular Biology (ISMB 1998), pp. 175–182. AAAI Press, Menlo Park (1998)

ESTATE: Strategy for Exploring Labeled Spatial Datasets Using Association Analysis

Tomasz F. Stepinski[1], Josue Salazar[1], Wei Ding[2], and Denis White[3]

[1] Lunar and Planetary Institute, Houston, TX 77058, USA
tom@lpi.usra.edu, salazar@lpi.usra.edu
[2] Department of Computer Science, University of Massachusetts Boston, Boston, MA 02125, USA
ding@cs.umb.edu
[3] US Environmental Protection Agency, Corvallis, OR 97333, USA
white.denis@epa.gov

Abstract. We propose an association analysis-based strategy for exploration of multi-attribute spatial datasets possessing naturally arising classification. Proposed strategy, ESTATE (**E**xploring **S**patial da**T**a **A**ssociation pat**TE**rns), inverts such classification by interpreting different classes found in the dataset in terms of sets of discriminative patterns of its attributes. It consists of several core steps including discriminative data mining, similarity between transactional patterns, and visualization. An algorithm for calculating similarity measure between patterns is the major original contribution that facilitates summarization of discovered information and makes the entire framework practical for real life applications. Detailed description of the ESTATE framework is followed by its application to the domain of ecology using a dataset that fuses the information on geographical distribution of biodiversity of bird species across the contiguous United States with distributions of 32 environmental variables across the same area.

Keywords: Spatial databases, association patterns, clustering, similarity measure, biodiversity.

1 Introduction

Advances in gathering spatial data and progress in Geographical Information Science (GIS) allow domain experts to monitor complex spatial systems in a quantitative fashion leading to collections of large, multi-attribute datasets. The complexity of such datasets hides domain knowledge that may be revealed through systematic exploration of the overall structure of the dataset. Often, datasets of interest either possess naturally present classification, or the classification is apparent from the character of the dataset and can be performed without resorting to machine learning. The purpose of this paper is to introduce a strategy for thorough exploration of such datasets. The goal is to discover all combinations of attributes that distinguish between the class of interest and the other classes in the dataset. The proposed strategy (ESTATE) is a tool for

B. Pfahringer, G. Holmes, and A. Hoffmann (Eds.): DS 2010, LNAI 6332, pp. 326–340, 2010.
© Springer-Verlag Berlin Heidelberg 2010

finding explanation and/or interpretations behind divisions that are observed in the dataset. Note that the aim of ESTATE is the reverse of the aim of classification/prediction tools; whereas a classifier starts from attributes of individual objects and outputs classes and their spatial extents, the ESTATE starts from the classes and their spatial extents and outputs the concise description of attribute patterns that best define the individuality of each class. The need for such classification-in-reverse tool arises in many domains, including cases that may influence economic and political decisions and have significant societal repercussions. For example, a fusion of election results with socio-economic indicators form an administrative region-based spatial dataset that can be explored using ESTATE to reveal a spatio-socio-economic makeup of electoral support for different office seekers [24]. The framework can be also utilized for analyzing a diversity of underlying drivers of change (temporal, spatial, and modal) in the spatial system. An expository example of spatial change analysis – pertaining to geographical distribution of biodiversity of bird species across the contiguous United States – is presented in this paper.

The ESTATE interprets the divisions within the dataset by exploring the structure of the dataset. The strategy is underpinned by the framework of association analysis [1,13,36] that assures that complex interactions between all attributes are accounted for in a model-free fashion. Specifically, we rely on the contrast data mining [10,2], a technique for identification of discriminative patterns – associative itemsets of attributes that are found frequently in the part of the dataset affiliated with the focus class but not in the remainder of the dataset. A collection of all discriminative patterns provides an exhaustive set of attribute dependencies found only in the focus class. These dependencies are interpreted as knowledge revealing what sets the focus class apart from the other classes. The set of dependencies for all classes is used to explain the divisions observed in the dataset.

The ESTATE framework consists of a number of independent modules; some of them are based on existing techniques while others represent original contributions. We present two original contributions to the field of data mining: 1) a novel similarity measure between itemsets that makes possible clustering of transactional patterns thus enabling effective summarization of thousands of discovered nuggets of knowledge, and 2) a strategy for disambiguation of class labels in datasets when classification is not naturally present and needs to be deduced from the character of the dataset.

2   Related Work

There is a vast literature devoted to classification/prediction techniques. In the context of spatial (especially, geospatial) datasets many broadly used predictors are based on the principle of regression, including multiple regression [26], logistic regression [32,7,15], Geographically Weighted Regression (GWR) [4,12], and kernel logistic regression [31]. These techniques are ill-suited for our stated purpose. A machine learning-based classifier could be constructed for the dataset in which all objects have prior labels (the entire dataset is the training set). Denoting a classifier function as $F : F(attributes) \rightarrow class$, its inverse $F^{-1} : F^{-1}(class) \rightarrow attributes$ would give a set of all of the objects (their attribute vectors) mapped to a given class. However, the outcome of F^{-1} would be of no help to our purpose because it does not provide any synthesis leading to the understanding the common characteristics of the objects belonging to a given class. The exception is the classification and regression tree (CART) classifier, whose hierarchical form of F allows interpretation of F^{-1}. Indeed, the use of regression trees was proposed [29] to map spatial divisions of class variable. However, for the CART classifier to work in this role, the number of terminal nodes needs to relatively small (because each node represent a cluster) but the nodes need to have high class label purity. These requirements are rarely fulfilled in real life applications. Moreover, the cluster description, as given by a series of conditions on features that define a specific terminal node in the tree, comments only on the small subset of the features and does not reveal interactions between different features. In short, CART works well as a classifier but it's less than ideal as a data exploration tool. Another potential alternative to ESTATE is to combine the GWR model with geovisual analytical exploration [8]. GWR is a regression model that yields local estimates of the regression parameters. Thus, the GWR assigns a vector of coefficients to each object in the dataset. Resulting multivariate space can be explored visually in order to find structure in the dataset. However, such strategy does not contrast the prior classes and the resulting clusters are grouping together objects on the basis similar regression models rather than similar attributes. ESTATE provides a natural, data-centric, model-free approach to dataset exploration that, by its very design, offers advantages over the approaches based on regression.

The possibility of using transactional patterns for exploration of spatial datasets received little attention. Application of association analysis to geospatial data was discussed in [11,23], and another application, to the land cover change was discussed in [20]. These studies did not utilized discriminative pattern mining. In addition, they lack any pattern synthesis techniques making the results difficult to interpret by domain scientists.

One of the major challenges of association analysis is the explosive number of identified patterns which leads to a need for pattern summarization. The two major approaches to pattern summarization are lossless and lossy representations. Lossless compression techniques include closed itemsets [21] and non-derivable itemsets [6]. In general, reduction in a number of patterns due to a lossless compression is insufficient to significantly improve interpretability of the results. More radical summarization is achieved via lossy compression techniques including maximal frequent patterns [3], top-k frequent patterns [14], top-k redundancy-aware patterns [27], profile patterns [34], δ-cover compressed patterns [33], and regression-based summarization [17].These techniques have been developed for categorical datasets where a notion of similarity between the items does not exist. The datasets we wish to explore with ESTATE are ordinal. We exploit the existence of an ordering information in the attributes of items to define a novel similarity between the itemsets. Our preliminary work on application of association analysis to exploration of spatial datasets is documented in [9,25].

3   ESTATE Framework

The ESTATE framework is applied to a dataset composed of spatial objects characterized by their geographical coordinates, attributes, and class labels. The spatial dataset can be in the form of a raster (objects are individual pixels), point data (objects are individual points), or shapefile (objects are polygons). Information in each object is structured as follows $o = \{x, y; f_1, f_2, ..., f_m; c\}$, where x and y are object's spatial coordinates, f_i, $i = 1, \ldots, m$, are values of m attributes as measured at (x, y), and c is the class label. From the point of view of association analysis, each object (after disregarding its spatial coordinates and its class label) is a transaction containing a set of exactly m items $\{f_1, f_2, ..., f_m\}$, which are assumed to have ordinal values. The entire spatial dataset can be viewed as a set of N fixed-length transactions, where N is the size of the dataset.

An itemset (hereafter also referred to as a pattern) is a set of items contained in a transaction. For example, assuming $m = 10$, $P = \{2, \_, \_, \_, 3, \_, \_, \_, \_, \_\}$ is a pattern indicating that $f_1 = 2$, $f_5 = 3$ while the values of all other attributes are not a part of this pattern. A transaction *supports* an itemset if the itemset is a subset of this transaction; the number of all transactions supporting a pattern is refereed to as a *support* of this pattern. For example, any transaction with $f_1 = 2$, $f_5 = 3$ "supports" pattern P regardless of the values of attributes in slots denoted by an underscore symbol in the representation of P given above. The support of pattern P is the number of transactions with $f_1 = 2$, $f_5 = 3$. Because transactions have spatial locations, there is also a spatial manifestation of support which we call a *footprint* of a pattern. For example a footprint of P is a set of spatial objects characterized by $f_1 = 2$, $f_5 = 3$.

The ESTATE framework consists of the following modules: (1) Mining for associative patterns that discriminate between two classes in the dataset (Section 3.1). (2) Disambiguating class labels so the divisions of objects into different classes coincide with footprints of discriminative patterns (Section 3.2). (3) Clustering all discriminative patterns into a small number of clusters representing diverse motifs of attributes associated with a contrast between the two classes (Section 3.3). (4) Visualizing the results in both attribute and spatial domains (see the case study in Section 4).

3.1   Mining for Discriminative Patterns

Without loss of generality we consider the case of the dataset with only two classes: $c = 1$ and $c = 0$. A *discriminating* pattern X is an itemset that has much larger support within a set of transactions $\mathcal{O}_p$ stemming from $c = 1$ objects than within a set of transactions $\mathcal{O}_n$ stemming from $c = 0$ objects. For a pattern X to be accepted as a discriminating pattern, its growth rate, $\frac{sup(X, \mathcal{O}_p)}{sup(X, \mathcal{O}_n)}$, must exceed a predefined threshold δ, where $sup(X, \mathcal{O})$ is the support of X in a dataset $\mathcal{O}$.

We mine for *closed* patterns that are *relatively* frequent in $\mathcal{O}_p^0$. A pattern is frequent if its support (in $\mathcal{O}_p^0$) is larger than a predefined threshold. Mining for frequent patterns reduces computational cost. Further significant reduction in computational cost is achieved by mining only for frequent closed patterns [22]. A closed pattern is a maximal set of items shared by a set of transactions. A closed pattern can be viewed as lossless compression of all non-closed patterns that can be derived from it. Mining only for closed patterns makes physical and computational sense inasmuch as closed patterns give the most detailed motifs of attributes associated with difference between the two classes.

3.2 Disambiguating Class Labels

In many (but not all) practical application, the class labels are implicit rather than explicit. For example, biodiversity index is continuously distributed across the United States without a naturally occurring boundary between "high bio-diversity" (class $c = 1$) and "not-high biodiversity" (class $c = 0$) objects. This introduces a question of what is the best way to partition the dataset into the two classes? One way is to divide the objects using distribution-deduced threshold on the class variable, another is to use the union of footprints of mined discriminative patterns. These two methods will result in different partitions of the dataset introducing potential ambiguity to class labels. We propose to disambiguate the labeling by iterating between the two definitions until the two partitions are as close to each other as possible.

We first calculate the initial $\mathcal{O}_p^0$–$\mathcal{O}_n^0$ partition using a threshold on the value of the class variable. Using this initial partition, our algorithm mines for discriminating patterns. We calculate a footprint of each pattern and the union of all footprints. The union of the footprints intersects, but is not identical to the footprint of $\mathcal{O}_p^0$. Second, we calculate the next iteration of the partition $\mathcal{O}_p^1$–$\mathcal{O}_n^1$ and the new set of discriminating patterns. The objects that were initially in $\mathcal{O}_n^0$ are added to $\mathcal{O}_p^1$ if they are in the union of footprints of the patterns calculated in first step, their values of class variable are "high enough", and they are neighbors of $\mathcal{O}_p^0$. Because of this last requirement, the second step is in itself an iterative procedure. The requirement that incorporated objects have "high enough" values of class variable is fulfilled by defining a buffer zone. The buffer zone is easily defined in a dataset of ordinal values; it consists of objects having a value one less than the minimum value allowed in $\mathcal{O}_p^0$. Finally, we repeat the second step calculating $\mathcal{O}_p^i$ and its corresponding set of discriminating patterns from the results of $i-1$ iteration until the iteration process converges. Note that convergence is assured by the design of the process. The result is the optimal $\mathcal{O}_p$–$\mathcal{O}_n$ partition and the optimal set of discriminating patterns.

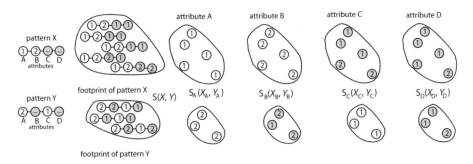

Fig. 1. Graphics illustrating the concept of similarity between two patterns. White items are part of the pattern, gray items are not the part of the pattern.

3.3   Pattern Similarity Measure

Despite considering only frequent closed discriminative patterns, the ESTATE finds thousands of patterns. A single pattern provides a specific combination of attribute values found in a specific subset of the $c = 1$ class of objects but nonexistent or rare among $c = 0$ class objects. The more specific (longer) the pattern the smaller is its footprint; patterns having larger spatial presence tend to be less specific (shorter). Because of this tradeoff there is not much we can learn about the global structure of the dataset from a single pattern; such pattern provides either little information on regional scale or a lot of information on local scale. In order to effectively explore the entire dataset we need to consider all mined patterns each covering only relatively small spatial patch, but together covering the entire domain of the $c = 1$ class. To enable such exploration we cluster the patterns into larger aggregates of similar patterns by taking advantage of ordering information contained in ordinal attributes of spatial objects. The clustering is made possible by the introduction of a similarity measure between the patterns. We propose to measure a similarity between two patterns as a similarity between their footprints. Hereafter we will continue to refer to the "pattern similarity measure" with the understanding that the term "pattern" is used as a shortcut for the set of objects in its footprint.

Fig. 1 illustrates the proposed concept of pattern similarity. In this simple example each object has four attributes denoted by A, B, C, and D, respectively. Each attribute has only one of two possible values: 1 or 2. Pattern $X = \{1, 2, \_, \_\}$ is supported by 5 objects and pattern $Y = \{2, \_, 1, \_\}$ is supported by 3 objects. The similarity between patterns X and Y is the similarity between the two sets of 4-dimensional vectors constructed from the values of items in transactions belonging to respective footprints. Similarity of each dimension (attribute) is calculated separately as a similarity between two sets of scalar entities. The total similarity is the weighted sum of the similarities of all attributes.

The similarity between patterns X and Y is $S(X, Y) = \sum_{i=1}^{m} w_i S_i(X_i, Y_i)$, where X_i, Y_i indicate the ith attribute, w_i indicates the ith weight (we use $w_i = 1$ in our calculations), and m is the number of attributes. The similarity between ith attribute in the two patterns $S_i(X_i, Y_i)$ is calculated using group average, a technique similar to the UPGMA (Unweighted Pair Group Method with Arithmetic mean) [19] method of calculating linkage in agglomerative clustering. The UPGMA method reduces to $S_i(X_i, Y_i) = s(x_i, y_i)$ for attributes which are present in both patterns (like an attribute A in an example shown in Fig. 1); here x_i and y_i are the values of attributes X_i and Y_i ($x_A = 1$ and $y_A = 2$ in the example on Fig. 1) and $s(x_i, y_i)$ is the similarity between those values (see below). If the ith attribute is present in the pattern Y but absent in the pattern X (like an attribute C in an example shown in Fig. 1) the UPGMA method reduces to

$$S(-, Y_i) = \sum_{k=1}^{n} P_X(x_k)s(z_k, y_i) \tag{1}$$

where $P_X(x_k)$ is the probability of ith attribute having the value x_k in all objects belonging to the footprint of X and n is the number of different values the ith attribute can have. The UPGMA reduces to an analogous formula if the ith attribute is present in the pattern X but it's absent in the pattern Y (like an attribute B in an example shown in Fig. 1). Finally, if the ith attribute is absent in both patterns (like an attribute D in an example shown in Fig. 1) the UPGMA gives

$$S(-_i, -_i) = \sum_{l=1}^{n} \sum_{k=1}^{n} P_X(x_l)P_Y(y_k)s(x_l, y_k) \tag{2}$$

We propose to calculate the similarity between the two values of ith attribute using a measure inspired by an earlier concept of measuring similarities between ordinal variables using information theory [18]. The similarity between two ordinal values of same attribute $s(x_i, y_i)$ is measured by the ratio between the amount of information needed to state the commonality between x_i and y_i, and the information needed to fully describe both x_i and y_i.

$$s(x_i, y_i) = \frac{2 \times \log \ P(x_i \vee z_1 \vee z_2 \ldots \vee z_k \vee y_i)}{\log \ P(x_i) + \log \ P(y_i)} \tag{3}$$

where $z_1, z_2, \ldots, z_k$ are ordinal values such that $z_1 = x_i + 1$ and $z_k = y_i - 1$. Probabilities, $P()$, are calculated using the known distribution of the values of ith attribute in $\mathcal{O}_\mathrm{p}$.

Using a measure of "distance" ($dist(X, Y) = \frac{1}{S(X,Y)} - 1$) between each pair of patterns in the set of discriminative patterns we construct a distance matrix. In order to gain insight into the structure of the set of discriminative patterns we visualize the distance matrix using clustering heat map [30]. The heat map is the distance matrix with its columns and rows rearranged to place rows and columns representing similar patterns near each other. We determine an appropriate order of rows and columns in the heat map by performing a hierarchical clustering (using an average linkage) of the set of discriminative patterns and sorting the rows and columns by the resultant dendrogram. The values of distances in the heat map are coded by a color gradient enabling the analyst to visually identify interesting clusters of patterns.

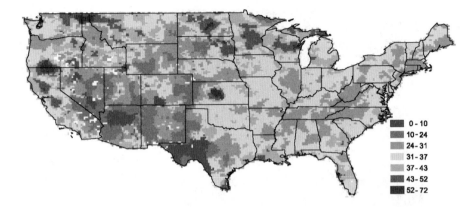

Fig. 2. Biodiversity of bird species across the contiguous United States. Two categories with the highest values of biodiversity (purple and red) are chosen as the initial high biodiversity region. Missing data regions are shown in white.

4   Case Study: Biodiversity of Bird Species

We apply the ESTATE framework to the case study pertaining to the discovery of associations between environmental factors and the spatial distribution of biodiversity across the contiguous United States. Roughly, biodiversity is a number of different species (of plants and/or animals) within a spatial region. A pressing problem in biodiversity studies is to find the optimal strategy for protecting the species given limited resources. In order to design such a strategy it is necessary to understand associations between environmental factors and the spatial distribution of biodiversity. In this context we apply ESTATE to discover existence of different environments (patterns or motifs of environmental factors) which associate with the high levels of biodiversity.

The database is composed of spatial accounting units resulting from tessellation of the US territory into equal area hexagons with center-to-center spacing of approximately 27 km. For each unit the measure of biodiversity (class variable) and the values of environmental variables (attributes) are given. The biodiversity measure is provided [35] by the number of species of birds exceeding a specific threshold of probability of occurrence in a given unit. Fig. 2 shows the distribution of biodiversity measure across the contiguous US. The environmental attributes [28] include terrain, climatic, landscape metric, land cover, and environmental stress variables that are hypothesized to influence biodiversity; we consider $m=32$ such attributes. The class variable and the attributes are discretized into up to seven ordinal categories (lowest, low, medium-low, medium, medium-high, high, highest) using the "natural breaks" method [16].

Because of the technical demands of the ESTATE label disambiguation module we have transformed the hexagon-based dataset into the square-based dataset. Each square unit (pixel) has a size of 22 × 22 km and there are $N=21039$ data-carrying pixels in the transformed dataset. The dataset does not have explicit

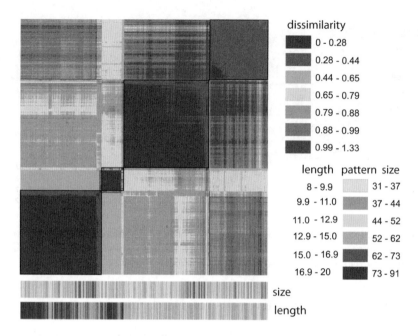

Fig. 3. Clustering heat map illustrating pairwise similarities between pairs of patterns in the set of 1503 discriminating patterns. The two bars below the heat map illustrate size of the pattern size and its length, respectively.

labels. Because we are interested in contrasting the region characterized by high biodiversity with the region characterized by not-high biodiversity we have partitioned the dataset into $\mathcal{O}_p$ corresponding to $c = 1$ class and consisting initially of the objects having high and highest categories of biodiversity and $\mathcal{O}_n$ corresponding to $c = 0$ class and consisting initially of the objects having lowest to medium-high categories of biodiversity. The label disambiguation module modifies the initial partition during the consecutive rounds of discriminative data mining.

We identify frequent closed patterns discriminating between $\mathcal{O}_p$ and $\mathcal{O}_n$ using an efficient depth-first search method [5]. We mine for patterns having growth rate ≥ 50 which are fulfilled by at least 2% of transactions (pixels) in $\mathcal{O}_p$. We also keep only the patterns that consist of eight or more attributes; shorter patterns are not specific enough to be of interest to us. We have found 1503 such patterns. The patterns have lengths between 8 and 20 attributes; the pattern length is broadly distributed with the maximum occurring at 12 attributes. Pattern size (support) varies from 31 to 91 pixels; the distribution of pattern size is skewed toward the high values and the maximum occurs at 40 pixels.

Fig. 3 shows a heat map constructed from a distance (dissimilarity) matrix calculated for all pairs of patterns in the set of 1503 patterns that discriminate between $\mathcal{O}_p$ and $\mathcal{O}_n$. The heat map is symmetric because distance between any two patterns is calculated twice. Deep purple and red colors indicate similar patterns whereas blue and green colors indicate dissimilar patterns. The heat map

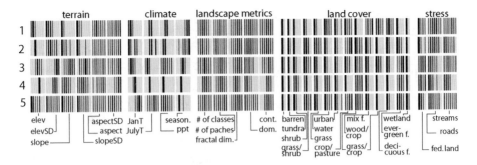

Fig. 4. Bar-code representation of the five regimes (clusters) of high biodiversity. See description in the main text.

clearly shows that the entire set of discriminative patterns naturally breakes into four clusters as indicated by purple and red color blocks on the map. Indeed, there are five top level clusters, but the fourth cluster, counting from the lower left corner, has only 4 patterns and is not visible in the heat map at the scale of Fig. 3. The patterns in each cluster identify similar combinations (motifs) of environmental attributes that are associated with the region of high biodiversity. The visual analysis of the heat map indicates that there are four (five if we count the small 4-pattern cluster) distinct motifs of environmental attributes associated with high levels of biodiversity. Potentially, these motifs indicate existence of multiple environmental regimes that differ from each other but are all conducive to high levels of biodiversity.

The clusters can be characterized and compared from two different perspectives. First, we can synthesis the information contained in all patterns belonging to each cluster; this will yield combinations of attributes that set apart the region associated with a given cluster from the not-high biodiversity region. Second, we can synthesize the information about prevailing attributes in the region associated with a given cluster; this will reveal a set of predominant environmental conditions associated with a given high biodiversity region (represented by a cluster). Because clusters are agglomerates of patterns and regions are agglomerates of transactions, they can be synthesized by their respective compositions. The biodiversity dataset has $m = 32$ attributes, thus each cluster (region) can be synthesized by 32 histograms, each corresponding to a composition of a particular attribute within a cluster (region). Our challenge is to present this large volume of information in a manner that is compact enough to facilitate immediate comparison between different clusters.

In this paper we restrict ourself to synthesizing and presenting the predominate environmental conditions associated with each of the five clusters identified in the heat map. Recall that the attributes are categorized into 7, 4, or 2 ordinal categories, thus a histogram representing a distribution of the values taken by an attribute in a given cluster consists of up to seven percentage-showing numbers. Altogether, 173 numbers, ranging in values from 0 (absence of a given attribute from cluster composition) to 1 (only a single value of a given attribute is present

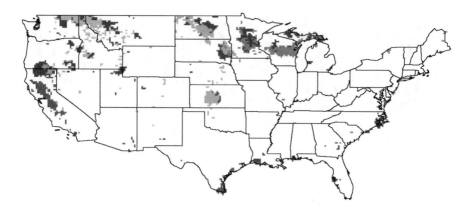

Fig. 5. Spatial footprints of five pattern clusters. White – not high biodiversity region; gray – high biodiversity region; purple (cluster #1), light green (cluster #2), yellow (clister #3), blue (cluster #4), and red (cluster #5) – footprints of the five clusters.

in a cluster) represents a summary of a cluster. We propose a bar-code representation of such summary. Such representation facilitates quick qualitative comparison between different clusters. Fig. 4 shows the bar-coded description for the five clusters corresponding to different biodiversity regimes. A cluster bar-code contains 32 fragments each describing a composition of a single attribute within a cluster. In Fig. 4 these fragments are grouped into five thematic categories: terrain (6 attributes), climate (4 attributes), landscape elements (5 attributes), land cover (14 attributes), and stress (3 attributes). Each fragment has up to seven vertical bars representing ordered categories of the attribute its represent. If a given category is absent within a cluster the bar is gray; black bars with increasing thickness denote categories with increasingly large presence in a cluster.

The five regimes of high biodiversity differs on the first four terrain attributes and all climate attributes. The landscape metrics attributes are similar except for regime #4. Many land cover attributes are similar indicating that a number of land cover types, such as, for example, tundra, barren land or urban are absent in all high biodiversity regimes. More in depth investigation of the bar codes reveals that the regime #1 is dominated by the crop/pasture cover, the regime #2 by the wood/crop cover, the regime #3 by the evergreen forest, and the regimes # 4 and #5 are not dominated by any particular land cover. Finally, environmental stress attributes are similar except for the federal land that is more abundant within the regions defined by the regimes #3 and #4.

Spatial manifestation of the five clusters identified in the heat map are shown in Fig. 5 where transactions (pixels) fulfilled by patterns belonging to different clusters are indicated by different colors. Interestingly, different environmental regimes (clusters) are located at distinct geographical locations. This geographical separation of the clusters is the result and not a build-in feature of our method. In principle, footprints of different discriminative patterns may overlap, and footprints of the entire clusters may overlap as well. It is a property of the biodiversity dataset that clusters of similar discriminative patterns have non-overlapping footprints.

Note that in our calculations the label disambiguation module did not achieve complete reconciliation between the region of high biodiversity and the union of support of all discriminating patterns. The gray pixels on Fig. 4 indicate trans-actions that are in the $\mathcal{O}_p$ but are not in the union of support of all the patterns. The ESTATE guarantees convergence of the disambiguation module but does not guarantee the complete reconciliation of the two regions. However, perfect correspondence is not required and, in fact, less than perfect correspondence provides some additional information. The gray areas on Fig. 4 represents atypical regions characterized by infrequent combinations of environmental attributes.

5   Discussion

A machine learning task of predicting labels of class variable using explanatory variables became an integral component of spatial analysis and is broadly utilized in many domains including geography, economy, and ecology. However, many interesting spatial datasets possesses natural labels, or their labels can be easily classified without resorting to machine-learning methods. We have developed the ESTATE framework in order to understand such naturally occurring divisions in terms of dataset attributes. In a broad sense, the purpose of ESTATE is reverse to the purpose of a classification.

ESTATE is a unique data exploration algorithm and thus it's output is not directly comparable to the outputs of other data exploration algorithms. As we mentioned in section 3, regression tree cartography [29] is the approach most similar in its goals to that of the ESTATE. The limited length of this paper prevents the detailed comparison of the two approaches, especially because the two methods yield different outputs that are not directly comparable. Classi-fication rather than regression tree needs to be used in the tree cartography method for closer correspondence to the ESTATE method. The methodology suffers from shortcomings described in section 3: it is unable to yield a relatively small number of label-pure terminal nodes. Thus, the two classes (for example, high biodiversity and not-high biodiversity) cannot be clearly separated. Each terminal node is characterized by a small number of conditions on few environ-mental attributes. These are the attributes that most clearly divide the data at any note of the tree. This makes sense for classification purposes, but not for data exploration purposes. The ESTATE provides reacher and more precise de-scription of the entire patterns of attributes that discriminate between the two classes.

Many real life problems analyzable by ESTATE may be formulated in terms of "spatial change" datasets (class labels change from one location to another). Other real life problems, analyzable by ESTATE, may be formulated as "tem-poral change" datasets (class labels indicate presence or absence of change in measurements taken at different times), or "modal change" datasets (class la-bels indicate agreement or disagreement between modeled and actual spatial system). An expository example given in Section 4 belongs to the spatial change dataset type. The biodiversity dataset has "natural" classes inasmuch as it can be divided into high and no-high biodiversity parts just on the basis of the distribution of biodiversity measure. Note that classes other than "high" can be as easily defined; for example, for a complete evaluation of the biodiversity dataset we would also define a "low" class. Other datasets (see, for example, [24]) have prior classes and require no additional pre-processing.

It is noted that ESTATE (like most other data discovery techniques) discovers associations and not causal relations. In the context of the biodiversity dataset it means that ESTATE has found five different environments that associate with high biodiversity but it does not proof actual causality between those environments and high levels of biodiversity. It is up to the domain experts to review the results and draw the conclusions. The causality is strongly suggested if the experts believe that the 32 attributes used in the calculation exhaust the set of viable controlling factors of biodiversity.

A crucial component of the ESTATE is the pattern similarity measure that enables clustering of similar patterns into agglomerates. We stress that our method does not use patterns to cluster objects, instead patterns themselves (more precisely their footprints) are the subject of clustering. This methodology can be applied outside of the ESTATE framework for summarization of any transactional patterns as long as their items consist of ordinal variables. Future research would address how to extend our similarity measure to categorical variables.

Acknowledgements

This work was partially supported by the National Science Foundation under Grant IIS-0812271.

References

1. Agrawal, R., Swami, A.N.: Fast algorithms for mining association rules. In: Proc. VLDB, pp. 487–499 (1994)
2. Bay, S.D., Pazzani, M.J.: Detecting change in categorical data: Mining contrast sets. In: Knowledge Discovery and Data Mining, pp. 302–306 (1999)
3. Bayardo Jr., R.J.: Efficiently mining long patterns from databases. In: SIGMOD 1998: Proceedings of the 1998 ACM SIGMOD International Conference on Management of Data, Seattle, Washington, United States, pp. 85–93 (1998)
4. Brunsdon, C.A., Fotheringham, A.S., Charlton, M.B.: Geographically weighted regression: a method for exploring spatial nonstationarity. Geographical Analysis 28, 281–298 (1996)
5. Burdick, D., Calimlim, M., Gehrke, J.: Mafia: a maximal frequent itemset algorithm for transactional databases. In: Proceedings of the 17th International Conference on Data Engineering, Heidelberg, Germany (2001)
6. Calders, T., Goethals, B.: Non-derivable itemset mining. Data Min. Knowl. Discov. 14(1), 171–206 (2007)

7. Cheng, J., Masser, I.: Urban growth pattern modeling: a case study of Wuhan City, PR China. Landscape and Urban Planning 62(4), 199–217 (2003)
8. Demar, U., Fotheringham, S.A., Charlton, M.: Combining geovisual analytics with spatial statistics: the example of Geographically Weighted Regression. The Cartographic Journal 45(3), 182–192 (2008)
9. Ding, W., Stepinski, T.F., Salazar, J.: Discovery of geospatial discriminating patterns from remote sensing datasets. In: Proceedings of SIAM International Conference on Data Mining (2009)
10. Dong, G., Li, J.: Efficient mining of emerging patterns: discovering trends and differences. In: KDD 1999: Proceedings of the Fifth ACM SIGKDD International Conference on Knowledge Discovery and Data Mining, San Diego, California, United States, pp. 43–52 (1999)
11. Dong, J., Perrizo, W., Ding, Q., Zhou, J.: The application of association rule mining to remotely sensed data. In: 345 (ed.) Proc. of the 2000 ACM Symposium on Applied Computing (2000)
12. Fotheringham, A.S., Brunsdon, C., Charlton, M.: Geographically Weighted Regression: the analysis of spatially varying relationships. Wiley, Chichester (2002)
13. Han, J., Pei, J., Yin, Y., Mao, R.: Mining frequent patterns without candidate generation: A frequent-pattern tree approach. Data Mining and Knowledge Discovery 8(1), 53–87 (2004)
14. Han, J., Wang, J., Lu, Y., Tzvetkov, P.: Mining top-k frequent closed patterns without minimum support. In: ICDM 2002: Proceedings of the 2002 IEEE International Conference on Data Mining, Washington, DC, USA, p. 211 (2002)
15. Hu, Z., Lo, C.: Modeling urban growth in Atlanta using logistic regression. Computers, Environment and Urban Systems 31(6), 667–688 (2007)
16. Jenks, G.F.: The data model concept in statistical mapping. International Yearbook of Cartography 7, 186–190 (1967)
17. Jin, R., Abu-Ata, M., Xiang, Y., Ruan, N.: Effective and efficient itemset pattern summarization: regression-based approaches. In: KDD 2008: Proceeding of the 14th ACM SIGKDD International Conference on Knowledge Discovery and Data Mining, Las Vegas, Nevada, USA, pp. 399–407 (2008)
18. Lin, D.: An information-theoretic definition of similarity. In: International Conference on Machine Learning, Madison, Wisconsin (July 1998)
19. McQuitty, L.: Similarity analysis by reciprocal pairs for discrete and continuous data. Educational and Psychological Measurement 26, 825–831 (1966)
20. Mennis, J., Liu, J.W.: Mining association rules in spatio-temporal data: An analysis of urban socioeconomic and land cover change. Transactions in GIS 9(1), 5–17 (2005)
21. Pasquier, N., Bastide, Y., Taouil, R., Lakhal, L.: Discovering frequent closed itemsets for association rules. In: Beeri, C., Bruneman, P. (eds.) ICDT 1999. LNCS, vol. 1540, pp. 398–416. Springer, Heidelberg (1998)
22. Pasquier, N., Bastide, Y., Taouil, R., Lakhal, L.: Discovering frequent closed itemsets for association rules. In: Beeri, C., Bruneman, P. (eds.) ICDT 1999. LNCS, vol. 1540, pp. 398–416. Springer, Heidelberg (1998)
23. Rajasekar, U., Weng, Q.: Application of association rule mining for exploring the relationship between urban land surface temperature and biophysical/social parameters. Photogrammetric Engineering & Remote Sensing 75(3), 385–396 (2009)
24. Stepinski, T., Salazar, J., Ding, W.: Discovering spatio-social motifs of electoral support using discriminative pattern mining. In: Proceedings of COM.geo. 2010 1st International Conference on Computing for Geospatial Reserch & Applications (2010)

25. Stepinski, T.F., Ding, W., Eick, C.F.: Controlling patterns of geospatial phenomena. submitted to Geoinformatica (2010)
26. Theobald, D.M., Hobbs, N.T.: Forecasting rural land use change: a comparison of regression and spatial transition-based models. Geographical and Environmental Modeling 2, 65–82 (1998)
27. Wang, C., Parthasarathy, S.: Summarizing itemset patterns using probabilistic models. In: KDD 2006: Proceedings of the 12th ACM SIGKDD International Conference on Knowledge Discovery and Data Mining, Philadelphia, PA, USA, pp. 730–735 (2006)
28. White, D., Preston, B., Freemark, K., Kiester, A.: A hierarchical framework for conserving biodiversity. In: Klopatek, J., Gardner, R. (eds.) Landscape Ecological Analysis: Issues and Applications, pp. 127–153. Springer, New York (1999)
29. White, D., Sifnenos, J.C.: Regression tree cartography. J. Computational and Graphical Statistics 11(3), 600–614 (2002)
30. Wilkinson, L., Friendly, M.: The history of the cluster heat map. The American Statistician 63(2), 179–184 (2009)
31. Wu, B., Huang, B., Fung, T.: Projection of land use change patterns using kernel logistic regression. Photogrammetric Engineering & Remote Sensing 75(8), 971–979 (2009)
32. Wu, F., Yeh, A.G.: Changing spatial distribution and determinants of land development in Chinese cities in the transition from a centrally planned economy to a socialist market economy: A case study of Guangzhou. Urban Studies 34(11), 1851–1879 (1997)
33. Xin, D., Han, J., Yan, X., Cheng, H.: Mining compressed frequent-pattern sets. In: VLDB 2005: Proceedings of the 31st International Conference on Very Large Data Bases, Trondheim, Norway, pp. 709–720 (2005)
34. Yan, X., Cheng, H., Han, J., Xin, D.: Summarizing itemset patterns: a profile-based approach. In: KDD 2005: Proceedings of the eleventh ACM SIGKDD International Conference on Knowledge Discovery in Data Mining, Chicago, Illinois, USA, pp. 314–323 (2005)
35. Yang, K., Carr, D., O'Connor, R.: Smoothing of breeding bird survey data to produce national biodiversity estimates. In: Proceeding of the 27th Symposium on the Interface Computing Science and Statistics, pp. 405–409 (1995)
36. Zaki, M., Ogihara, M.: Theoretical foundations of association rules. In: 3rd ACM SIGMOD Workshop on Research Issues in Data Mining and Knowledge Discovery (1998)

Adapted Transfer of Distance Measures for Quantitative Structure-Activity Relationships

Ulrich Rückert[1], Tobias Girschick[2], Fabian Buchwald[2], and Stefan Kramer[2]

[1] International Computer Science Institute,
Berkeley, USA
`rueckert@eecs.berkeley.edu`
[2] Technische Universität München,
Institut für Informatik/I12,
85748 Garching b. München, Germany
`{tobias.girschick,fabian.buchwald,stefan.kramer}@in.tum.de`

Abstract. Quantitative structure-activity relationships (QSARs) are regression models relating chemical structure to biological activity. Such models allow to make predictions for toxicologically or pharmacologically relevant endpoints, which constitute the target outcomes of trials or experiments. The task is often tackled by instance-based methods (like k-nearest neighbors), which are all based on the notion of chemical (dis-)similarity. Our starting point is the observation by Raymond and Willett that the two big families of chemical distance measures, fingerprint-based and maximum common subgaph based measures, provide orthogonal information about chemical similarity. The paper presents a novel method for finding suitable combinations of them, called adapted transfer, which adapts a distance measure learned on another, related dataset to a given dataset. Adapted transfer thus combines distance learning and transfer learning in a novel manner. In a set of experiments, we compare adapted transfer with distance learning on the target dataset itself and inductive transfer without adaptations. In our experiments, we visualize the performance of the methods by learning curves (i.e., depending on training set size) and present a quantitative comparison for 10% and 100% of the maximum training set size.

1 Introduction

Quantitative structure-activity relationships (QSARs) are models quantitatively correlating chemical structure with biological activity or chemical reactivity. In technical and statistical terms, QSARs are often regression models on graphs (molecular structures being modeled as graphs). QSARs and small molecules are subject of very active research in data mining [1,2]. The task is often tackled by instance-based and distance-based methods, which predict biological activity based on the similarity of structures. As the success of those methods critically depends on the availability of a suitable distance measure, it would be desirable to automatically determine a measure that works well for a given dataset and

B. Pfahringer, G. Holmes, and A. Hoffmann (Eds.): DS 2010, LNAI 6332, pp. 341–355, 2010.

endpoint. Recently proposed solutions for other, related problems (general classification problems instead of domain-specific regression problems as discussed here) include distance learning methods [3] and methods from inductive transfer [4]. In distance learning, the distance measures (e.g., parameterized distances like the Mahalanobis distance) are directly learned from labeled training distances. Inductive transfer is concerned with transferring the bias of one learning task to another, related task.

In this paper, we propose adapted transfer, a combination of distance learning and inductive transfer. We learn the contributions of the distances on a task related to our problem and then transfer them to our learning task at hand. The approach is evaluated specifically for QSAR problems (regression on graphs). In the experiments, we investigate how adapted transfer performs compared to distance learning or inductive transfer alone, depending on the size of the available training set. These questions are studied using five pairs of distinct datasets, each consisting of two datasets of related problems.

For the distance measures, our starting point is the observation by Raymond and Willett [5] that maximum common subgraph (MCS) based measures and fingerprint-based measures provide orthogonal information and thus should be considered as complementary. The reason for this may be that MCS-based measures aim to quantify the *global* similarity of structures, whereas fingerprint-based measures rather quantify *local* similarity in terms of smaller, common substructures. We devised an approach that optimally combines the contributions of the two types of measures, and thus balances the importance of global and local similarity for chemical structures.

This paper is organized as follows: In the next section, we present the technical details of learning and adapting distance measures for QSAR problems. Then the datasets, preprocessing steps and the experimental setup are explained before we present the results of the experimental evaluation. We relate the approach to existing work before we give conclusions in the last section.

2   Distance Learning, Inductive Transfer and Adapted Transfer

We frame the learning problem as follows. We are given a set of n labeled examples $X := \{(x_1, y_1), \ldots, (x_n, y_n)\}$, where the examples $x_i \in \mathcal{X}$ are arbitrary objects taken from an instance space $\mathcal{X}$ and the $y_i \in \mathbb{R}$ are real-valued target labels. For the learning setting, we aim at finding a regression function $r : \mathcal{X} \to \mathbb{R}$ that predicts the target label well on new unseen data. We measure the accuracy of a predictor, by taking the squared difference between the predicted target label y' and the true target label y. In other words, we evaluate a prediction using the *squared loss* $l_2(y, y') = (y - y')^2$. We also assume that we have a *distance function* $d : \mathcal{X} \times \mathcal{X} \to \mathbb{R}$ at our disposal, which quantifies the distance between two instances. More precisely, we demand that $d(x, x) = 0$ and that $d(x_1, x_2) < d(x_1, x_3)$, if x_2 is more similar to x_1 than x_3. For ease of notation, we store the distances between all training examples in one $n \times n$ matrix D, so that $D = [d_{ij}] = d(x_i, x_j)$. One well known way to perform regression with distance functions is the *k-nearest neighbor* rule. Given an unlabeled example x, one determines the k nearest neighbors in the training data according to the distance function and then predicts the average over the k neighbors. Let $Y := (y_1, \ldots, y_n)^T$ denote the target label vector and let $W = [w_{ij}]$ be a $n \times n$ *neighbor matrix* that has $w_{ij} = \frac{1}{k}$ if x_j is among the k nearest neighbors of x_i and $w_{ij} = 0$ otherwise. With this, the vector of predicted target labels of the training instances is simply $\hat{Y} := WY$.

Our main contribution, the adapted transfer, is based on the two building blocks: distance learning and inductive transfer. Therefore, we introduce the building blocks first, and describe our main contribution subsequently. In the following we deal with settings, where we have more than one distance function to rate the distance between examples. Rather than restricting ourselves to one fixed function, we would like to use all available information for the prediction by combining the m distance functions $d_1, \ldots, d_m$. One simple way to do so is to take the average: $\hat{Y} = \frac{1}{m} \sum_{i=1}^{m} W_i Y$. In practice, however, one will often encounter settings, in which some distances provide better information than the others. In such settings it makes sense to use a *weighted average* $\hat{Y} = \sum_{i=1}^{m} \alpha_i W_i Y$, where the weight vector $\alpha = (\alpha_1, \ldots, \alpha_m)^T \in \mathbb{R}^m$ with $\sum_{i=1}^{m} \alpha_i = 1$ specifies to which extent each distance function contributes to the prediction. If we aim at low empirical error on the training set, we can determine the optimal α by minimizing the squared error on the training set:

$$\alpha^* := \operatorname*{argmin}_{\alpha} \left\| \sum_{i=1}^{m} \alpha_i W_i Y - Y \right\|^2 \tag{1}$$

$$\text{subject to } \sum_{i=1}^{m} \alpha_i = 1$$

$$0 \le \alpha_i \le 1 \text{ for } i = 1, \ldots, m$$

This is a standard quadratic program with linear constraints and can be solved efficiently by standard convex optimization software.

To extend this setting, we use a different nearest neighbor matrix W than that of the standard k-nearest neighbor approach. In the original definition, this matrix makes a hard cut: the first k neighbors contribute equally to the prediction, whereas the remaining examples are ignored. This appears to be a somewhat arbitrary choice and one can envision many more fine-grained and less restrictive prediction schemes. In principle, a matrix W must fulfill two properties in order to lead to reasonable predictions: its rows must sum to one and it must assign larger weights to more similar instances. In our experiments we used a nearest neighbor approach with a distance threshold. Instead of choosing a fixed number k of nearest neighbors, one selects a distance threshold t and determines the set T of all neighbors whose distance to the test example is less than t. Each example in T influences the prediction with weight $\frac{1}{|T|}$.

Our second building block is *inductive transfer*. Inductive transfer is suitable for settings where the amount of available training data is too small to determine a good weight vector. Instead of learning a completely new weight vector α from the (limited) target training data, we make use of an additional dataset, which is assumed to have similar characteristics as the target data. We call this additional dataset the *source dataset* to distinguish it from the *target training set*, so that the inductive transfer takes place from source to target. In the *Simple Transfer* setting, one induces a weight vector β only from the source data (by solving (1) for the source dataset) and uses this β without modification for the actual prediction. The actual training data provides the neighbors for the prediction, but is not used for the computation of α.

Enhancing this *Simple Transfer* setting, the *Adapted Transfer* setting allows for the transferred weight vector β to be adapted to the target training data. This can be done in two ways:

- *Bounded Adaptation.* One induces a weight vector β from the source data, but adapts it in a second step slightly to the target training data. For the adaptation step, we would like to avoid overfitting on the (limited) training data. Thus, we extend the optimization criterion (1) with the additional criterion that the α may not differ too much from the transferred β. More precisely, for a fixed $\epsilon > 0$ we compute

$$\alpha^* := \operatorname*{argmin}_{\alpha} \left\| \sum_{i=1}^{m} \alpha_i W_i Y - Y \right\|^2 \tag{2}$$

$$\text{subject to } \sum_{i=1}^{m} \alpha_i = 1$$

$$|\alpha_i - \beta_i| \leq \epsilon \text{ for } i = 1, \ldots, m$$

$$0 \leq \alpha_i \leq 1 \text{ for } i = 1, \ldots, m$$

- *Penalized Adaptation.* In this approach, we also adapt the weight vector β induced from the source data. Instead of limiting the interval from which the α can be taken, we add a regularization term to the optimization criterion that penalizes αs that deviate too much from β. Formally, for $C > 0$, we solve

$$\alpha^* := \operatorname*{argmin}_{\alpha} \left\| \sum_{i=1}^{m} \alpha_i W_i Y - Y \right\|^2 + C\|\alpha - \beta\|^2 \tag{3}$$

$$\text{subject to } \sum_{i=1}^{m} \alpha_i = 1$$

$$0 \leq \alpha_i \leq 1 \text{ for } i = 1, \ldots, m$$

In the following sections, these variants will be evaluated and tested experimentally.

3   Data and Experimental Setup

In this section we give an overview of the datasets, toxicological endpoints and similarity measures used in the study, and we describe the experimental setup.

3.1   Data

All of the datasets for our study were taken from the data section of the cheminformatics web repository[1]. Since we are interested in adapted transfer between different datasets, we put a special focus on finding pairs of datasets with similar or identical endpoints.[2] Note that due to the wealth of data produced in all areas of science and industry today, the existence of related datasets is frequently occurring and thus practically relevant. In fact, even in computational chemistry, the five pairs used in this paper are just a selection from a wider range of possibilities.

For the first pair of datasets [6,7] abbreviated DHFR_4q (361 compounds) and DHFR_S. (673), the goal is to predict the dihydrofolate reductase inhibition of compounds as measured by the pIC_{50} value, that indicates how much of a given substance is needed to block a biological activity by half. We had to remove a number of instances, which were considered to be inactive in the original publication and marked with default values. Overall the compounds in this pair of datasets share a high similarity. Consequently, there often are only local changes to the molecular graph structure and the graphs are very similar on a global level. The second pair, CPDB_m (444, mouse) and CPDB_r (580, rat) are generated from data obtained by the carcinogenic potency project[3]. The compounds' carcinogenicity is rated according to the molar TD_{50} value TD_{50}^m, where a low value indicates a potent carcinogen. The two datasets contained several instances where the actual structure of the compound was missing. If the molecule could be identified uniquely we downloaded the structure from the NCBI PubChem database[4]. If this was not possible, the molecule was removed from the set. The third pair of datasets [7], ER_TOX (410) and ER_LIT (381), measure the logarithmized relative binding affinities (RBA) of compounds to the estrogen receptor with respect to β-estradiol. All inactive compounds were removed from the datasets as they all have the same value. The fourth pair, ISS_m (318, mouse) and ISS_r (376, rat)[8], is similar to the second pair. The target value under consideration is again the carcinogenic potency of a compound as measured by the molar TD_{50} value. The two datasets contained several instances where the actual structure of the compound was missing. If the molecule could be identified uniquely via the given CAS number, we downloaded the structure from the NCBI PubChem database. If this was not possible, the molecule was removed from the set. The fifth and last pair of datasets [7,6], COX2_4q (282) and COX2_S. (414), are used to predict the cyclooxygenase-2 inhibition of compounds as measured by the pIC_{50} value. We had to remove a number of instances, which were considered to be inactive in the original publication and marked with default values. As in the first pair of datasets, the compounds contained in this

---

[1] http://www.cheminformatics.org

[2] In pharmacology and toxicology, an endpoint constitutes the target outcome of a trial or experiment.

[3] http://potency.berkeley.edu/chemicalsummary.html

[4] http://pubchem.ncbi.nlm.nih.gov/

dataset pair are highly similar. The preprocessed and cleaned datasets used in our experiments are available for download on the authors' website[5].

3.2 Distances

For all ten datasets, we generated three different distance matrices. The first and the second matrix are based on the Tanimoto distance metric for binary fingerprints. The first set of fingerprints are occurrence fingerprints for frequently occurring substructures. The substructures are closed free trees. The free trees were calculated with the Free Tree Miner (FTM) [9] software so that approximately 1000 free trees were found. The resulting set was further reduced to the set of closed trees, i.e., trees occurring in the same set of structures as one of the supertrees were removed. The second fingerprint set is built of pharmacophoric (binary) fingerprints containing more than 50 chemical descriptors computed with the cheminformatics library JOELIB2[6]. The third distance matrix is based on a Tanimoto-like maximum common subgraph (MCS) based distance measure:

$$d_{mcs}(x, y) = 1 - \left(\frac{|mcs(x, y)|}{|x| + |y| - |mcs(x, y)|} \right), \tag{4}$$

where $|\cdot|$ gives the number of vertices in a graph, and $mcs(x, y)$ calculates the MCS of molecules x and y. JChem Java classes were used for computing the maximum common subgraph (MCS), JChem 5.1.3_2, 2008, ChemAxon (http://www.chemaxon.com).

4   Experiments

We consider a QSAR learning task given by a *target* training and test set. Additionally, we assume that we can transfer information from a *source* dataset containing related training data. The task is to induce a predictor from the target training set and the source dataset, which features good predictive accuracy on the test set. To solve this task we propose a strategy called *adapted transfer*. This approach combines adaptation and inductive transfer, as outlined in the second section. We start with identifying the weight vector α that optimizes (1) on the source dataset. Instead of using this weight vector directly, we adapt it to better match with the target training data. This is done either within an ϵ-environment by optimizing (2) or with a quadratic distance penalty by solving (3). The resulting weight vector is then applied with the target training data in a nearest-neighbor classifier.

To get reliable results, we repeat our experiments one hundred times, where each run consists of a ten-fold cross-validation. This means we estimate the methods' success on one thousand different configurations of training- and test-folds. To quantify predictive accuracy, we choose mean squared error, a standard measure in regression settings. We evaluate the adapted transfer approaches against three baseline strategies:

---

[5] http://wwwkramer.in.tum.de/research

[6] http://www-ra.informatik.uni-tuebingen.de/software/joelib

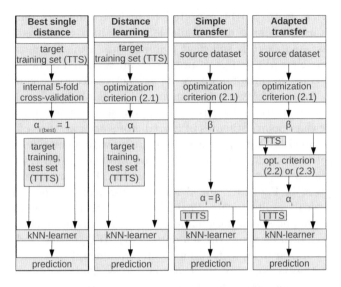

Fig. 1. Graphical overview of the four strategies used in the experiments. Abbreviations: opt. = optimization, $\alpha_{i(best)} = \alpha_i$ for best single distance.

- **Best single distance.** We perform an internal 5-fold cross validation on the target training set to determine the best of the three distances. This distance is then used to predict the target values for the target test set. The source dataset is not used.
- **Distance learning.** We compute the solution to the optimization problem (1) to determine the best linear combination α on the target training set. The weighted combination of distances is then used as new distance for the prediction on the test data. Again, the source dataset is not used.
- **Simple transfer.** Here, we optimize (1) on the source dataset instead of the target training data. The weighted combination of distances is then used as new distance for the prediction on the test data. Here, the target training data is only used for the nearest-neighbor classifier, not for the adaptation of the distance measure.

All four strategies are illustrated in Figure 1. All algorithms and methods were implemented in MATLAB Version 7.4.0.336 (R2007a). We applied the MOSEK[7] Optimization Software (Version 5.0.0.60) that is designed to solve large-scale mathematical optimization problems.

4.1   Learning Curves

At their core, distance adaptation and inductive transfer methods are approaches to improve predictive accuracy by fine-tuning the learning bias of a machine

---

[7] MOSEK ApS, Denmark. http://www.mosek.com

learning scheme. Both can be expected to make a difference only if there is not enough target training data available to obtain a good predictor. If this is not the case and there is sufficient training data available, most reasonable learning approaches will find good predictors anyway, and distance adaptation or inductive transfer cannot improve its predictive accuracy significantly. To evaluate this trade-off between the amount of available training data and the applicability of transfer and adaptation approaches, we first present *learning curves* rather than point estimates of a predictor's accuracy for a fixed training set size. More precisely, we repeat each experiment with increasing subsets of the original target training data. We start by using only the first 10%, then 20%, and so on until the complete training data is available. The corresponding learning curves are given in Figures 2 and 3 for one representative parameter setting producing typical results (nearest neighbor with distance threshold $t = 0.2$, $\epsilon = 0.2$ for the bounded adaptation and $C = 10.0$ for the penalized adaptation). The plots are shown for six of the ten datasets (DHFR, CPDB and ER). While the differences appear to level off for increasing training set sizes, there are clearly differences at the beginning of the learning curves. The single best distance is outperformed by other methods (outside the scale of the y-axis for CPDB), and distance learning does not work well yet for small training set sizes. For a more principled comparison, we now examine under which circumstances one approach outperforms another significantly.

4.2   Comparison of Approaches

To investigate whether our adapted transfer strategy outperforms the presented baseline approaches we first evaluate how the baseline approaches perform compared to each other. The corresponding results are shown in Table 1. Second, we investigate how the adapted transfer strategy competes (results shown in Table 2). We conducted one hundred runs of tenfold cross-validation. For each run, we noted whether the first or the second method performed better. We tested if the resulting set of predictive accuracies were statistically significant improvements or deteriorations at a significance level of 5%.

As a first experiment, we would like to address the question whether distance learning improves predictive accuracy. More precisely, we compare whether having a linear combination of the distances' contributions optimizing criterion (1) (distance learning strategy) outperforms the single best distance. The results are given in Table 1. It shows that distance adaptation using a linear combination significantly outperforms the single best distance in nine out of ten cases for 10% and eight out of ten cases for 100%. Thus, there is some empirical evidence that a learning method, which adjusts its bias to better accommodate for the underlying data, is more successful than an approach with a single fixed bias (i.e. distance). The result also supports the observation by Raymond and Willett [5] that maximum common subgraph based measures and fingerprint-based measures provide orthogonal information.

For the second experiment we would like to investigate whether inductive transfer improves predictive accuracy. To do so, we compare the best single

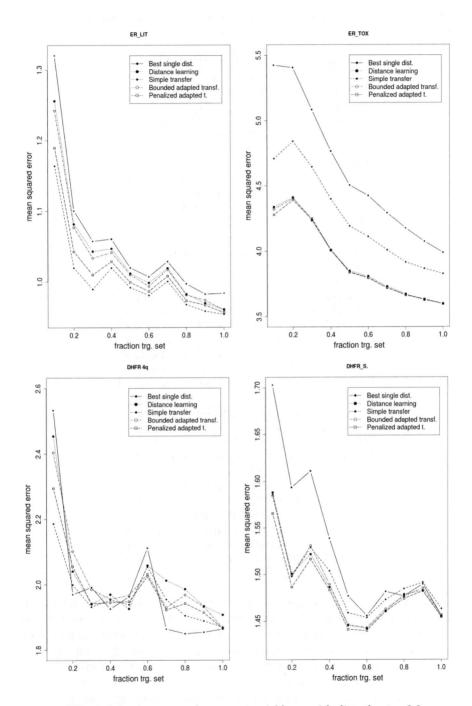

Fig. 2. Learning curves for nearest neighbour with dist. thr. $t = 0.2$

Table 1. A vs. B: • = A significantly better than B, ○ = A significantly worse than B; w = how often out of 100 times A "wins" against B

Distance learning vs. Best single distance				
frac. trg. set	10%		100%	
Short-hand	w	p-value	w	p-value
$DHFR\_4q$	39	0.2445	17	8.9928e-15 ○
$DHFR\_S.$	74	7.3586e-10 •	50	0.9916
$CPDB\_m$	100	3.8397e-50 •	100	5.0502e-62 •
$CPDB\_r$	100	5.1532e-51 •	100	3.1704e-67 •
ER_TOX	96	3.1118e-30 •	99	8.1943e-37 •
ER_LIT	69	2.3297e-06 •	66	5.3560e-06 •
ISS_m	100	1.4792e-50 •	100	8.8650e-76 •
ISS_r	100	1.5046e-46 •	100	3.5386e-44 •
$COX2\_4q$	83	6.5791e-10 •	59	1.4442e-05 •
$COX2\_S.$	56	0.0070 •	93	3.3759e-24 •
Simple transfer vs. Best single distance				
$DHFR\_4q$	68	1.9913e-06 •	40	0.1496
$DHFR\_S.$	72	5.6524e-09 •	47	0.1856
$CPDB\_m$	100	1.7442e-59 •	100	2.4231e-61 •
$CPDB\_r$	100	3.0277e-56 •	100	6.3283e-69 •
ER_TOX	80	1.8104e-12 •	79	1.5431e-10 •
ER_LIT	92	1.7692e-22 •	80	1.7956e-06 •
ISS_m	99	1.4792e-50 •	100	4.7435e-26 •
ISS_r	97	1.5046e-46 •	55	1.3118e-22 •
$COX2\_4q$	91	6.5791e-10 •	34	7.5261e-04 ○
$COX2\_S.$	44	0.0070 •	78	0.0050 •
Distance learning vs. Simple transfer				
$DHFR\_4q$	14	2.9153e-17 ○	21	1.2097e-11 ○
$DHFR\_S.$	44	0.8970	55	0.1030
$CPDB\_m$	13	1.5582e-38 ○	69	8.3637e-04 •
$CPDB\_r$	2	2.4788e-19 ○	26	1.1582e-04 ○
ER_TOX	73	5.3856e-09 •	84	1.0532e-12 •
ER_LIT	12	6.5350e-20 ○	44	0.0067 ○
ISS_m	29	8.8116e-07 ○	100	1.4392e-42 •
ISS_r	85	1.4217e-19 •	100	4.6961e-78 •
$COX2\_4q$	15	6.0041e-13 ○	85	4.6009e-18 •
$COX2\_S.$	52	0.0178 •	87	5.5511e-16 •

distance strategy with the simple transfer where the weights for the linear combination are computed on the source dataset rather than the target training data. Our experiments indicate (Table 1) that inductive transfer using a linear combination significantly outperforms the single best distance in all cases for 10% and seven out of ten cases for 100%. Apparently, inductive transfer has the same effect as distance learning.

Since both building blocks of our adapted transfer strategy, distance learning and inductive transfer, improve predictive accuracy, the next experiment deals with the question under which circumstances one approach outperforms the other. The experiments indicate (Table 1) that inductive transfer is significantly better than distance learning, if only few training data are available, but the opposite is true, if all the available training data is used. Hence, one can say that one should resort to inductive transfer methods, whenever there is comparably few training data available and when the source data for the transfer is of sufficiently good quality (as appears to be the case on all datasets except for ER_TOX, ISS_r and COX2_Sutherland). Unfortunately, it is often hard to tell in advance, whether the source data is good enough for successful transfer and how the size of the available target data compares to the size of source data of unknown quality.

In order to avoid this problem, we introduced a "mixed strategy", which transfers weights from the source dataset, but ensures that the actual weights differ not too much from the ones, which can be obtained by distance adaptation on the target data. We now compare the penalized adapted transfer approach to its two building block baseline methods. Table 2 shows that adapted transfer outperforms distance learning on small training data, but leads to no further improvement, if there is sufficient amount of training data. On the other hand, the mixed strategy performs better than simple transfer in settings with large amounts of training data. When only few training data is present, its performance is sometimes better and sometimes worse than the simple transfer (see Table 2), possibly depending on the quality of the source data and the representativeness of the few training examples. In summary, these results indicate that adapted transfer is a good compromise, which keeps the high predictive accuracy of distance adaptation on small and large training datasets, and improves on simple transfer in settings with large amounts of training data. This holds for both variants of the adapted transfer (bounded and penalized) which perform comparably with a slight advantage for the penalized version.

4.3   Analysis of Optimized Weights

Figure 4 shows horizontally stacked bar-plots of the weights α_i optimized in the distance learning approach and of the weights α_i^p optimized in the penalized adaptation approach (mean over the hundred repetitions of ten fold cross-validation). The weights α_1 based on the sub-structural features are shown in white, the pharmacophoric fingerprint based weights α_2 in grey and the MCS-based α_3 in black. A general observation is that the strength of the adaptation of the α_is is consistent with the learning curves in Figures 2 and 3. Strong adaptation can, for example, be seen, e.g., in the $DHFR\_4q$ dataset at 10% and at 100%. This effect can clearly be seen in the learning curves. Especially notable is that the MCS weights α_3 (black) are significantly lower for the DHFR and COX2 datasets. This reflects very nicely the fact that the compounds in those four datasets are much less diverse. Less diverse compounds can be distinguished more easily with local than with global differences as represented by the MCS-based weights α_3.

Table 2. A vs. B: •/o = A significantly better/worse than B; w = "wins" of A

		Penalized adapted transfer vs. Distance learning		
$DHFR\_4q$	75	3.6350e-11 •	84	1.7402e-18 •
$DHFR\_S.$	61	0.0363 •	54	0.6914
$CPDB\_m$	94	3.5594e-29 •	39	0.0160
$CPDB\_r$	50	0.9865	69	4.7849e-04 o
ER_TOX	49	0.0244 •	48	0.8390
ER_LIT	76	6.8167e-11 •	55	0.1088
ISS_m	69	1.3269e-05 •	58	0.1663
ISS_r	30	4.2098e-05 o	81	2.6523e-13 •
$COX2\_4q$	87	3.5753e-12 •	7	5.2729e-29 o
$COX2\_S.$	55	0.4762	22	7.1220e-11 o
		Penalized adapted transfer vs. Simple transfer		
$DHFR\_4q$	21	2.1467e-06 o	54	0.1907
$DHFR\_S.$	61	0.0014 •	54	0.0229 •
$CPDB\_m$	17	9.5657e-13 o	52	0.2353
$CPDB\_r$	5	1.3033e-28 o	42	0.3879
ER_TOX	82	8.5636e-12 •	81	3.5649e-13 •
ER_LIT	28	5.6761e-06 o	44	0.3060
ISS_m	41	0.1454	100	1.1372e-41 •
ISS_r	81	3.9968e-15 •	100	1.3256e-79 •
$COX2\_4q$	21	5.9618e-09 o	60	0.0033 •
$COX2\_S.$	75	3.4792e-08 •	64	7.0540e-05 •

5   Related Work

A related approach for classification instead of regression has been proposed by Woznica *et al.* [10]. The authors combine distances for different complex representations of a given learning problem, and use the learned distance for kNN classification. Hillel and Weinshall [11] learned distance functions by coding similarity in the context of image retrieval and graph based clustering. In the study of Raymond and Willett [5] it was shown that graph-based (MCS) and fingerprint-based measures of structural similarity are complementary; we followed up on this observation in our study.

Weinberger and Tesauro [12] is another approach learning a distance metric for regression problems. While this distance learning method (and many others) optimize a parametrized distance measure (typically the Mahalanobis metric), the approach presented in this paper optimizes contributions from various given distance measures in the form of a neighbourhood matrix. Another major difference is that we explicitly aim for learning schemes for molecular structures and not for general feature vectors. As we want to use an MCS-based distance measure (which is not based on features) as an input, a meaningful experimental comparison is not possible.

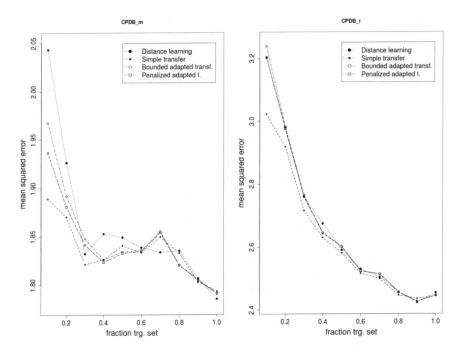

Fig. 3. Learning curves for nearest neighbour with dist. thr. $t = 0.2$

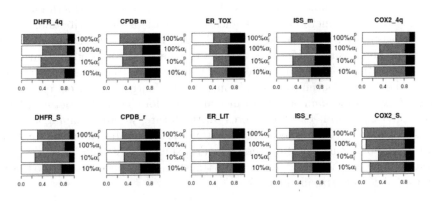

Fig. 4. Graphical representation of the α_i and α_i^p at 10% and 100% of the training data. α_1 (cFTs) = white, α_2 (joelib) = grey and α_3 (MCS) = black

In 1995, Baxter [13] proposed to learn internal representations sampled from many similar learning problems. This internal representation is then used to bias the learner's hypothesis space for the learning of future tasks stemming from the same environment. A related topic in machine learning research is multi-task learning. For example, Evgeniou *et al.* [14] study the problem of learning many related tasks simultaneously using kernel methods and regularization. Our approach differs in that it does not learn multiple tasks but only one. The related tasks are used more in the sense of background knowledge that is used to bias the generalisation problem, especially in cases where there is insufficient data on the given learning task.

The presented approach is also related to multiple kernel learning [15], but differs in two important points: First, one of the requirements of our application domain is the use of a maximum common subgraph based similarity measure. Although there are attempts to define positive semi-definite kernels on the basis of maximum common subgraphs [16], there is no convincing approach yet to achieve this in a straightforward manner. Second, it is important to note that we do not combine distance measures themselves linearly, but their contributions in the form of neighbouring instances. Therefore, there is yet another level of indirection between the distance measures and the way they are used and combined. Simple linear combinations did not work well in preliminary experiments.

A recent approach by Zha *et al.* [17] learns distance metrics from training data and auxiliary knowledge, in particular, auxiliary metrics possibly learned from related datasets. The approach was defined for classification, not regression, and tested on image data. Finally, the work by Erhal *et al.* [18] is similar in spirit, but technically different, and also has a less comprehensive experimental evaluation.

6   Conclusion

In the paper, we proposed adapted transfer, a method combining inductive transfer and distance learning, and evaluated its use for quantitative structure-activity relationships. The method derives linear combinations of contributions of distance measures for chemical structures. Compared to inductive transfer and distance learning alone, the method appears to be a good compromise that works well both with large and small amounts of training data. Technically, the method is based on convex optimization and combines the contributions from representatives of two distinct families of distance measures for chemical structures, MCS-based and fingerprint-based measures. In the future, we are planning to test if related learning tasks can be detected automatically, not based on the similarity of the endpoint, but based on the structural similarity of their instances. This would resemble other work on inductive transfer [4], but incorporate those ideas in a more domain-specific manner.

Acknowledgements. This work was partially supported by the EU FP7 project (HEALTH-F5-2008-200787) OpenTox (http://www.opentox.org).

References

1. Horváth, T., Gärtner, T., Wrobel, S.: Cyclic pattern kernels for predictive graph mining. In: Proc. of KDD 2004, pp. 158–167. ACM Press, New York (2004)
2. Shervashidze, N., Vishwanathan, S., Petri, T., Mehlhorn, K., Borgwardt, K.: Efficient Graphlet Kernels for Large Graph Comparison. In: Proc. of AISTATS 2009 (2009)

3. Goldberger, J., Roweis, S., Hinton, G., Salakhutdinov, R.: Neighborhood Component Analysis. In: Proc. of NIPS 2004, pp. 513–520 (2005)
4. Eaton, E., Desjardins, M., Lane, T.: Modeling transfer relationships between learning tasks for improved inductive transfer. In: Proc. of ECML PKDD 2008, pp. 317–332. Springer, Heidelberg (2008)
5. Raymond, J.W., Willett, P.: Effectiveness of graph-based and fingerprint-based similarity measures for virtual screening of 2D chemical structure databases. JCAMD, 59–71 (January 2002)
6. Sutherland, J.J., O'Brien, L.A., Weaver, D.F.: Spline-fitting with a genetic algorithm: A method for developing classification structure-activity relationships. J. Chem. Inf. Model 43(6), 1906–1915 (2003)
7. Sutherland, J.J., O'Brien, L.A., Weaver, D.F.: A comparison of methods for modeling quantitative structure-activity relationships. J. Med. Chem. 47(22), 5541–5554 (2004)
8. Benigni, R., Bossa, C., Vari, M.R.: Chemical carcinogens: Structures and experimental data,
   http://www.iss.it/binary/ampp/cont/ISSCANv2aEn.1134647480.pdf
9. Rückert, U., Kramer, S.: Frequent free tree discovery in graph data. In: SAC 2004, pp. 564–570. ACM Press, New York (2004)
10. Woznica, A., Kalousis, A., Hilario, M.: Learning to combine distances for complex representations. In: Proc. of ICML 2007, pp. 1031–1038. ACM Press, New York (2007)
11. Hillel, A.B., Weinshall, D.: Learning distance function by coding similarity. In: Proc. of ICML 2007, pp. 65–72. ACM Press, New York (2007)
12. Weinberger, K.Q., Tesauro, G.: Metric learning for kernel regression. In: Proc. of AISTATS 2007 (2007)
13. Baxter, J.: Learning Internal Representations. In: Proc. COLT 1995, pp. 311–320. ACM Press, New York (1995)
14. Evgeniou, T., Micchelli, C.A., Pontil, M.: Learning multiple tasks with kernel methods. J. Mach. Learn. Res. 6, 615–637 (2005)
15. Sonnenburg, S., Rätsch, G., Schäfer, C., Schölkopf, B.: Large scale multiple kernel learning. J. Mach. Lear. Res. 7, 1531–1565 (2006)
16. Neuhaus, M., Bunke, H.: Bridging the Gap Between Graph Edit Distance and Kernel Machines. World Scientific Publishing Co., Inc, Singapore (2007)
17. Zha, Z.J., Mei, T., Wang, M., Wang, Z., Hua, X.S.: Robust distance metric learning with auxiliary knowledge. In: Proc. of IJCAI 2009, pp. 1327–1332 (2009)
18. Erhan, D., Bengio, Y., L'Heureux, P.J., Yue, S.Y.: Generalizing to a zero-data task: a computational chemistry case study. Technical Report 1286, Département d'informatique et recherche opérationnelle, University of Montreal (2006)

Incremental Mining of Closed Frequent Subtrees

Viet Anh Nguyen and Akihiro Yamamoto

Graduate School of Informatics, Kyoto University
Yoshida Honmachi, Sakyo-ku, Kyoto, 606-8501, Japan
vietanh@iip.ist.i.kyoto-u.ac.jp,
akihiro@i.kyoto-u.ac.jp

Abstract. We study the problem of mining closed frequent subtrees
from tree databases that are updated regularly over time. Closed frequent subtrees provide condensed and complete information for all frequent subtrees in the database. Although mining closed frequent subtrees
is in general faster than mining all frequent subtrees, this is still a very
time consuming process, and thus it is undesirable to mine from scratch
when the change to the database is small. The set of previous mined
closed subtrees should be reused as much as possible to compute new
emerging subtrees. We propose, in this paper, a novel and efficient incremental mining algorithm for closed frequent labeled ordered trees. We
adopt a divide-and-conquer strategy and apply different mining techniques in different parts of the mining process. The proposed algorithm
requires no additional scan of the whole database while its memory usage is reasonable. Our experimental study on both synthetic and real-life
datasets demonstrates the efficiency and scalability of our algorithm.

1 Introduction and Motivations

In this paper, we examine a mining problem, called incremental tree mining. Tree
mining, an extension of the itemset mining paradigm [2], refers to the extraction
of all frequent trees in tree-structured databases. Due to its useful applications
in various areas such as bioinformatics [9], management of the web and XML
data [1], and marketing channel management [13], tree mining has attracted
noticeable attention from researchers in the data mining community. There have
been several efficient algorithms proposed in the previous works, see for example,
[3], [7], [11], [12]. These algorithms mine the entire database and output the set
of results. This *static fashion*, however, has limitation in application fields where
the data increasing continuously as the time moves on. For example, an online
shopping website where browsing behavior of each user is produced in the form of
trees may be browsed by hundreds of people every minute. In such situations, it
is undesirable to mine the incremented databases from scratch, and so naturally,
an incremental mining algorithm which reuses the results of past minings to
minimize the computation time must be adopted.

This paper develops an incremental mining algorithm for closed frequent labeled ordered trees. Closed subtrees are the maximal ones among each equivalent
class that consists of all frequent subtrees with the same transaction sets in a

B. Pfahringer, G. Holmes, and A. Hoffmann (Eds.): DS 2010, LNAI 6332, pp. 356–370, 2010.

tree database. In many cases, we only care about closed subtrees; given the set of all closed frequent subtrees, we can always derive the set of all frequent subtrees with their support. After a careful survey of the previous work, we found that the essential technique used in many existing algorithms is to buffer the set of *semi-frequent* patterns, those that are likely to become frequent in the new iteration, in addition to the set of all frequent patterns, see for example [4], [6], [10]. For the case of closed tree mining, this technique could not be applied effectively because even when all the new frequent patterns are found, filtering out those that are closed is still expensive. We, instead, adopt a divide-and-conquer strategy and treat each kind of new emerging closed tree patterns differently based on their own characteristics. The key point of our algorithm is that it requires no additional scan of the whole database. Besides, the memory for keeping information from previous mining iterations is still reasonable even for very low minimum support thresholds.

Related Work. There is very few work on incremental mining for tree data up to now. The reason may partly be the complexity of this problem. The main challenge here is how to achieve fast response time, even real-time response in some strict scenarios, with reasonable memory space usage. Asai et al. [4] introduced an online algorithm for mining tree data streams, but their problem setting in which data arrive continuously node-by-node may not be realistic because in the real world, the most common case is that data is acquired in small batches over the time. Further, only approximate answers are returned and no accuracy guarantee on the result set is provided by the algorithm. Hsieh et al. [10] proposed a complete algorithm but the problem setting is still the same as that of Asai et al.. Bifet and Gavalda [5] presented an incremental mining algorithm for frequent unlabeled rooted trees. After computes the set of new closed subtrees of the additional batch, the algorithm checks the subsumption of every subtree of the new detected closed subtrees against all closed subtrees mined in the previous run. In mining labeled rooted trees, the number of closed subtrees is often many times bigger than that of the unlabeled cases and the subsumption checking for labeled trees is also often much more expensive than the subsumption checking for unlabeled trees. These obstacles may make the method propose in [5] not suitable for labeled trees.

The rest of the paper is organized as follows. Section 2 formally defines the problem of incremental closed subtree mining. Section 3 describes the new incremental algorithm. Section 4 reports the experimental results. Section 5 concludes the paper.

2   Preliminaries

In this section, we introduce the notations and concepts used in the remainder of the paper .

Rooted Labeled Trees. Let $\Sigma = \{l_1, \ldots, l_m\}$ be a set of labels. A *rooted labeled tree* $T = (V, E, r, L)$ on Σ is an acyclic connected graph, where $V = \{v_1, \ldots, v_n\}$

is a finite set of nodes, $E \subseteq V \times V$ is a finite set of edges, $r \in V$ is a distinguished node that has no entering edges, called the *root*, and L is a labelling function $L : V \to \Sigma$ assigning a label $L(v)$ to each node v of T.

A *rooted ordered tree* is a rooted tree that has a predefined left-to-right ordering among the children of each node. In this paper, a tree means a rooted ordered labeled tree. Let u, v be nodes in V. If u is on the path from the root to v then u is an *ancestor* of v and v is a *descendant* of u. If $(u, v) \in E$, then u is the *parent* of v and v is a *child* of u. Two nodes with the same parent are *siblings* to each other. If a node v has no child, then v is called a *leaf*.

Tree Inclusion. Let $S = (V_S, E_S, r_S, L_S)$ and $T = (V_T, E_T, r_T, L_T)$ be two trees on Σ. Then S is included in T, denoted by $S \preceq T$, if there exists an injective mapping $\varphi : V_S \to V_T$ such that

1. φ preserves the node labels: $\forall v \in V_S$ $L_S(v) = L_T(\varphi(v))$.
2. φ preserves the parent-child relation: $(v_1, v_2) \in E_S$ iff $(\varphi(v_1), \varphi(v_2)) \in E_T$.
3. The left-to-right ordering among the children of a node v in S is a subordering of the children of $\varphi(v)$ in T.

If $S \preceq T$ holds, then we also say that S *occurs* in T, S is a *subtree* of T, T is a *supertree* of S, and T *contains* S. The subtree defined above is called an *induced subtree* [8]. In this paper, a subtree is sometimes called a *pattern*, and unless otherwise specified, a subtree means an induced subtree.

Suppose V_S has k nodes which are indexed with numbers $1, 2, \ldots, k$ as $v_1, \ldots, v_k$. The *total occurrence* of S in T w.r.t. φ is the list $Total(\varphi) = (\varphi(v_1), \ldots, \varphi(v_k)) \in (V_T)^k$. The *root occurrence* of S in T w.r.t. φ is the node $root(\varphi) = (\varphi(r_S)) \in V_T$. A root occurrence of S in T is also called an occurrence of S in T.

Support of a Subtree and Frequent Subtrees. Let D denote a database of a set of transactions, where each transaction $s \in D$ is a tree. For a given pattern t, let $\sigma_t(s) = 1$ if t is a subtree of s, and 0 otherwise. The *support* of a pattern t in the database D is defined as $support_D(t) = \sum_{s \in D} \sigma_t(s)$. A pattern t is called *frequent* in D if $support_D(t)$ is greater than or equal to a support threshold $minsup_D$. Let $0 \leq \sigma \leq 1$ be a non-negative number called the *minimum support* specified by a user, we have $minsup_D = \sigma \cdot |D|$, where $|D|$ is the size of D.

Just as the item set mining, the *monotone property* holds that, for two patterns s and t, $s \preceq t$ implies $support_D(s) \geq support_D(t)$.

Closed Subtrees. A subtree t is *closed* if and only if no proper supertree of t has the same support that t has.

Incremental Mining Problem. Suppose a set of new tree transactions, Δ, is to be added to the database D. The database D is referred to as the *original database*, the database Δ as an *increment database*, and the database $U = D \cup \Delta$ as the *updated database* where $\cup$ denotes the union of two sets. The incremental mining problem which we treat is to find all closed frequent subtrees in the database U given D, Δ, and the minimum support σ. We denote the set of all closed frequent subtrees mined from a database D as $C(D)$.

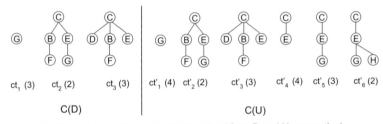

(a) A database D, an increment database Δ and their union

ct_1 (3) ct_2 (2) ct_3 (3) ct'_1 (4) ct'_2 (2) ct'_3 (3) ct'_4 (4) ct'_5 (3) ct'_6 (2)

C(D)   C(U)

(b) Sets of closed frequent subtrees mined from D and U, respectively

Fig. 1. Running Example

Figure 1(a) shows the original tree database D comprised of three transactions T_1, T_2, and T_3, the incremental database Δ comprised of T_4, and the updated database U comprised of T_1, T_2, T_3, and T_4. Figure 1(b) shows the sets of closed frequent subtrees, namely $C(D)$ and $C(U)$, for the databases D and U respectively, when σ is set to 50%. The number in brackets denotes the support of a closed subtree. This example is used as a running example in our paper.

3   The Mining Algorithm

In this section we discuss a divide-and-conquer method for incremental mining of closed subtrees, where the set of closed subtrees will be divided into different non-overlapping subsets, each of which can be obtained independently using its own stored information from the previous mining iteration.

3.1   A Divide-and-Conquer Solution

The insertion of new transactions in Δ may give rise to new closed subtrees which never appear before. Some existing closed subtrees mined from D may have support count increased by Δ. There also are old closed subtrees which remain completely unchanged by the insertion of Δ. In general, we can divide $C(U)$, the set of all closed subtrees mined from $U = D \cup \Delta$, into two subsets, $C(U) = C(U)|_\Delta \cup C(U)|_s$, where

$$C(U)|_\Delta = \{t \in C(U) \mid \exists \text{ a transaction } T \in \Delta \text{ s.t. } t \preceq T\},$$
$$C(U)|_s = \{t \in C(D) \mid support_U(t) = support_D(t) \text{ and } support_U(t) \geq minsup_U\}.$$

The sets $C(U)|_\Delta$ and $C(U)|_s$ are respectively the set of closed subtrees in $C(U)$ that occur in Δ, and that of closed subtrees in $C(U)$ which are also closed subtrees in $C(D)$ with supports stay unchanged.

Proposition 1. *It holds that* $C(U) = C(U)|_\Delta \cup C(U)|_s$.

Proof. By definitions, we have $C(U)|_\Delta \subseteq C(U)$ and $C(U)|_s \subseteq C(U)$. Therefore, $C(U)|_\Delta \cup C(U)|_s \subseteq C(U)$. Suppose that there exists a closed subtree c s.t. $c \in C(U)$ and $c \notin C(U)|_\Delta \cup C(U)|_s$. We have $c \notin C(U)|_\Delta$ and $c \notin C(U)|_s$. Because $c \notin C(U)|_\Delta$, then all support of c is from D, which means c remains closed in U as in D, thus we have $c \in C(U)|_s$, this contradicts with $c \notin C(U)|_s$, then the proposition follows. $\square$

For the running example, we have $C(U)|_\Delta = \{ct'_1, ct'_4, ct'_5, ct'_6\}$, and $C(U)|_s = \{ct'_2, ct'_3\}$.

We can compute $C(U)$ by computing $C(U)|_\Delta$ and $C(U)|_s$ separately. However, the computation is still expensive because the whole database U still need to be scanned. For this reason, we try to further divide $C(U)|_\Delta$ into smaller subsets which can be computed efficiently with minimized access to the database. The first subset is the one that contains closed subtrees that occur in $C(D)$. Closed subtrees in $C(U)|_\Delta$ which do not occur in $C(D)$ are divided into two parts: those whose prefixes occur in $C(D)$ and those whose none of their prefixes occur in $C(D)$. The subsets are formally defined as follows:

$$C(U)|_{o\Delta} = \{t \in C(U)|_\Delta \mid t \vdash C(D)\},$$
$$C(U)|_{p\Delta} = \{t \in C(U)|_\Delta \mid \exists i, i < size(t) \ \forall j \leq i \ prefix(t, j) \vdash C(D) \text{ and}$$
$$prefix(t, j) \not\vdash C(D)\},$$
$$C(U)|_{r\Delta} = \{t \in C(U)|_\Delta \mid \forall i \leq size(t) \ prefix(t, i) \not\vdash C(D)\},$$

where i, j are positive integers, $t \vdash C(D)$ means that there exists a closed subtree $c \in C(D)$ s.t. $t \preceq c$, $t \not\vdash C(D)$ if otherwise, and $prefix(t, i)$ is the subtree containing first i nodes in the pre-order traversal of t.

In summary, we have $C(U) = C(U)|_{o\Delta} \cup C(U)|_{p\Delta} \cup C(U)|_{r\Delta} \cup C(U)|_s$. In fact, we do not need to find $C(U)|_s$ since it can be obtained together with $C(U)|_{o\Delta}$. From the example of Figure 1, we have $C(U)|_{o\Delta} = \{ct'_1, ct'_4, ct'_5\}$, $C(U)|_{p\Delta} = \{ct'_6\}$ where ct'_6 is the prefix with length 3 of $ct_2 \in C(D)$, and $C(U)|_{r\Delta} = \emptyset$.

3.2   Enumerating Frequent Subtrees

We adopt the *rightmost extension* technique which was originally proposed by Asai et al. [3] and Zaki [14]. A subtree t of size k is extended to a tree of size $k+1$ by adding a new node only to a node on the *rightmost path* of t. The rightmost path of a tree t is the unique path from the root to the rightmost node of t and the node to be added is called a *rightmost extension* of t.

Figure 2 shows the basic idea of the rightmost extension technique. The subtree t is extended in a bottom-up manner, starting from the rightmost node up to the root of t. By systematically enumerating all frequent subtrees using

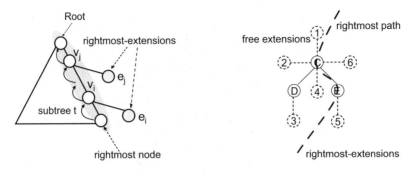

Fig. 2. The rightmost extension technique   **Fig. 3.** Extensions of a subtree

the rightmost extension technique which starts from the empty tree ϵ, an *enumeration tree* is formed. Figure 4 shows the enumeration tree for the database $U = D \cup \Delta$ if we mine U from scratch. To apply the rightmost extension technique we have to record the occurrences of all nodes on the rightmost path of t. However, in fact, only occurrences of the rightmost node of t need to be stored because the occurrences of all other nodes on the rightmost path can be easily computed by taking the parent of its child.

Free extensions, rightmost extensions and immediate supertrees. Here we introduce some important notions that will be used for the purpose of pruning and closure checking in the later parts of the paper. Given a subtree t, we call a supertree t' of t that has one more node than t an *immediate supertree* of t. The additional node in t' that is not in t representing the extension from t to t'. Please notice that the extension represents not only the node label but also the position. Based on their positions, extensions are parted into two categories: *rightmost* extensions and *free* extensions as shown in Figure 3.

Delta frequent rightmost extensions. If we mine $C(U)$ from scratch, for each subtree t, we have to consider every possible rightmost frequent extension of t. However, to mine $C(U)|_\Delta$, the number of possible rightmost extensions of a subtree t can be greatly reduced since we have to care for those occurring in Δ only. We call a node e an Δ-*frequent* rightmost extension of t if it satisfies the following conditions: (i) e is a rightmost extension of t, (ii) t' occurs in Δ where t' is the new subtree obtained by attaching e to t, and (iii) $support_U(t') \geq minsup_U$. Here, t' is called a candidate immediate supertree of t.

The Δ-frequent rightmost extensions of ϵ are frequent nodes that occur in Δ. In Figure 4, Δ-frequent rightmost extensions of ϵ are nodes with lables C, E, G, and H (t_3, t_5, t_7, and t_8 in the figure 4 respectively). Subtrees t_1, t_2, t_4, t_6, t_{10} and t_{11} together with their descendants can be safely removed from the enumeration tree.

Nodes in Δ that are infrequent in D but become frequent under the insertion of Δ are those that will lead to $C(U)|_{r\Delta}$. Other nodes in Δ whose support is

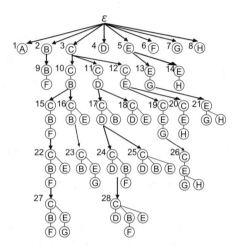

Fig. 4. The enumeration tree for the database $U = D \cup \Delta$

increased under the insertion of Δ will lead to $C(U)|_{o\Delta}$ and/or $C(U)|_{p\Delta}$. To compute Δ-frequent rightmost extensions of ϵ, we store the support counts of all nodes, both frequent and infrequent, of the original database D. Each time when the increment database Δ is received, those support counts will be updated.

The computing of subsets of $C(U)|_\Delta$ are explained in following subsections. We assume constant-time access to all parts of the database. More specifically speaking, given an occurrence of a node v, the occurrence of its parent, its first child, or its next sibling can be computed in constant time.

3.3   Computing $C(U)|_{r\Delta}$

Since neither c, a closed subtree $c \in C(U)|_{r\Delta}$, nor any of its prefixes are contained in a closed subtree in $C(D)$, the information of $C(D)$ is useless to mine $C(U)|_{r\Delta}$. We deal with this problem as follows. When mining $C(D)$ in the original database D, we store the occurrence lists of all infrequent nodes in D. When the increment database Δ is received, it is scanned once to update all those occurrence lists. New frequent nodes (nodes that are infrequent in D but become frequent in U) now have enough information to grow until all the closed subtrees rooted at that nodes are discovered.

Pruning. The rightmost extension technique enumerates all frequent subtrees, however, not all of frequent subtrees are closed, and many of them can be efficiently pruned under the conditions as shown in the following proposition.

Proposition 2. *For a frequent subtree t, if there exists a free extension e_f of t such that e_f occurs at all occurrences of t, then t together with all the descendants of t in the enumeration tree can be pruned.*

Function $Extend(t, D, \Delta)$
 1: $E \leftarrow \emptyset$; $stopFlag \leftarrow$ false;
 2: **for each** node v on the rightmost path of t, bottom-up, **do**
 3:     **for each** Δ-frequent rightmost extension e of t **do**
 4:         $t' \leftarrow t$ plus e, with v as e's parent;
 5:         $E \leftarrow E \cup t'$;
 6:         **if** e occurs at all occurrences of t **then** $stopFlag \leftarrow$ true;
 7:     **if** $stopFlag$ **then break**;
 8: **return** E;

Fig. 5. Compute candidate immediate supertrees of t

Proof. First, since e_f is not a rightmost extension of t, it will never occur in any of t's descendants. Next, since e_f occurs at all occurrences of t, it also occurs at all occurrences of every descendant of t in the enumeration tree. Therefore, t and all the descendants of t in the enumeration tree can not be closed because we can always add e_f to t (or to the descendant of t) to form an immediate supertree with the same support.   □

We can early prune some rightmost extensions of a subtree t as follows. Let us look at Figure 2 again. Both e_i and e_j are rightmost extensions of t with v_i and v_j are the parents of e_i and e_j respectively. Let t_j be the subtree obtained by attaching e_j to t. Since v_j is a proper ancestor of v_i, e_i becomes a free extension of t_j. If e_i occurs at all occurrences of t then e_i occurs at all occurrences of t_j. By Proposition 2, t_j together with all the descendants of t_j in the enumeration tree can be pruned, in other words, there is no need to extend the subtree t with the extension v_j in this case. Figure 5 shows the function for obtaining candidate supertrees of a frequent subtree t with early pruning of some rightmost extensions. This function is also used in mining $C(U)|_{o\Delta}$ and $C(U)|_{p\Delta}$ with some modification.

Closure Checking. A frequent subtree t which is not pruned can still be not closed. The following proposition, which also utilizes the notion of extensions, can be used to determine whether a frequent subtree t is closed or not.

Proposition 3. *For a frequent subtree t, if there exists an extension e of t such that e occurs at all transactions of t, then t is not closed, otherwise t is closed.*

Proof. Let t' be the immediate supertree obtained by adding e to t. If e occurs at all transactions of t then $support(t') = support(t)$, and thus t is not closed. If there does not exist such an extension e then there does not exist a supertree of t that has the same support with t, and thus t is closed.   □

In our running example, no closed subtree of $C(U)|_{r\Delta}$ is generated because the 1-subtree t_8 is pruned by Proposition 2 (at every occurrence of t_8, there is a root extension with label E). Note that the pruning technique of Proposition 2 is equivalent to the pruning techniques based on the notions of *blankets* proposed by Chi et al. [7].

3.4   Computing $C(U)|_{o\Delta}$

For a frequent subtree t such that $t \vdash C(D)$, we denote by $C(D)|^t$ the set of closed subtrees in $C(D)$ that contain t. All frequent extensions, including frequent rightmost extensions and frequent free extensions of t in D are preserved in $C(D)|^t$. Moreover, the support of t in D can be identified by $C(D)|^t$ because there always exists a closed subtree in $C(D)|^t$ that occurs in the same transaction set of t. This means that, we can grow the subtree t by using $C(D)|^t$ and Δ only, without looking at the original database D. For example, we can use ct_2, ct_3 and T_4 to grow the subtree t_3 in Figure 4. We have $support_U(t_3) = support_D(ct_3) + support_\Delta(t_3) = 3 + 1 = 4$. Similarly, to grow t_{13}, only ct_2 and T_4 are enough.

Pruning. We can still apply the pruning technique of Proposition 2. If there exists a free extension e of t in $C(D)|^t \cup \Delta$ such that e occurs at all occurrences of t in $C(D)|^t \cup \Delta$ then e occurs at all occurrences of t in D, by Proposition 2 we can safely prune t and all descendants of t in the enumeration tree. For example, the subtree t_5 and its descendants t_{13}, t_{14} are pruned because at every occurrence of t_5 in ct_2, ct_3 and T_4, there is a root extension with label C. In general, the number of closed subtrees in $C(D)|^t$ is much smaller than the number of transactions in D that contain t, and thus, the check for pruning is fast.

Closure Checking. Let $C(D)|_s^t$ be the set of all closed subtrees in $C(D)|^t$ that are supertrees of t and occur in every transaction of t. If e is an extension of t in any closed subtree in $C(D)|_s^t$ then e occurs at all transactions of t in D. We have the following proposition which is derived from Proposition 3.

Proposition 4. *For a frequent subtree t, $t \vdash C(D)$, if there exists an extension of e of t such that e occurs at all transactions of t in Δ and that e occurs in a closed subtree $c \in C(D)|_s^t$ then t is not closed, otherwise t is closed.*

In the example of Figure 4, we have t_7, t_{12}, and t_{19} are closed subtrees that belong to $C(U)|_{o\Delta}$. The number of closed subtrees in $C(D)|_s^t$ is very small (often 1), making the closure checking very fast compared with using the original database D.

3.5   Computing $C(U)|_{p\Delta}$

Only $C(D)$ and Δ are not enough to compute $C(U)|_{p\Delta}$. Our solution is to use $C(D)$ as a bridge to the original database D by keeping some additional information together with $C(D)$. By this, when growing a subtree t that occurs in $C(D)$, we are able to detect delta frequent rightmost extensions of t that occur in D but not in $C(D)$. For a node v of a closed subtree $c \in C(D)$, the following additional informations are kept:

1. $CE(v)$: the list of occurrences of v in the database D, such that at each occurrence $o \in CE(v)$, v has some child node w at some tree $T \in D$ and that w is not a child node of v in the closed subtree c.

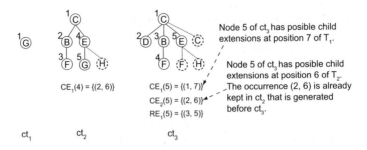

Fig. 6. Complement information of $C(D)$

2. $RE(v)$: the list of occurrences of v in the database D such that at each occurrence $o \in CE(v)$, v has some right sibling node w at some tree $T \in D$ and that w is not a sibling node of v in the closed subtree c.

Note that we do not store all the possible extensions themselves which can be very huge but only occurrences of nodes in closed subtrees which have potential extensions. However, an occurrence of a node may occurs in more than one closed subtrees that may lead to redundant computation. To avoid redundant, we use the following simple but efficient solution.

Suppose that closed subtrees in $C(D)$ are ordered by their generation times. For a node v of a closed subtree $c \in C(D)$, the list $CE(v)$ is split into two parts, $CE_1(v)$ and $CE_2(v)$, where $CE_1(v)$ contains occurrences that are detected for the first time for c, and $CE_2(v)$ contains occurrences that are already kept in some closed subtrees generated before c. Now, suppose that we are growing a subtree t with $C(D)|^t = \{c_{i_1}, c_{i_2}, \ldots, c_{i_k}\}$ where c_{i_j} is generated before c_{i_k} for $j < k$. Then $c_{i_1}.CE_2(v) \cup c_{i_1}.CE_1(v) \cup c_{i_2}.CE_1(v) \cup \ldots \cup c_{i_k}.CE_1(v)$ contains all occurrence of v in D at which v has possible child extensions and there is no duplication among the groups. Here $c_{i_1}.CE_2(v)$ means the list CE_2 of the node v of the closed subtree c_{i_1}. In figure 6, the additional information of the node with label E is given. We do the similar for $RE(v)$.

Suppose t' is a child of a subtree t in the enumeration tree where t occurs in $C(D)$ and t' does not. Once the occurrences of the rightmost node of t' in U are known, the occurrences of all nodes on the rightmost path of t' can be easily computed. This means that we have enough information to grow t' using the rightmost extension technique.

Pruning. We can still apply Proposition 2 to check whether t' or any descendant of t' can be pruned because this proposition does not require having the occurrences of all nodes of t'. This is called *partly pruning* since we can compute free extensions for those nodes whose occurrences are known only.

Closure Checking. Proposition 3 can be used to check if t' (or a descendant of t') is not closed. If there exists an extension (at some nodes whose occurrences

Algorithm $ICTreeMiner(D, \Delta, \sigma, C(D))$

Input: an original database D, an increment database Δ, a minimum support σ, and the set of closed subtrees $C(D)$.

Output: all closed frequent subtrees of $D \cup \Delta$.

1: $CS \leftarrow \emptyset$; $DS \leftarrow \emptyset$;
2: Scan Δ once, find
        C_1: set of new frequent nodes;
        C_2: set of frequent nodes whose support increased by Δ;
3: **for each** $t \in C_1$ **do** $Grow\_rDelta(t, CS, D, \Delta)$;
4: Scan $C(D)$ once, for every subtree $t \in C_2$ find occurrences of t in $C(D)$, and find
        $C(D)|^t$: set of closed trees that contain t;
        $C(D)|_s^t$: set of closed trees that contain t and have the same support as t;
        $C(D)|_i^t$: set of closed trees that are identical with t;
5: **for each** $t \in C_2$ **do** $Grow\_oDelta(t, CS, DS, C(D)|^t, C(D)|_s^t, C(D)|_i^t, D, \Delta)$;
6: Delete closed subtrees in $C(D)$ whose support is less than $minsup_U$;
7: **return** $C(D) \setminus DS \cup CS$;

Fig. 7. The ICTreeMiner Algorithm

Procedure $Grow\_rDelta(t, CS, D, \Delta)$

1: **if** $\exists$ a free extension e_f of t s.t. e_f occurs at all occurrences of t **then return**;
2: $E \leftarrow Extend(t, D, \Delta)$;
3: **for each** $t' \in E$ **do** $Grow\_rDelta(t', CS, D, \Delta)$;
4: **if** $\nexists$ an extension e of t s.t. e occurs at all transactions of t **then** $CS \leftarrow CS \cup t$;
5: **return**;

Fig. 8. The procedure for growing frequent subtrees that lead to $C(U)|_{r\Delta}$

are known) that occurs at every transaction of t' then t' is not closed. However, if t' survives the test of Proposition 3 it still can be not closed and must have to pass the last test as follows.

For a subtree $t \in C(U)|_{p\Delta}$, by the definition of closed subtrees, t is closed if and only if no proper supertree of t has the same support that t has. If s is a proper supertree of t then we have $s \notin C(D)$ and $s \notin C(U)|_{o\Delta}$ because $t \notin C(D)$ and $t \notin C(U)|_{o\Delta}$ itself, thus s can only occur in $C(U)|_{p\Delta}$ or $C(U)|_{r\Delta}$. However, if $s \in C(U)|_{r\Delta}$ and $support(s) = support(t)$ then there will exist a *root extension* (for example, the free extension number 1 in Figure 3) of t that occurs at every transaction of t. In this case, t is not closed by Proposition 3. So we need to consider the case $s \in C(U)|_{p\Delta}$ only. If s is not a descendant of t in the enumeration tree then s is always generated before t due to the rightmost extension method. We use a hash-based technique to check the closedness of t. When a closed subtree c in $C(U)|_{p\Delta}$ is detected, it is put into the hash table $HT(k)$, where k is the support of the subtree c. To check the subtree t, we simply look up in the hash table $HT(support_U(t))$ to see if there exists a supertree of t in the hash table; if so then t is not closed. This simple hash-based method is efficient because the number of closed subtrees in $C(U)|_{p\Delta}$ is often very small as will be shown the experimental section.

Procedure $Grow\_oDelta(t, CS, DS, C(D)|^t, C(D)|^t_s, C(D)|^t_i, D, \Delta)$
1: **if** $\exists$ a free extension e_f of t s.t. e_f occurs at all occurrences of t in $C(D)|^t \cup \Delta$
   then return;
2: $E_1 \leftarrow Extend(t, C(D)|^t, \Delta)$;
3: Find for each $t' \in E_1$
   $C(D)|^{t'}$: set of closed trees that contain t';
   $C(D)|^{t'}_s$: set of closed trees that contain t' and have the same support as t';
   $C(D)|^{t'}_i$: set of closed trees that are identical with t';
4: $E_2 \leftarrow Extend\_p(t, C(D)|^t, \Delta)$;
5: **for each** $t' \in E_1$ **do** $Grow\_oDelta(t', CS, DS, C(D)|^{t'}, C(D)|^{t'}_s, C(D)|^{t'}_i, D, \Delta)$;
6: **for each** $t' \in E_2$ **do** $Grow\_pDelta(t', CS, D, \Delta)$;
7: **if** $\nexists$ an extension e of t s.t. e occurs at all transactions of t in Δ and e is found in
   one closed subtree in $C(D)|^t_s$ **then**
   $CS \leftarrow CS \cup t$;
   $DS \leftarrow DS \cup C(D)|^t_i$;
8: **return**;

Fig. 9. The procedure for growing frequent subtrees that lead to $C(U)|_{o\Delta}$ and $C(U)|_{p\Delta}$

Procedure $Grow\_pDelta(t, CS, D, \Delta)$
1: **if** $\exists$ a free extension e_f of t s.t. e_f occurs at all occurrences of t **then return**;
2: $E \leftarrow Extend(t, D, \Delta)$;
3: **for each** $t' \in E$ **do** $Grow\_pDelta(t', CS, D, \Delta)$;
4: **if** $\exists$ an extension e of t s.t. e occurs at all transactions of t **then return**;
5: **if** $\nexists$ a supertree of t in $HT(support_U(t))$ **then**
6:    $CS \leftarrow CS \cup t$;
7:    insert t into $HT(support_U(t))$;
8: **return**;

Fig. 10. The procedure for growing frequent subtrees that lead to $C(U)|_{p\Delta}$

In the example of Figure 4, t_{20} is pruned and t_{26} is detected as a closed subtree of $C(U)|_{p\Delta}$. The final algorithm, named *ICTreeMiner* (Incremental Closed Tree Miner) is summarized in Figures 7, 8, 9, and 10. In the algorithm, DS is the set of closed frequent subtrees in $C(D)$ whose support increased by the insertion of Δ. In the example $DS = \{t_7\}$. The function *Extend_p* (line 4, Figure 9) computes candidate intermidiate supertrees of a subtree t using the complement information of $C(D)|^t$.

4   Experiments

In this section, we present experimental results on synthetic and real-life datasets. All experiments are measured on a 2.4GHz Intel Core 2 Duo CPU with 2GB of RAM, running Windows XP. The algorithm is implemented in C++. For synthetic dataset, we use the $T1M$ dataset generated by the tree generation program provided by Zaki [14]. For $T1M$, we set the number of distinct node labels

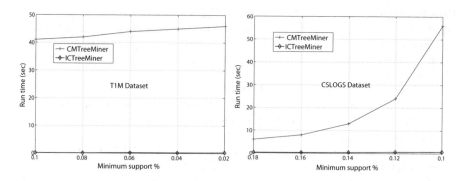

Fig. 11. Effect of varying the minimum support

$N = 100$, the total number of nodes $M = 10,000$, the maximal depth $D = 10$, the maximum fanout $F = 10$ and the total number of trees $T = 1,000,000$. For real-life dataset, we use the $CSLOGS$ dataset[1] which contains web logs collected over a month in the Computer Science Department at the Rensselaer Polytechnic Institute. This dataset consists of $59,690$ trees with total $772,188$ nodes. The average depth and average fan-out of this dataset are 4 and 2.5, respectively.

We compare ICTreeMiner with CMTreeMiner, an efficient static algorithm for mining closed frequent subtrees. First, we evaluate the response times by varying the minimum support. For each value of the minimum support, we take the average value of 20 runtimes of ICTreeMiner. For $T1M$, we take its first $900,000$ transactions as the original database. For $CSLOGS$, the original database consists of its first $50,000$ transactions. The incremental database Δ for $T1M$ consists of $10,000$ transactions taken arbitrarily from the rest of the $T1M$ dataset whereas the incremental database Δ for $CSLOGS$ has 500 transactions for each test. The performance results are shown in Figure 11.

We can see that the proposed incremental algorithm ICTreeMiner is much more efficient than the *static* algorithm that mines from scratch. This is because when the size of the incremental database is much less than that of the original database, the set $C(U)|_{o\Delta}$ will contribute most of new frequent closed subtrees of $C(U)|_{\Delta}$ and can be mined very efficiently. Table 1 shows the statistic values of ICTreeMiner for $T1M$ and $CSLOGS$. The second to last column shows the number of subtrees checked by the subsumption checking when mining $C(U)|_{p\Delta}$ (line 5 in Figure 10). The last column shows the memory usage for storing complement information of $C(D)$. Here we can see that the memory usage is not so big even for very low of minimum supports. However, the memory consumption could be very high if the number of closed subtrees in $C(D)$ is big and the support of each closed subtree is high.

Secondly, we evaluate the scalability of our algorithm by varying the size of the database and fixing the minimum support value. The values of the minimum support are 0.12% and 0.1% for $T1M$ and $CSLOGS$, respectively. Each time

---

[1] http://www.cs.rpi.edu/~zaki/software/

Table 1. Some statistics of running ICTreeMiner on T1M and CSLOGS

Data set	$\sigma(\%)$	$\|C_U\|$	$\|C_{U\|_\Delta}\|$	$\|C_{U\|_i\Delta}\|$	#subtrees checked	Memory (MB)
T1M	0.10	254	136	136	0	29.46
	0.08	307	164	163	0	30.23
	0.06	392	208	206	0	31.88
	0.04	652	289	285	1	34.84
	0.02	1,793	465	461	1	39.79
CSLOGS	0.18	4,273	1,664	1,653	10	17.22
	0.16	5,591	2,003	1,988	25	24.69
	0.14	7,537	2,309	2,289	28	44.51
	0.12	10,724	2,553	2,530	42	118.58
	0.10	17,707	2,977	2,952	43	365.92

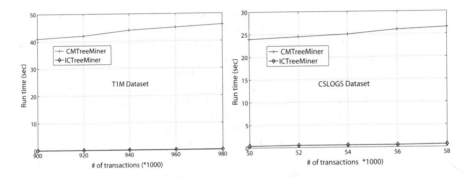

Fig. 12. Scalability testing

the database is updated, we run CMTreeMiner from scratch. The performance results are shown in Figure 12. Its shows that both algorithms scale well with the database size, but the incremental mining algorithm still slightly outperforms the static one.

5   Conclusion

In this paper we have proposed an algorithm for mining closed frequent tree patterns from tree databases that are updated regularly over time. The proposed algorithm which adopted the divide-and-conquer technique outperforms the closed subtree mining algorithms of the static manner especially when the change to the database is small. For next work, we would like to test our algorithm on more real-world datasets. We would also like to extend the ICTreeMiner algorithm to handle databases of labeled rooted unordered trees and to mine embedded closed subtrees as well.

Acknowledgment

This work was partly supported by Kyoto University Global COE "Information Education and Research Center for Knowledge-Circulating Society".

References

1. Abiteboul, S., Buneman, P., Suciu, D.: Data on the Web. Morgan Kaufmann, San Francisco (2000)
2. Agrawal, R., Imielinski, T., Swami, A.: Mining association rules between sets of items in large databases. In: Proc. the ACM SIGMOD Intl. Conf. on Management of Data, pp. 207–216 (1993)
3. Asai, T., Abe, K., Kawasoe, S., Arimura, H., Sakamoto, H., Arikawa, S.: Efficient Substructure Discovery From Large Semi-structured Data. In: Proc. the Second SIAM International Conference on Data Mining (SDM 2002), pp. 158–174 (2002)
4. Asai, T., Arimura, H., Abe, K., Kawasoe, S., Arikawa, S.: Online Algorithms for Mining Semi-structured Data Stream. In: Proc. IEEE International Conference on Data Mining (ICDM 2002), pp. 27–34 (2002)
5. Bifet, A., Gavalda, R.: Mining Adaptively Frequent Closed Unlabeled Rooted Trees in Data Streams. In: Proc. the 14th ACM SIGKDD International Conference on Knowledge Discovery and Data Mining (KDD-08), pp. 34–42 (2008)
6. Cheng, H., Yan, X., Han, J.: IncSpan: Incremental Mining of Sequential Patterns in Large Database. In: Proc. the 10th ACM SIGKDD International Conference on Knowledge Discovery and Data Mining (KDD-04), pp. 527–532 (2004)
7. Chi, Y., Yang, Y., Xia, Y., Muntz, R.R.: CMTreeMiner: Mining Both Closed and Maximal Frequent Subtrees. In: Dai, H., Srikant, R., Zhang, C. (eds.) PAKDD 2004. LNCS (LNAI), vol. 3056, pp. 63–73. Springer, Heidelberg (2004)
8. Chi, Y., Muntz, R.R., Nijssen, S., Kok, J.N.: Frequent Subtree Mining - An Overview. In: Fundamenta Informaticae, vol. 66, pp. 161–198 (2005)
9. Hashimoto, K., Takigawa, I., Shiga, M., Kanehisa, M., Mamitsuka, H.: Mining Significant Tree Patterns in Carbohydrate Sugar Chains. In: Proc. the 7th European Conference on Computational Biology, pp. 167–173 (2008)
10. Hsieh, M., Wu, Y., Chen, A.: Discovering frequent tree patterns over data streams, in Proc. SIAM International Conference on Data Mining (SDM 2006), pp. 629-633 (2006)
11. Nijssen, S., Kok, J.N.: Efficient Discovery of Frequent Unordered Trees. In: Proc. the First International Workshop on Mining Graphs, Trees and Sequences (MGTS2003), in conjunction with ECML/PKDD 2003, pp. 55-64 (2003)
12. Termier, A., Rousset, M.C., Sebag, M.: Dryade: A New Approach for Discovering Closed Frequent Trees in Heterogeneous Tree Databases. In: Perner, P. (ed.) ICDM 2004. LNCS (LNAI), vol. 3275, pp. 543–546. Springer, Heidelberg (2004)
13. Wang, D., Peng, G.: A New Marketing Channel Management Strategy Based on Frequent Subtree Mining. Communications of the IIMA 7(1), 49–54 (2007)
14. Zaki, M.J.: Efficiently Mining Frequent Trees in a Forest. In: Proc. the Eighth ACM SIGKDD International Conference on Knowledge Discovery and Data Mining (KDD 2002), pp. 71–80 (2002)

Optimal Online Prediction in Adversarial Environments

Peter L. Bartlett

Computer Science Division and Department of Statistics, University of California at Berkeley, Berkeley CA 94720, USA
bartlett@cs.berkeley.edu

In many prediction problems, including those that arise in computer security and computational finance, the process generating the data is best modelled as an adversary with whom the predictor competes. Even decision problems that are not inherently adversarial can be usefully modeled in this way, since the assumptions are sufficiently weak that effective prediction strategies for adversarial settings are very widely applicable.

The first part of the talk is concerned with the regret of an optimal strategy for a general online repeated decision problem: At round t, the strategy chooses an action (possibly random) a_t from a set $\mathcal{A}$, then the world reveals a function ℓ_t from a set $\mathcal{L}$, and the strategy incurs a loss $\mathbf{E}\ell_t(a_t)$. The aim of the strategy is to ensure that the regret, that is, $\mathbf{E}\sum_t \ell_t(a_t) - \inf_{a\in\mathcal{A}} \sum_t \ell_t(a)$ is small. The results we present [1] are closely related to finite sample analyses of prediction strategies for probabilistic settings, where the data are chosen iid from an unknown probability distribution. In particular, we relate the optimal regret to a measure of complexity of the comparison class that is a generalization of the Rademacher averages that have been studied in the iid setting.

Many learning problems can be cast as online convex optimization, a special case of online repeated decision problems in which the action set $\mathcal{A}$ and the loss functions ℓ are convex. The second part of the talk considers optimal strategies for online convex optimization [2,3]. We present the explicit minimax strategy for several games of this kind, under a variety of constraints on the convexity of the loss functions and the action set $\mathcal{A}$. The key factor is the convexity of the loss functions: curved loss functions make the decision problem easier. We also demonstrate a strategy that can adapt to the difficulty of the game, that is, the strength of the convexity of the loss functions, achieving almost the same regret that would be possible if the strategy had known this in advance.

References

1. Abernethy, J., Agarwal, A., Bartlett, P.L., Rakhlin, A.: A stochastic view of optimal regret through minimax duality. arXiv:0903.5328v1 [cs.LG] (2009)
2. Abernethy, J., Bartlett, P.L., Rakhlin, A., Tewari, A.: Optimal strategies and minimax lower bounds for online convex games. UC Berkeley EECS Technical Report EECS-2008-19 (2008)
3. Bartlett, P.L., Hazan, E., Rakhlin, A.: Adaptive online gradient descent. UC Berkeley EECS Technical Report EECS-2007-82 (2007)

B. Pfahringer, G. Holmes, and A. Hoffmann (Eds.): DS 2010, LNAI 6332, p. 371, 2010.
© Springer-Verlag Berlin Heidelberg 2010

Discovery of Abstract Concepts by a Robot

Ivan Bratko

University of Ljubljana, Faculty of Computer and Information Sc.,
Tržaška 25, 1000 Ljubljana, Slovenia
ivan.bratko@fri.uni-lj.si

Abstract. This paper reviews experiments with an approach to discovery through robot's experimentation in its environment. In addition to discovering laws that enable predictions, we are particularly interested in the mechanisms that enable the discovery of *abstract* concepts that are not explicitly observable in the measured data, such as the notions of a tool or stability. The approach is based on the use of Inductive Logic Programming. Examples of actually discovered abstract concepts in the experiments include the concepts of a movable object, an obstacle and a tool.

Keywords: Autonomous discovery, robot learning, discovery of abstract concepts, inductive logic programming.

1 Introduction

In this paper we look at an approach to autonomous discovery through experiments in an agent's environment. Our experimental domain is the robot's physical world, and the subject of discovery are various quantitative or qualitative laws in this world. Discovery of such laws of (possibly naive) physics enables the robot to make predictions about the results of its actions, and thus enable the robot to construct plans that would achieve the robot's goals. This was roughly the scientific goal of the European project XPERO (www.xpero.org).

In addition to discovering laws that directly enable predictions, we are in this paper particularly interested in the mechanisms that enable the discovery of *abstract* concepts. In XPERO, such abstract concepts are called "insights". By an insight we mean something conceptually more general than a law. One possible definition of an insight in the spirit of XPERO is the following: an insight is a new piece of knowledge that makes it possible to simplify the current agent's theory about its environment. So an insight may enhance the agent's description language, and thus it should also make further discovery easier because the hypothesis language becomes more powerful and suitable for the domain of application.

What may count as an insight in the XPERO sense? Suppose the robot is exploring its physical environment and trying to make sense of the measured data. Assume the robot has no prior theory of the physical world, nor any knowledge of relevant mathematics. But it can use one, or several, machine learning methods that may use a general logic as a hypothesis language. Then, examples of insights would be the discoveries of notions like absolute coordinate system, arithmetic operations, notion of

B. Pfahringer, G. Holmes, and A. Hoffmann (Eds.): DS 2010, LNAI 6332, pp. 372–379, 2010.

gravity, notion of support between objects, etc. These concepts were never explicitly observed in the robot's measured data. They are made up by the robot, possibly through predicate invention, as useful abstract concepts that cannot be directly observed. An insight would thus ideally be a new concept that makes the current domain theory more flexible and enables more efficient reasoning about the domain.

The expected effect of an insight is illustrated in Figure 1. Initially, the robot starts with a small theory, which will become better, and also larger, after first experiments and learning steps. Newly discovered relations and laws of the domain are added to the theory, so the theory keeps growing. Then, when an insight occurs, the insight may enable a simplification of the theory, so the theory shrinks. A simplification is possible because an insight gives rise to a better representation language which facilitates a more compact representation of the current knowledge. Then it may start growing again. This process is reminiscent of the evolution of scientific theories.

In this paper I outline an approach to the discovery of abstract concepts by a robot using Inductive Logic Programming (ILP), and give examples of actually discovered abstract concepts in experiments. These include the concepts of movable object, an obstacle, and a tool.

It should be noted that our scientific goals of discovering abstract concepts are considerably different from typical goals in robotics. In a typical robotics project, the goal may be to improve the robot's performance at carrying out some physical task. To this end, any relevant methods, as powerful as possible, will be applied. In contrast to this, here we are less interested in improving the robot's performance at some specific task, but in making the robot improve its theory and "understanding" of the world. We are interested in finding mechanisms, as generic as possible, that enable the gaining of insights. For such a mechanism to be generic, it has to make only a few rather basic assumptions about the agent's prior knowledge. We are interested in minimizing such "innate knowledge" because we would like to demonstrate how discovery and gaining insights may come about from only a minimal set of "first principles". Not all machine learning methods are appropriate. Our aim requires that the induced insights can be interpreted and understood, flexibly used in robot's reasoning, and are not only useful for making direct predictions.

2 The Experimental Setting

We assume that the robot's discovery process takes the form of an indefinite "experimental loop". The robot starts with some initial knowledge (possibly zero). This is the robot's initial theory of the domain. Then it repeats the steps:

1. Perform experiments and collect observation data
2. Apply a ML method to the data, which results in a new theory
3. Design new experiments aiming at collecting most informative new data
4. Plan the execution of these experiments using the current theory of the domain
5. Go to step 1 to repeat the loop.

When designing next experiments, the robot has to estimate the potential benefits of possible experiments. This can be done by using methods of active learning, but there may be an additional difficulty of finding a plan to execute these experiments. A new experiment is defined by measurements to be taken in some situation, e.g. at some robot's position with respect to an object of interest. So each experiment defines a goal for the robot to achieve. To achieve such goals, the robot has to plan sequences of actions to be carried out. However, when looking for such plans, the robot uses the current theory – nothing better is available at the moment. The current theory is an imperfect model of the environment and may thus lead the robot to false or impossible plans.

The above described experimental loop is somewhat similar to that of reinforcement learning. However, in our case the only reward is improved knowledge. So the loop is driven by the robot's "hunger for knowledge". Also, we are interested in the robot acquiring explicit symbolic knowledge that enables understanding and symbolic reasoning.

Experiments with this experimental scenario, using a number of learning methods at step 2 of the loop, are described in [1]. [2] is a comparison of the used ML techniques w.r.t. a collection of learning tasks and criteria relevant for autonomous robot discovery. ML methods used in these experiments include regression trees, decision trees (implementations in Orange [3]), induction of qualitative trees with QUIN [4], induction of equations with Goldhorn [5], Inductive Logic Programming with Aleph [6] and Hyper [7], and statistical relational learning with Alchemy [8].

3   Gaining Insights through Predicate Invention in ILP

In this section I describe an approach to and experiments with the discovery of abstract concepts using the mechanism of predicate invention in ILP. The experimental scenario consisted of real or simulated mobile robot(s) pushing blocks in a plane. Figure 2 shows two examples. In these experiments, the concepts of a movable object, an obstacle, and a tool were discovered from measured data. These notions were expressed as new predicates.

First, let us consider how the concept of a movable object emerged. When given commands to move specified objects by given distances, the robot was able to actually move some of the blocks, but some of the blocks could not be moved. After some time, the robot had collected a number of experimental data recorded as ground facts about predicates:

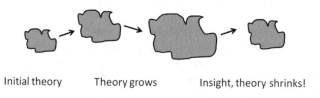

Initial theory   Theory grows   Insight, theory shrinks!

Fig. 1. Evolution of a theory during execution of experimental loop

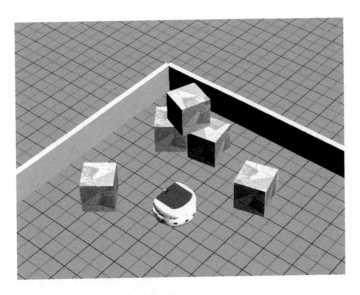

Fig. 2. Two of various robotic experimental settings used in the experiments: a simulated Khepera robot, and a real humanoid robot Nao [9]

- **at(Obj,T,P)**, meaning object Obj was observed at position P at time T;
- **move(Obj,P1,D,P2)**, meaning command "move Obj from P1 by distance D" resulted in Obj at P2.

Here all positions are two-dimensional vectors. The robot's prior knowledge (communicated to the ILP program as "background knowledge") consisted of the predicates:

- **different(X,Y)**, meaning X and Y are not approximately equal;
- **approx_equal(X, Y)**, meaning $X \approx Y$;
- **add(X, Y, Z)**, meaning $Z \approx X + Y$.

These relations are defined as approximations so that they are useful in spite of noise in numerical data.

It should be noted that neither the observations nor the prior knowledge contain the concept of mobility of objects, or any mention of it. There are no examples given of movable and immovable objects. The ILP program Hyper [7] was used on this learning problem to induce a theory of moving in this world (that is learn predicate **move/4** which would for a given command "move Obj from position P1 by distance D" predict the position P2 of Obj after the command has been executed). The induced theory by Hyper was stated in logic by the following Prolog clauses:

> **move(Obj,Pos1,Dist,Pos2):-**
> **approx_equal(Pos1, Pos2),**
> **not p(Obj).**

> **move(Ob,Pos1,Dist,Pos2):-**
> **add(Pos1,Dist,Pos2),**
> **p(Obj).**

> **p(Obj):-**
> **at(Obj,T1,Pos2),**
> **at(Obj,T2,Pos2),**
> **different(Pos1,Pos2).**

In the clauses above, the variables were renamed for easier reading. The first clause deals with immovable objects (after the move command, the position of the object remains unchanged). The second clause handles movable objects. The point of interest is that HYPER invented a new predicate, p(Object) which is true for objects that can be moved. At least the "intention" is to define the movability property of an object. The definition above says that an object is movable if it has been observed at two different positions in time, which is not quite correct in general. The robot has come up with a new concept never mentioned in the data or problem definition. The new concept p(Object) enabled the learning system to divide the problem in the two cases. A meaningful name of the newly invented predicate p(Object) would be movable(Object).

In another experiment where many objects were present in the scene, the robot invented another predicate which corresponds to the notion of obstacle. If an immovable object appears on the trajectory of a moving object then the stationary object impedes the movement. The ILP learner found it useful to introduce a new predicate whose definition corresponded to such an object – an obstacle. Again, the notion of obstacle was never mentioned in the problem definition, nor in the learning data. The learner just figured it out that such a new notion was useful for explaining the behavior of objects in the robot's world.

In another experiment, two robots, one stronger and one weaker, were experimenting with block pushing. The concept of movability was this time relative to the particular robot. There were blocks that the stronger robot could move and the weaker could not. In this case the new invented predicate was:

```
p(R, Obj) :-
   moving(R, T),
   contact(R, Obj, T).
```

This can be interpreted as: an object Obj is movable by robot R if there is time T when R was observed to be moving while it was in contact with Obj. The touch sensor is on when there is a force contact between the robot and the object, so in the case of contact the robot must have been pushing the object.

The essential learning mechanism in these experiments was predicate invention in ILP [12]. This capability further increases computational complexity of ILP, but it enables a way of gaining insights. Details of these experiments are described by Leban et al. [10]. That paper also explains a method for generating negative examples needed by HYPER. In brief, negative examples are based on a kind of closed-world assumption, namely that the results of robot's actions are a function the action's parameters. This idea however requires a refinement to produce *critical* negative examples. First, negative examples are generated from positive examples by random perturbations of the "function argument". The learning from the given positive examples and these negative examples typically results in an overly general intermediate theory. When this theory is tested by prediction made on a positive example, typically more than one answer is produced. But since we know the correct predictions for these cases, the spurious predictions are easily recognised and added as additional negative examples. Another, improved theory is then induced from the positive examples and the so enhanced set of negative examples.

Further, more complex experiments led to the discovery of the notions of tool and obstacle. In these experiments, a robot was carrying out block moving tasks whose goals were of the kind: at(Block, Pos). These tasks required the robot's planning of sequences of actions that resulted in the specified goals. Figure 3 shows an example. The robot's planner constructed plans for solving a collection of such tasks. The plans were constructed by search using CLP(R) to handle constraints that ensured collision-free robot's trajectories and align the robot with a block to be pushed. The plans were also "explained" by a means-ends planner in terms of the goals that each action is supposed to achieve. Macro operators were then learned from this collection of concrete plans as generalised subsequences of actions. The generalisation was accomplished by replacing block names by variables whenever possible. The robot then induced a classification of macro operators in terms of logic definitions that discriminated between macro operators. At this stage two new concepts were invented that can be interpreted as definitions of the concept of a tool and of an obstacle. For example, the following definition defines that an object Obj has the role of a tool in a macro operator:

```
tool( MacroOp, Obj) :-
   object( Obj),
   member( Action --> Goals, MacroOp),      % Purpose of Action is Goals
   argument( Obj, Action),      % Obj appears as an argument in Action
   not argument( Obj, Goals).   % Obj does not appear as argument in Goals
```

Essentially, this definition says that Obj has the role of a tool in a macro operator if there is an action Action in the operator such that the purpose of Actions is Goals, and Obj appears as one of the arguments that describe Action, and it does not appear as an argument of Goals that Action achieves. Details on these experiments are documented in [11].

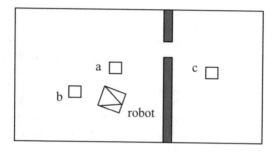

Fig. 3. A blocks moving task; the goal is to move block a to the right of the wall (shaded area represents the wall). The robot cannot squeeze itself through the door, so it has to use block b as a tool with which block a can be pushed through the door.

4 Concluding Remarks

In the paper, some experiments in the discovery of abstract concepts through predicate invention in ILP in a simple robotic domain were reviewed. The approach is based on enabling Hyper to invent new predicates by including initial "dummy" definitions of auxiliary predicates for which there are no examples given. We showed examples where Hyper was able to invent such auxiliary predicates that represent the notion of the object's movability and the notion of a tool. In these experiments, the number of examples sufficient to induce sensible definitions of such concepts was typically relatively small, in the order of tens or hundreds. The data that was used for learning was collected using both a simulated or a real robot and contained noise.

One critical question concerning the significance of the experimental results is of course that of background knowledge: how much help was conveyed to the system by providing useful and carefully selected background predicates? Although the question of what is reasonably acceptable background knowledge is often raised, it needs principled, in-depth study.

For the task of abstract discovery considered in this paper, in experiments with a number of ML methods, ILP proved to be the only approach with potential of success. This was analysed in [2]. The hypothesis language of ILP is typically predicate logic. This is sufficiently expressive to enable, at least in principle, the applicability also to tasks where the other approaches seem to be insufficient. In particular, such tasks are the discovery of aggregate and functional notions. For example, the key mechanism in discovery of aggregate and functional notions is that of predicate invention. In the discovery and general handling of aggregate notions, the recursion facility is essential. Also, the use of logic theories as background knowledge allows very natural transition between the learning of theories in increasingly complex worlds.

The general problem with ILP is its high computational complexity. Predicate invention in ILP [12] further increases complexity. Another deficiency is lack of quantitative facilities in logic. The latter are typically added to the basic logic formalism of ILP as background knowledge, most elegantly as CLP(R) (constraint logic programming with real numbers). However, computational complexity which arises from the main strength of ILP, that is the expressiveness of ILP's hypothesis language, is the critical limitation and it seems the situation in this respect is unlikely to change considerably with more efficient algorithms for ILP learning. So, improvements can mainly be expected from fundamentally enhancing the ways of applying this approach in concrete situations. For example, one possibility to considerably reduce the complexity is through the introduction of hierarchy of learning problems.

Acknowledgements

Research described in this paper was supported by the European Commission, 6th Framework project XPERO, and the Slovenian research agency ARRS, Research program Artificial Intelligence and Intelligent Systems. A number of people contributed to the related experimental work, including G. Leban and J. Žabkar.

References

1. Bratko, I., Šuc, D., Awaad, I., Demšar, J., Gemeiner, P., Guid, M., Leon, B., Mestnik, M., Prankl, J., Prassler, E., Vincze, M., Žabkar, J.: Initial experiments in robot discovery in XPERO. In: ICRA'07 Workshop Concept Learning for Embodied Agents, Rome (2007)
2. Bratko, I.: An Assessment of Machine Learning Methods for Robotic Discovery. Journal of Computing and Information Technology – CIT 16, 247–254 (2008)
3. Demšar, J., Zupan, B.: Orange: Data Mining Fruitful & Fun - From Experimental Machine Learning to Interactive Data Mining (2006), http://www.ailab.si/orange
4. Šuc, D.: Machine Reconstruction of Human Control Strategies, Frontiers Artificial Intelligence Appl., vol. 99. IOS Press, Amsterdam (2003)
5. Križman, V.: Automatic Discovery of the Structure of Dynamic System Models. PhD thesis, Faculty of Computer and Information Sciences, University of Ljubljana (1998)
6. Srinivasan, A.: The Aleph Manual. Technical Report, Computing Laboratory, Oxford University (2000),
   http://web.comlab.ox.ac.uk/oucl/research/areas/machlearn/Aleph/
7. Bratko, I.: Prolog Programming for Artificial Intelligence, 3rd edn. Addison-Wesley/Pearson (2001)
8. Richardson, M., Domingos, P.: Markov Logic Networks. Machine Learning 62, 107–136 (2006)
9. Aldebaran robotics – Nao (2010), http://www.aldebaran-robotics.com/eng/index.php
10. Leban, G., Žabkar, J., Bratko, I.: An experiment in robot discovery with ILP. In: Železný, F., Lavrač, N. (eds.) ILP 2008. LNCS (LNAI), vol. 5194, pp. 77–90. Springer, Heidelberg (2008)
11. Košmerlj, A., Leban, G., Žabkar, J., Bratko, I.: Gaining Insights About Objects Functions, Properties and Interactions, XPERO Report D4.3. Univ. of Ljubljana, Faculty of Computer and Info. Sc. (2009)
12. Stahl, I.: Predicate invention in Inductive Logic Programming. In: De Raedt, L. (ed.) Advances in Inductive Logic Programming, pp. 34–47. IOS Press, Amsterdam (1996)

Contrast Pattern Mining and Its Application for Building Robust Classifiers

Kotagiri Ramamohanarao

Department of Computer Science and Software Engineering
The University of Melbourne
Kotagiri@unimelb.edu.au

Abstract. The ability to distinguish, differentiate and contrast between different data sets is a key objective in data mining. Such ability can assist domain experts to understand their data and can help in building classification models. This presentation will introduce the techniques for contrasting data sets. It will also focus on some important real world applications that illustrate how contrast patterns can be applied effectively for building robust classifiers.

B. Pfahringer, G. Holmes, and A. Hoffmann (Eds.): DS 2010, LNAI 6332, p. 380, 2010.
© Springer-Verlag Berlin Heidelberg 2010

Towards General Algorithms for Grammatical Inference[*]

Alexander Clark

Department of Computer Science
Royal Holloway, University of London
alexc@cs.rhul.ac.uk

Abstract. Many algorithms for grammatical inference can be viewed as instances of a more general algorithm which maintains a set of primitive elements, which distributionally define sets of strings, and a set of features or tests that constrain various inference rules. Using this general framework, which we cast as a process of logical inference, we re-analyse Angluin's famous LSTAR algorithm and several recent algorithms for the inference of context-free grammars and multiple context-free grammars. Finally, to illustrate the advantages of this approach, we extend it to the inference of functional transductions from positive data only, and we present a new algorithm for the inference of finite state transducers.

---

[*] The full version of this paper is published in the Proceedings of the 21th International Conference on Algorithmic Learning Theory, Lecture Notes in Artificial Intelligence Vol. 6331.

B. Pfahringer, G. Holmes, and A. Hoffmann (Eds.): DS 2010, LNAI 6332, p. 381, 2010.

The Blessing and the Curse
of the Multiplicative Updates

Manfred K. Warmuth[*]

Computer Science Department
University of California, Santa Cruz
CA 95064, U.S.A.
manfred@cse.ucsc.edu

Abstract. Multiplicative updates multiply the parameters by nonnegative factors. These updates are motivated by a Maximum Entropy Principle and they are prevalent in evolutionary processes where the parameters are for example concentrations of species and the factors are survival rates. The simplest such update is Bayes rule and we give an in vitro selection algorithm for RNA strands that implements this rule in the test tube where each RNA strand represents a different model. In one liter of the RNA "soup" there are approximately 10^{20} different strands and therefore this is a rather high-dimensional implementation of Bayes rule.

We investigate multiplicative updates for the purpose of learning online while processing a stream of examples. The "blessing" of these updates is that they learn very fast because the good parameters grow exponentially. However their "curse" is that they learn too fast and wipe out parameters too quickly. We describe a number of methods developed in the realm of online learning that ameliorate the curse of these updates. The methods make the algorithm robust against data that changes over time and prevent the currently good parameters from taking over. We also discuss how the curse is circumvented by nature. Some of nature's methods parallel the ones developed in Machine Learning, but nature also has some additional tricks.

[*] Supported by NSF grant IIS-0917397.

B. Pfahringer, G. Holmes, and A. Hoffmann (Eds.): DS 2010, LNAI 6332, p. 382, 2010.

Author Index

Aiguzhinov, Artur 16
Ariki, Yasuo 87

Baba, Teruyuki 221
Bannai, Hideo 132
Bartlett, Peter L. 371
Bifet, Albert 1
Boley, Mario 57
Bratko, Ivan 372
Buchwald, Fabian 341

Clark, Alexander 381

Ding, Wei 326
Drew, Mark S. 236

Enomoto, Nobuyuki 221

Foulds, James R. 102
Frank, Eibe 1, 102
Fürnkranz, Johannes 266

Ganzert, Steven 296
Girschick, Tobias 341
Grosskreutz, Henrik 57
Grčar, Miha 174
Gunn, Steve R. 42
Guttmann, Josef 296

Hamrouni, Tarek 189
Hatano, Kohei 132
Honda, Yuya 72
Hopf, Thomas 311

Imada, Keita 117
Inokuchi, Akihiro 205

Jones, Gareth 42
Juršič, Matjaž 174

Kami, Nobuharu 221
Kashihara, Kazuaki 132
Kimura, Masahiro 144

Kivinen, Jyrki 251
Kramer, Stefan 159, 296, 311, 341
Krause-Traudes, Maike 57

Lavrač, Nada 174
Lovell, Chris 42
Luosto, Panu 251

Mannila, Heikki 251
Motoda, Hiroshi 144
Mueller, Marianne 159

Nagano, Shinya 72
Nagy, Iulia 87
Nakamura, Katsuhiko 117
Nguyen, Viet Anh 356

Ohara, Kouzou 144
Ontañón, Santiago 281

Park, Sang-Hyeun 266
Plaza, Enric 281
Podpečan, Vid 174

Qin, ZhiGuang 27

Ramamohanarao, Kotagiri 380
Rückert, Ulrich 341

Saito, Kazumi 144
Salazar, Josue 326
Schulte, Oliver 236
Seki, Hirohisa 72
Serra, Ana Paula 16
Soares, Carlos 16
Steinmann, Daniel 296
Stepinski, Tomasz F. 326
Suzuki, Einoshin 27

Takeda, Masayuki 132
Tanaka, Katsuyuki 87
Tong, Bin 27

Vinh, Nguyen Duy   205

Warmuth, Manfred K.   382
Washio, Takashi   205
Weizsäcker, Lorenz   266
White, Denis   326

Yahia, Sadok Ben   189
Yamamoto, Akihiro   356
Yoshikawa, Takashi   221
Younes, Nassima Ben   189

Zauner, Klaus-Peter   42

Printing: Mercedes-Druck, Berlin
Binding: Stein+Lehmann, Berlin